Electrician's

EXAM PREP

Charles R. Miller

NFPA®

nec®

JONES AND BARTLETT PUBLISHERS

Sudbury, Massachusetts

BOSTON TORONTO LONDON SINGAPORE

World Headquarters
Jones and Bartlett Publishers
40 Tall Pine Drive
Sudbury, MA 01776
978-443-5000
info@jbpub.com
www.jbpub.com

Jones and Bartlett Publishers
Canada
6339 Ormindale Way
Mississauga, Ontario L5V 1J2
Canada

Jones and Bartlett Publishers
International
Barb House, Barb Mews
London W6 7PA
United Kingdom

National Fire Protection
Association
1 Batterymarch Park
Quincy, MA 02169-7471
www.NFPA.org

Jones and Bartlett's books and products are available through most bookstores and online booksellers. To contact Jones and Bartlett Publishers directly, call 800-832-0034, fax 978-443-8000, or visit our website www.jbpub.com.

Production Credits
Chief Executive Officer: Clayton Jones
Chief Operating Officer: Don W. Jones, Jr.
President, Higher Education and Professional Publishing: Robert W. Holland, Jr.
V.P., Sales and Marketing: William J. Kane
V.P., Design and Production: Anne Spencer
V.P., Manufacturing and Inventory Control: Therese Connell
Publisher—Public Safety Group: Kimberly Brophy
Acquisitions Editor—Electrical: Martin Schumacher
Associate Editor: Amanda Brandt
Senior Production Editor: Karen C. Ferreira
Director of Marketing: Alisha Weisman
Marketing Associate: Meagan Norlund
Composition: Shepherd, Inc.
Text and Cover Design: Anne Spencer
Printing and Binding: Courier Stoughton
Cover Printing: Courier Stoughton

Library of Congress Cataloging-in-Publication Data
Miller, Charles R.
NFPA : electrician's exam prep / Charles R. Miller.
p. cm.
Includes index.
ISBN 978-0-7637-5118-0
1. Electric engineering—Examinations, questions, etc. I. Title.
TK169.M45 2008
621.319'24076—dc22
2007051991

6048
Printed in the United States of America
13 12 11 10 09 10 9 8 7 6 5 4 3 2

Contents

About the Author

Charles R. Miller owned and operated a successful commercial electrical contracting company, Lighthouse Electric Co., Inc., in Nashville, Tennessee, for eighteen years. While working in the electrical industry, he discovered that electricians needed help in passing their electrical exams. This discovery led Charles to design a custom seminar that would give electricians that help, and in 1988 he began operating a second company, Lighthouse Educational Services. In designing the seminar, Charles could not find the ideal textbook, so he wrote and illustrated his own. He traveled to many locations taking electrical exams to gain a greater understanding of the types of questions found on electrical exams. This book, *Electrician's Exam Prep,* has a proven track record of helping electricians pass electrical exams. *Electrician's Exam Prep* is also a great resource for instructors teaching electrical exam prep classes.

In 1998 Charles started work on another electrical book, the *Illustrated Guide to the National Electrical Code.* This book focuses on interpreting the *NEC* through clearly understandable wording and illustrations. In 1999, Charles began writing and illustrating the "Code in Focus" column for *Electrical Contractor* magazine. His detailed illustrations and writing style make this one of the top-read columns in the monthly magazine. In 2003, Charles began writing books for the NFPA. Titles include *Pocket Guide to Residential Electrical Installations, Pocket Guide to Commercial and Industrial Electrical Installations, NFPA's Electrical References,* and *NFPA's Pocket Electrical References.* Besides teaching, writing, and illustrating, Charles cohosted a home improvement radio talk show in Nashville, Tennessee, for more than three years.

Charles continues to be a student of the *National Electrical Code,* always seeking new experience on topics that will broaden his knowledge base of expertise. For more information on Charles and Lighthouse Educational Services, visit www.charlesRmiller.com.

Acknowledgments

I would like to say thank you to my wife, Linda, for all your love and support as I spend long hours writing and illustrating. I would also like to say thank you to the National Fire Protection Association (NFPA) for the opportunity to work with such a great organization.

Introduction

Electrician's Exam Prep was written out of a desire to start an electrician's exam prep class and not being able to find the book I wanted to use in class. In 1988 I started looking for a book that met certain qualifications: 1) it must be easy to read and easy to understand, 2) it must not be overwhelming, 3) it must show methods for solving the types of calculations found on electrical exams, 4) it must have sample calculation questions with answers showing how the correct answer is determined, 5) it must have sample general knowledge questions with answers, and 6) it must have sample *National Electrical Code®* questions with answers showing *Code®* references. I could not find the book I was looking for and therefore decided to write and illustrate the book myself. While this manual has been updated because of changes in the *National Electrical Code* and changes in electrical exams, it remains true to the standards upon which it was written.

This manual covers the 2005 edition of the *National Electrical Code,* but it is not limited to that edition. The calculations covered in Part I: Exam Topics have changed very little over the years. Although section numbers in Articles may change, calculation procedures seldom change. Part II contains general electrical questions, which will rarely change.

Why one manual for both journeyman and master electrician license exams? Because with most exams, there is very little difference between the journeyman and master (or contractor) electrician license exams. One difference is the type of occupancy used in a feeder or service load calculation. Load calculation questions on journeyman exams usually pertain to dwelling units and not to commercial occupancies. Another difference is the voltage system in an exam question. While master exams will probably have questions with both single- and three-phase voltage systems, journeyman exam questions usually do not have questions with three-phase voltage systems. Considering the importance of knowing commercial load calculations and knowing how to perform calculations with three-phase voltage, why not study these types of calculations regardless of the exam? Besides, there is no guarantee that these types of questions will not be on a journeyman electrical exam.

Studying this Manual

This manual is divided into three main parts. Part I: Exam Topics covers a wide variety of calculations typically found on electrical exams. Of all the elements, calculations can be the most intimidating when studying for an electrical exam, because math is required. That's why the first chapter is Math and the Electrician. This chapter covers the types of math problems required throughout the rest of Part I. All the chapters in Part I are easy to read and easy to understand. Each chapter contains detailed explanations and sample calculation-type problems found on electrical exams. Each problem is broken down into a step-by-step procedure. In the back of each chapter are more sample problems. The sample problems test the student's understanding of the material presented. Solutions to the sample problems not only give the correct answer but show how the answer was established.

In Part II: General Exam Questions, only a few of the sample questions come from the *Code*; the ones that do should be memorized. Most of these questions pertain to electrical theory. Originally, many electrical exams contained a *closed book* portion of the exam. Questions from Part II are the types of questions typically found during the closed book portion of the exam. Today, not many exams contain closed book questions, but you should not overlook the importance of these questions. These types of questions are usually scattered throughout the exam. Part II is divided into seven sample exams, each exam containing fifty questions. The time limit on each exam is one hour. Since few general exam questions come from the *Code*, answering all fifty questions in the time allotted should not be a problem. It is even advisable to memorize as many of the general exam questions as you can.

Sample questions from Part III: *NEC* Exam Questions come from anywhere and everywhere in the *Code*. Pay particular attention to exceptions and fine print notes (FPN). Part III is divided into sixteen sample exams, and each exam contains fifty questions. The time limit on each exam is two hours. Unlike the exams in Part II, answering all fifty questions in the time allotted is very difficult. Do not get frustrated with this part of the manual. For the first few exams, do not pay attention to the clock. Before developing speed, you must learn how to use the index and table of contents. You must also learn how to select a key word or key words out of the question. If you have trouble locating the section in the *Code* that provides the answer, look in the back of this manual and find the *Code* reference for that question. Look up the reference and look at the question again. Now try to pick out a key word that could have led you to this reference. Again, the first few exams will probably take longer (and maybe much longer) than two hours. As you spend time in the *Code* and working with these questions, your time will improve. By the time you complete most of the *National Electrical Code* exams, you should be able to complete an exam in less than two hours. After completing all sample exams and answering all sample exam questions, it is time to highlight your *Code* book.

The best advice for passing any test: study, study, study.

Test Taking Tips

There are different schools of thought on strategies for taking exams. The multipass method is one popular method. The idea behind this method is to read through all the exam questions quickly and answer only the ones you know without looking up anything. With a second pass, answer questions that can be found or calculated in a short amount of time. Third and fourth passes may be necessary to answer all the questions. Using this method can be dangerous if time runs short and doesn't allow for review of questions not answered on the first or second pass. The single-pass method is another strategy for taking an exam. This method implies that all questions are answered on the first pass through the exam. Even if you must guess at the answer, it is better than taking a chance on leaving the question unanswered. Most computerized exams will allow exam takers to bookmark a question for easy review. The benefit to this method is a reduction in time for the overall exam by relying on the exam takers' knowledge and intuition. Both methods offer some tips and tricks. Select the ones that fit well with your style of exam taking and choose that method.

With both methods, it is essential to keep up with the remaining time compared to the remaining questions. Computerized tests usually have a countdown clock somewhere on the screen. Before starting the exam, know how many questions must be answered each hour. For example, if there are 100 questions and the time limit is four hours, at least 25 questions must be answered each hour. If at the end of two hours you are answering question fifty-seven (and have not skipped any), then you are managing your time well and are set to finish early.

Do not leave the exam early. You paid to take the exam, so use all the time allowed. If time remains, look over the exam again—especially questions that you bookmarked. If there are no bookmarked questions, reread and rework the calculation-type questions. Do not change an answer unless you find you have missed something in the question or have made a mistake. Studies have shown that changing answers on multiple-choice exams often results in incorrect answers.

Sometimes it will be possible to narrow the choice of answers to two or three that could be correct. For example:

What is the efficiency of a 20-hp, 240-volt, three-phase motor that is drawing 48 amperes? The circuit has a power factor of 97%.
a) 133.7%
b) 103%
c) 86%
d) 77.09%

Answers a) and b) can be eliminated immediately because efficiency is always less than 100%.

As is true of most standardized exams, a question may appear to have two correct answers; however, only one answer will be correct. Of course you should select the best or closest answer. For example:

Disconnects shall be installed in a(n) _____ position.
a) diagonal
b) vertical
c) horizontal
d) upright

Although "vertical" and "upright" are synonyms, the *Code* uses the word "vertical" and not "upright." As specified in 240.33, "enclosures for overcurrent devices shall be mounted in a vertical position unless that is shown to be impracticable." Answer b) is correct.

Sometimes it is possible to have an idea of the correct answer before calculating. For example:

What is the approximate length of a conductor that is connecting the service to the load? The conductor is a 6-AWG THW bare uncoated copper conductor with a resistance of 0.035 ohms.
a) 0.071′
b) 0.491′
c) 71′
d) 1296′

This question is asking for the approximate length of a conductor. The first answer is unrealistic. The length of the conductor in answer a) is slightly less than 7⁄8 inch. The second answer is also unrealistic. The length of the conductor in answer b) is a little less than six inches. Since the length of a conductor is not often greater than one thousand feet, answer d) is probably not the correct answer. Therefore, before calculating the problem, it looks like answer c) may be the one. See Part I, Chapter 5 for instructions on calculating this type of problem.

Questions throughout an exam can use a mixture of terms, phrases, and symbols. For example, voltage systems can be expressed as single-phase voltage, 1-phase voltage, single-Ø voltage, 1-Ø voltage, three-phase voltage, 3-phase voltage, three-Ø voltage, and 3-Ø voltage.

Don't spend too much time on any one question. If you can't find the answer quickly, bookmark the question (or write down the number) and move to the next question. Keep that question in the back of your mind because sometimes you run across the right *Code* section while looking for another answer. Keep in mind also that some answers will come from exceptions and fine print notes. Always read the question carefully, especially calculation questions. Not reading the question carefully could result in a wrong answer. Finally, always answer every question, even if you have to guess at the answer.

PART I

Exam Topics

CHAPTER 1 Math and the Electrician

Basic math is an essential tool when taking a Master Electrician's or Journeyman Electrician's Examination. Most electrical examinations have one to three hours of calculations, so it is important to have an understanding of basic math principles. This first chapter provides a quick review of these principles.

Being fully prepared is a crucial part of success when taking examinations. Having the right size and type calculator is an important part of being prepared. The keys on the calculator need to be large enough to be pressed quickly and accurately. The calculator could stop working during the test, so it would be wise to carry an extra calculator or spare batteries to the examination. It would also be helpful for the calculator to have a square root key.

Working with Decimals

In most cases it will be necessary to carry out the calculations to only two or three numbers behind the decimal point. In this book, however, calculations will be carried out to at least four decimal places. This can be done by setting the calculator to a floating decimal. (Most handheld calculators are automatically set to a floating decimal.)

Numbers will be rounded off by the following rule: 5 and above will be rounded up; 4 and below will be rounded down.

EXAMPLES

0.816 = 0.82

Because 6 is more than 5, 1 can be rounded up to 2.

16.72 = 16.7

Because 2 is less than 5, it can be dropped.

It is important to be able to add, subtract, multiply, and divide numbers containing decimals. While working with the following problems, be sure to place the decimal in the correct position.

PROBLEMS

1. 485.773 + 13.667 = ______________
2. 2471 - 238.8642 = ______________
3. 0.491 × 0.300 = ______________
4. 13,225 ÷ 115 = ______________

Working with Fractions

There are different ways to work with fractions. The easiest way for the electrician is to change all fractions into decimals.

What is the decimal equivalent of $\frac{1}{2}$?

What is the decimal equivalent of $\frac{1}{4}$?

What is the decimal equivalent of $\frac{1}{10}$?

These figures are fractions that can quickly be changed into decimals. From these numbers a formula can be made to change other fractions into decimals.

To change a fraction into a decimal, divide the numerator (top number) by the denominator (bottom number).

Top Number divided by Bottom Number = Decimal Equivalent

EXAMPLES

$1/2$ = 0.5 (or 1 divided by 2 = 0.5)

$1/4$ = 0.25 (or 1 divided by 4 = 0.25)

$1/10$ = 0.1 (or 1 divided by 10 = 0.1)

EXAMPLES

$1/3 + 3/8$ = _____

Step 1. In each fraction, **divide** the top number by the bottom number.

$1/3 = 1 \div 3 = 0.3333$

$3/8 = 3 \div 8 = 0.375$

Step 2. Now **add** the decimals.

$0.3333 + 0.375 = 0.7083$

$1/8 \times 1/4$ = _____

Step 1. In each fraction, **divide** the top number by the bottom number.

$1/8 = 1 \div 8 = 0.125$

$1/4 = 1 \div 4 = 0.25$

Step 2. Now **multiply** the decimals.

$0.125 \times 0.25 = 0.0313$

$5/32 - 1/16$ = _____

Step 1. In each fraction, **divide** the top number by the bottom number.

$5/32 = 5 \div 32 = 0.1563$

$1/16 = 1 \div 16 = 0.0625$

Step 2. Now **subtract** the decimals.

$0.1563 - 0.0625 = 0.0938$

$5/4 \div 4/5$ = _____

Step 1. In each fraction, **divide** the top number by the bottom number.

$5/4 = 5 \div 4 = 1.25$

$4/5 = 4 \div 5 = 0.8$

Step 2. Now **divide** the decimals.

$1.25 \div 0.8 = 1.5625$

Working with Percentages

Percent (%) means per hundred. Many electrical examinations require a passing score of 70 percent, which means that 70 of 100 questions have to be answered correctly. Percentages are found easily with the aid of a calculator and by understanding a few basic math skills. As with fractions, a percent figure must first be changed to decimal form to perform mathematical operations. Although the percent key on a calculator performs this change automatically, it is important to have a working knowledge of these operations.

Changing a Percentage to a Decimal

Simply position the decimal point two places to the left and drop the percentage sign (%).

EXAMPLES

25% = 0.25

80% = 0.80 or 0.8

250% = 2.50 or 2.5

Changing a Decimal to a Percentage

Simply move the decimal point two places to the right and add the percentage sign (%).

EXAMPLES

0.30 = 30%

0.75 = 75%

1.25 = 125%

Increasing a Number by a Percentage

EXAMPLE

Increase 25 by 10 percent.

Step 1. Change 10% to a decimal.

10% = 0.10

Step 2. Multiply the numbers.

$25 \times 0.10 = 2.5$

Step 3. Add the answer to the original number.

$25 + 2.5 = 27.5$

Decreasing a Number by a Percentage

EXAMPLE

Decrease 25 by 10 percent.

Step 1. Change 10% to a decimal.

10% = 0.10

Step 2. Multiply the numbers.

$25 \times 0.10 = 2.5$

Step 3. Subtract the answer from the original number.

$25 - 2.5 = 22.5$

Working with Squares and Square Roots

Squaring a Number

The square of a number is the product of a number multiplied by itself.

EXAMPLES

$3^2 = 3 \times 3 = 9$

$12^2 = 12 \times 12 = 144$

$115^2 = 115 \times 115 = 13{,}225$

To square a number on most calculators, simply enter the number, then press the × key. Finally, press the = key. To square the number 3, press 3, press ×, then press =. The answer is 9. Some calculators have a function key, x^2, that performs this operation automatically. To square the number 3 with this type of calculator, simply press 3 and then press the x^2 key.

NOTE

A common mistake in squaring a number is to multiply the number by 2. This is incorrect except for 2^2, which is 2×2.

EXAMPLES

3^2 is *not* 3×2

12^2 is *not* 12×2

115^2 is *not* 115×2

Finding the Square Root of a Number

To find the square root of a number, look for a product of that number that, when multiplied by itself, matches the original number under the square root sign, $\sqrt{\ }$.

EXAMPLES

$\sqrt{9} = 3$ because $3 \times 3 = 9$

$\sqrt{100} = 10$ because $10 \times 10 = 100$

The best way to find the square root of any problem is to use a calculator with a square root key, $\sqrt{\ }$.

NOTE

In the electrical trade, there is one square root that should be memorized:

$\sqrt{3} = 1.732$

This number will be used in most three-phase calculations.

Additional Helpful Facts

There are different ways to express numbers that need to be multiplied.

EXAMPLE

$2 \times 4 \times 6$ can also be expressed as

$(2)(4)(6)$

Letters that are used in formulas have to be replaced by actual numbers. When letters are side by side they are to be multiplied.

EXAMPLES

$IR = I \times R$

$2KLI = 2 \times K \times L \times I$

Numbers or letters in parentheses or brackets must be calculated first.

EXAMPLE

12 × 3 × (1.5 × 8) = _____

Step 1. Multiply the numbers in the parentheses.

(1.5 × 8) = 12

Step 2. Replace the numbers in the parentheses with the total.

12 × 3 × **12** = _____

Step 3. Finish the problem.

12 × 3 × 12 = 432

Continue working with these basic math skills until you are comfortable with all applications. Remember, exam writers are looking for the most correct answer.

NOTE

In some instances there may be more than one correct choice, but always choose the *most correct* answer.

EXAMPLE

The volt-amps of a 10-hp, 230-volt, 3-phase motor are closest to _____.

a) 11,150 **b)** 10,000 **c)** 7460 **d)** 6440

Step 1. Select the correct Ohm's law formula.

VA = E × I.

Step 2. Find the motor's full load current from *NEC*® Table 430.250.

28 amperes.

Step 3. Solve the problem.

230 × 1.732 × 28 = 11,154.

Step 4. Choose a) 11,150, because it is closest to your answer.

Practical Applications

Taking one step at a time makes electrical formulas easier to calculate.

EXAMPLE

What is the resistance of a 1440-watt load that is drawing 12 amperes?

Step 1. Find the correct Ohm's law formula that solves for resistance (R).

$R = W \div I^2$

Step 2. Replace the letters with the given numbers.

$R = 1440 \div 12^2$

Step 3. Find the square of 12 and put it in the equation.

R = 1440 ÷ (12 × 12)

R = 1440 ÷ 144

Step 4. Solve for R (resistance) by finishing the calculation.

R = 10 ohms

When working with square roots, first calculate the numbers under the sign, and then find the square root.

EXAMPLE

What is the voltage of a circuit that has a resistance of 132.25 ohms and a total wattage of 100 watts?

Step 1. Find the correct Ohm's law formula that solves for voltage (E).

$$E = \sqrt{W \times R}$$

Step 2. Replace the letters with the given numbers.

$$E = \sqrt{100 \times 132.25}$$

Step 3. Multiply the numbers under the square root sign.

$$E = \sqrt{13{,}225}$$

Step 4. Find the square root of 13,225.

E = 115 volts

An exam question may ask for the neutral current on a three-phase, 4-wire, wye-connected panel. This can be done by using a formula and a few known factors. The phase current of each line is needed to calculate the neutral current. The formula containing a square root is shown below.

I_A, I_B, and I_C represent the current draw of each phase. It may also be expressed as the current draw of L_1 (Line 1), L_2, and L_3.

$$I_N = \sqrt{I_A^2 + I_B^2 + I_C^2 - (I_A \times I_B) - (I_B \times I_C) - (I_A \times I_C)}$$

Even this calculation can be solved easily by doing *one step at a time.*

EXAMPLE

What is the unbalanced neutral current of a 100-amp three-phase, 4-wire, wye-connected load? The load on Phase A is 40 amps, the load on Phase B is 40 amps, and the load on Phase C is 20 amps.

Step 1. Replace I_A, I_B, and I_C with the currents given in the problem.

$$I_N = \sqrt{40^2 + 40^2 + 20^2 - (40 \times 40) - (40 \times 20) - (40 \times 20)}$$

Step 2. Square the current draw of Phases A, B, and C.

$I_A^2 = 40^2 = 40 \times 40 = 1600$

$I_B^2 = 40^2 = 40 \times 40 = 1600$

$I_C^2 = 20^2 = 20 \times 20 = 400$

Step 3. Multiply the numbers in the parentheses.

$I_A \times I_B = 40 \times 40 = 1600$

$I_B \times I_C = 40 \times 20 = 800$

$I_A \times I_C = 40 \times 20 = 800$

Step 4. Add and subtract the numbers under the square root sign.

$$I_N = \sqrt{1600 + 1600 + 400 - 1600 - 800 - 800}$$

$$I_N = \sqrt{400}$$

Step 5. Solve for the square root of 400.

$$I_N = \sqrt{400} = 20 \text{ amps}$$

The unbalanced current on the neutral is 20 amperes.

END-OF-CHAPTER PROBLEMS

1. 4500 ÷ 18.75 = _______
   a) 2.4
   b) 24
   c) 240
   d) 84,375
2. 746 × 25 = _______
   a) 29.84
   b) 186.5
   c) 1865
   d) 18,650
3. 230 × 230 ÷ 10.58 = _______
   a) 21.74
   b) 50
   c) 5000
   d) 52,900
4. 80% of 20 = _______
   a) 4
   b) 16
   c) 25
   d) 1600
5. The decimal equivalency of ½ = _______
   a) 0.25
   b) 0.5
   c) 2
   d) 50
6. 15 × 80% = _______
   a) 12
   b) 18.75
   c) 120
   d) 1200
7. Increase 8000 by 10% = _______
   a) 800
   b) 7200
   c) 8000
   d) 8800
8. 3% of 120 = _______
   a) 3.6
   b) 4
   c) 36
   d) 40

9. $13{,}225 \div 115 =$ ________

a) 11.5

b) 115

c) 1150

d) 11,500

10. $\dfrac{2 \times 12.9 \times 100 \times 10.8}{3.6} =$ ________

a) 774

b) 7740

c) 27,864

d) 77,400

11. $10^2 =$ ________

a) 10

b) 20

c) 100

d) 1000

12. $0.491 \times 0.300 =$ ________

a) 0.1473

b) 1.473

c) 14.73

d) 147.3

13. $\dfrac{26{,}240 \times 0.491}{1000} =$ ________

a) 0.1288384

b) 12.88384

c) 1288.384

d) 1,288,384

14. $40 \times 0.90 \times 0.80 \times 80\% =$ ________

a) 23.04

b) 28.8

c) 36

d) 2304

15. $\dfrac{12 \times 125\%}{0.75 \times 0.80} =$ ________

a) 9

b) 15

c) 20

d) 25

Ohm's Law

CHAPTER 2

Ohm's law is the relationship between voltage (E), current (I), and resistance (R). The formula for Ohm's law is I = E ÷ R. Below are the Ohm's law definitions.

Definitions

Volt

E is the pressure required to force one ampere through a resistance of one ohm. The force or pressure necessary to cause electrons to flow through a conductor is known as electromotive force (EMF) or voltage. The basic unit of measurement in electricity is the volt. Voltage may be expressed by any of the following terms or symbols:

- E (volt)
- EMF (electromotive force)
- RMS (root mean squared)
- Difference of potential
- Effective voltage

Ampere

I refers to the electric current that will flow through one ohm under a pressure of one volt, the intensity of the flow of electrons through a conductor. Current is measured in amperes and may be expressed by either of the following terms or symbols:

- I (intensity)
- A (ampere or amp)

Resistance

R refers to the opposition that a device or material offers to the flow of current. Resistance depends on four factors: material, length, temperature, and cross-sectional area. Resistance is measured in ohms and may be expressed by either of the following terms or symbols:

- R (resistance)
- Ω (ohms)

Power

W refers to the rate at which electrical energy is delivered and consumed. Power is measured in watts. The *NEC* expresses power as VA or volt-amps. Power may be expressed by any of the following terms or symbols:

- P (power)
- W (watts)
- VA (volt-amps)
- kW (kilowatts)
- kVA (kilo-volt-amps)

Proportionality

Another property of Ohm's law states that the current that flows in a circuit is directly proportional to the applied voltage of that circuit. The current is inversely proportional to the resistance of that circuit.

If the voltage is increased and the resistance of a circuit remains the same, then the current will be increased proportionally.

If the resistance is increased and the voltage of a circuit remains the same, then the current will be decreased proportionally.

EXAMPLE

A circuit has a voltage of 120 volts, a current draw of 10 amperes, and a resistance of 12 ohms.

If the voltage in this circuit is doubled to 240 volts and the resistance remains at 12 ohms, the current will double to 20 amperes.

If the resistance of the circuit is doubled to 24 ohms and the voltage remains at 120 volts, the current will be cut in half to 5 amperes.

Use the following chart to find a formula for any value of the terms discussed in Ohm's law. Formulas in the chart are expressed with a power factor of 100% (unity).

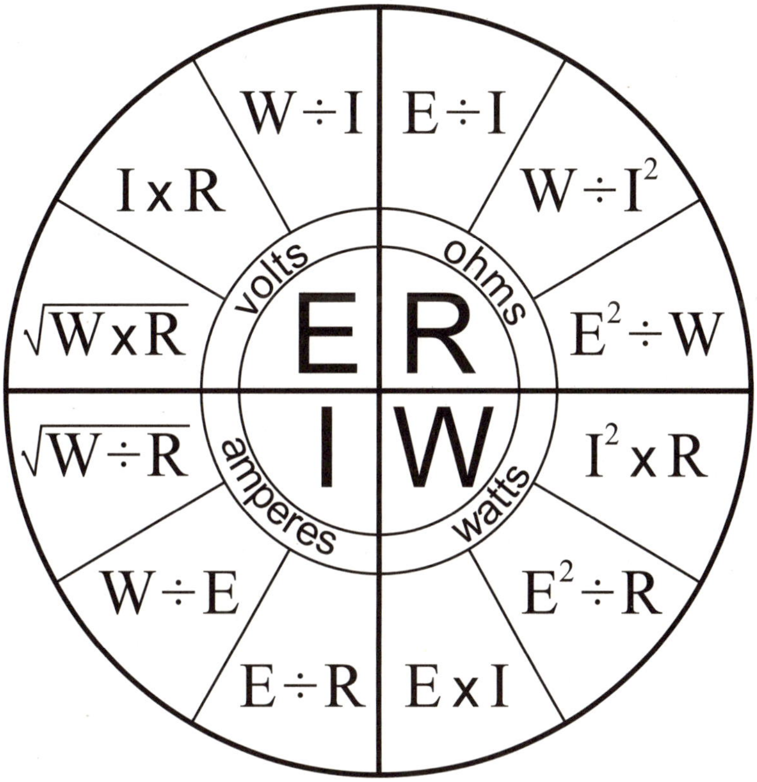

FIGURE 2-1

When given the values of any two factors, a third value may be calculated from a formula derived from the Ohm's law chart.

Understanding the question is part of the problem when taking examinations. Read the question carefully and determine exactly what the question is asking. Find all of the information necessary and solve the question.

A scratch sheet of paper, used for calculations, should contain the entire math necessary to solve the question. The question number should be written on the scratch sheet beside the information used in finding the answer. At the end of the test, if there is time remaining, all questions should be double-checked to be sure the answer is correct. Double-checking the answers would be very difficult if the scratch sheet really looked like a scratched-up sheet of paper!

Before choosing the answer, read the question again to be sure the answer chosen is exactly what the question is asking. For example, the question could be asking for the neutral conductor instead of the service conductor. Or, a question may be asking for kW instead of wattage, but both answers are given.

EXAMPLE

A 4 AWG TW conductor is good for how many kW at 120 volts?

a) 8.4 b) 70 c) 120 d) 8400

Step 1. The question is asking for kilowatts (1000 watts). A formula from the Ohm's law chart is needed to solve for wattage.

$W = E \times I$

Step 2. Voltage and current can be used to solve for wattage. Voltage is directly given in the question but current is not. The allowable ampacity of a 4 AWG TW conductor can be found in Table 310.16 of the *National Electrical Code*. A 4 AWG TW conductor is good for 70 amperes.

$W = E \times I = 120 \times 70 = 8400$ watts

Step 3. Answer d) is 8400 watts, **but the question is asking for kW** (kilowatts). 8400 kW is actually 8,400,000 watts.

Step 4. The 8400 watts must be divided by 1000 to solve for kW.

$8400 \div 1000 = 8.4$ kW

The correct answer is a).

The formula for power in the Ohm's law chart is based on a single-phase circuit with unity power factor (PF).

On exams, all line voltages and line currents will be calculated with a power factor of 100% (unity) unless otherwise specified.

Power factor is calculated only in ac circuits, not in dc circuits.

EXAMPLES

What is the current draw of an incandescent light bulb that has a rating of 100 watts at 115 volts? The light bulb is connected to a source voltage of **115 volts.**

Step 1. Select the correct Ohm's law formula.

$I = W \div E$

Step 2. Insert the known factors and solve the equation.

$I = 100 \div 115$

$I = 0.8696$ amperes

What is the current draw of an incandescent light bulb that has a rating of 100 watts at 115 volts? The light bulb is connected to a source voltage of **120 volts.**

Step 1. Find the resistance of the light bulb.

$R = E^2 \div W$ (Use the voltage rating of the bulb, not the source voltage.)

$R = 115 \times 115 \div 100$

$R = 132.25$ ohms

Step 2. Find the current by using the resistance of the light bulb and the source voltage.

$I = E \div R$

$I = 120 \div 132.25$

$I = 0.9074$ amperes

The resistance of the light bulb is the only factor that remains the same when the voltage is changed. The current will change proportionally to the change in voltage. Because the voltage source is different from the rating, the wattage will also be different.

How much power is being consumed in the problem above?

Step 1. Select the correct Ohm's law formula.

$W = E^2 \div R$

Step 2. Insert the known factors and solve the equation.

$W = 120 \times 120 \div 132.25$

$W = 108.8847$ watts

Alternative Ohm's Law Wheels

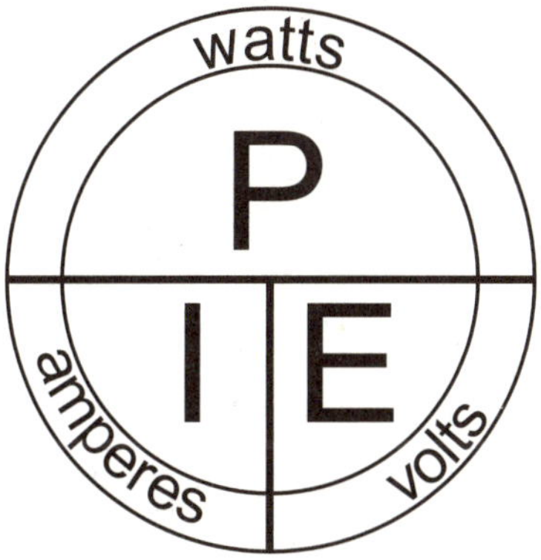

FIGURE 2-2

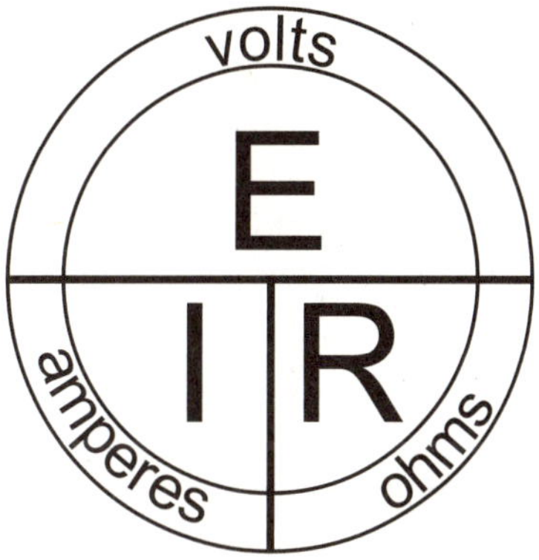

FIGURE 2-3

NOTE

The W has been replaced by P to help in memorizing it as the PIE chart.

The PIE and EIR charts shown above may be easier to remember than the Ohm's law wheel containing twelve formulas. By using one or a combination of both wheels, most of the unknown factors can be calculated from the two known factors. To find the formula, simply cover the unknown factor. Multiply the known factors if the two remaining letters are side by side. Divide the top known factor by the bottom known factor if one letter is above and one below.

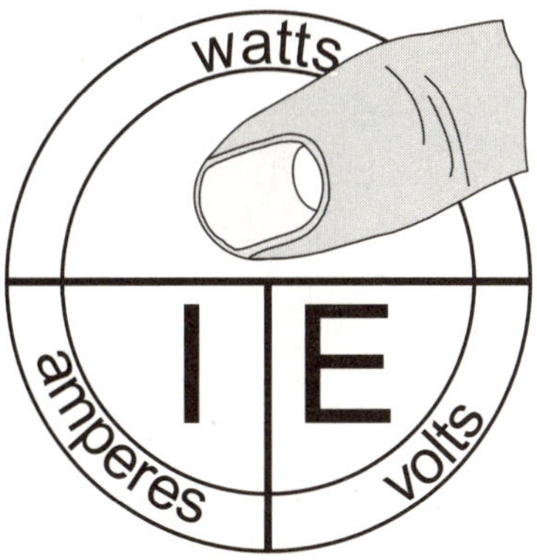

I x E

FIGURE 2-4A

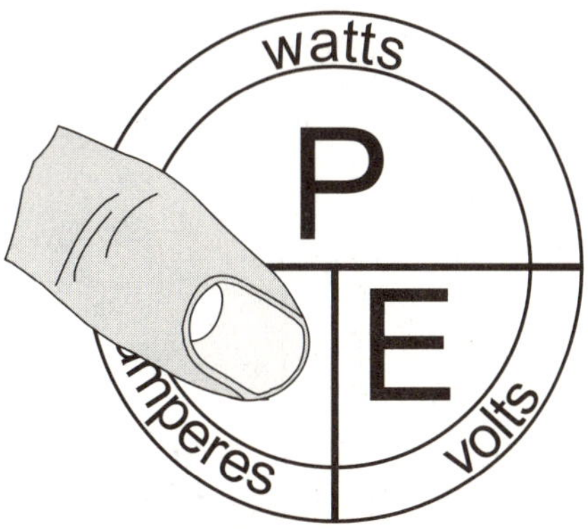

W ÷ E

FIGURE 2-4B

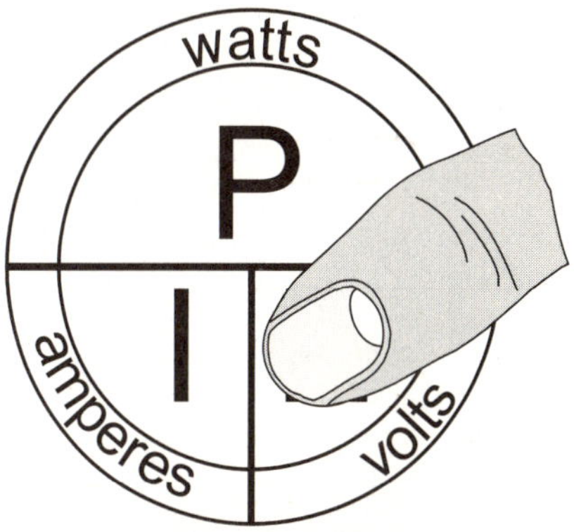

W ÷ I

FIGURE 2-4C

EXAMPLE

How much power is being consumed by a 120-volt circuit that draws 10 amperes?

Step 1. As watts are the unknown factor, cover the letter P. As the two uncovered letters are side by side, the known factors must be multiplied.

P = I × E

Step 2. Change the P to a W.

W = I × E

Step 3. Insert the two known factors and solve the equation.

W = I × E

Watts = Amps × Volts

W = 10 × 120 = 1200 watts

A circuit drawing 10 amperes at 120 volts consumes 1200 watts.

FIGURE 2-5A

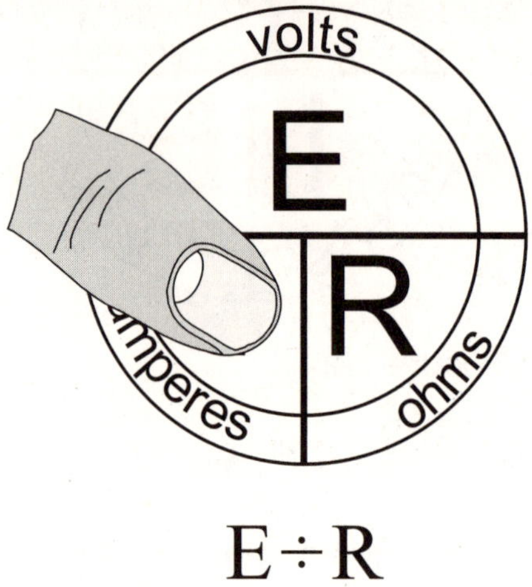

FIGURE 2-5B

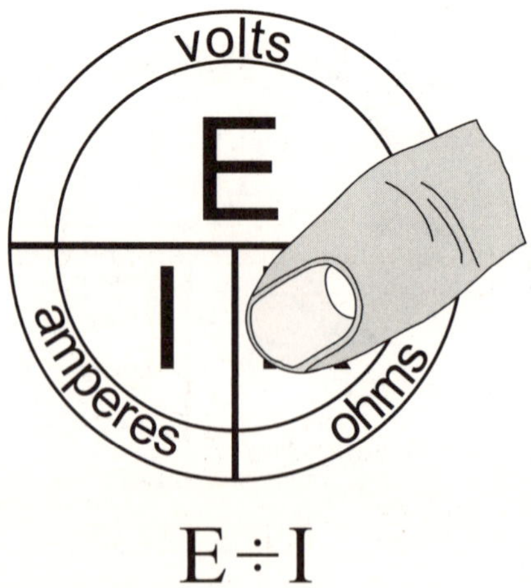

FIGURE 2-5C

EXAMPLE

If the current is 10 amperes and the resistance is 12 ohms, what is the voltage?

Step 1. Cover the unknown (voltage) factor and multiply the two known factors.

$E = I \times R$

Volts = Amps × Ohms

$E = 10 \times 12 = 120$ volts

If the current is 10 amperes and the resistance is 12 ohms, the voltage will be 120 volts.

Each of the smaller wheels provides three formulas. By using a combination of both wheels, additional unknown factors can be calculated.

EXAMPLE

What is the resistance of an electric heater rated 1200 watts at 120 volts?

Step 1. Although the question wants the resistance, start by finding current with the PIE chart. Divide watts by volts.

$$I = \frac{W}{E}$$

$$\text{Amps} = \frac{\text{Watts}}{\text{Volts}}$$

$$I = \frac{1200}{120} = 10 \text{ amps}$$

Step 2. Now that current and volts are known, solve for resistance. Divide volts by amps.

$$R = \frac{E}{I}$$

$$\text{Ohms} = \frac{\text{Volts}}{\text{Amps}}$$

$$R = \frac{120}{10} = 12 \text{ ohms}$$

The resistance of an electric heater rated 1200 watts at 120 volts is 12 ohms.

Series Circuits

In a series circuit, current flows from a voltage source through every part of the electrical device in a single path before returning to the voltage source.

The properties of a series circuit are

- Resistance is additive.
- Current remains the same throughout the circuit.
- Voltage is additive.
- Voltage source is the sum of the voltage drop of all resistors.
- Power is additive.

The formula to find the total resistance in a series circuit is

$$R_T = R_1 + R_2 + R_3$$

EXAMPLE

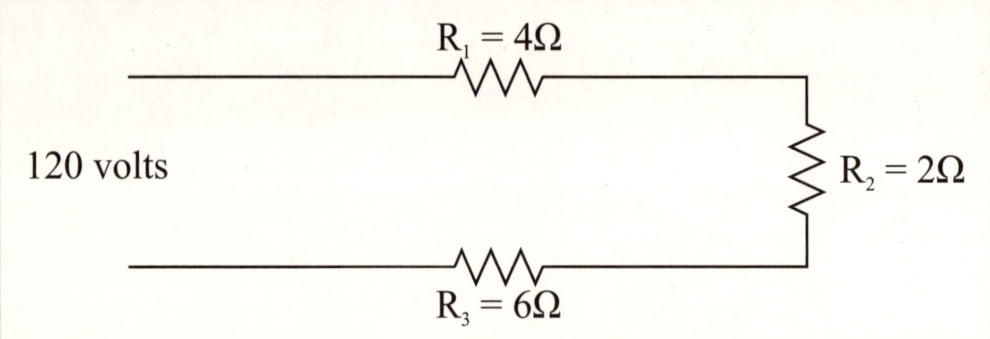

FIGURE 2-6

Resistance is additive:

$R_T = R_1 + R_2 + R_3$

$R_T = 4 + 2 + 6$

$R_T = 12$ ohms

The current remains the same:

$I_T = E \div R$

$I_T = 120 \div 12$

$I_T = 10$ amperes

Voltage is additive:

$E = I \times R$

Voltage drop across R_1 is $E_{Drop} = 10 \times 4 = 40$ volts

Voltage drop across R_2 is $E_{Drop} = 10 \times 2 = 20$ volts

Voltage drop across R_3 is $E_{Drop} = 10 \times 6 = 60$ volts

Voltage source is the sum of the voltage drop of all resistors:

$E_{Total} = R_1$ volts $+ R_2$ volts $+ R_3$ volts

$E_{Total} = 40 + 20 + 60$

$E_{Total} = 120$ volts

Power is additive:

$W = E \times I$

Power consumed by R_1 is $W = 40 \times 10 = 400$ watts

Power consumed by R_2 is $W = 20 \times 10 = 200$ watts

Power consumed by R_3 is $W = 60 \times 10 = 600$ watts

$W_{Total} = R_1$ watts $+ R_2$ watts $+ R_3$ watts

$W_{Total} = 400 + 200 + 600$

$W_{Total} = 1200$ watts

PROBLEMS

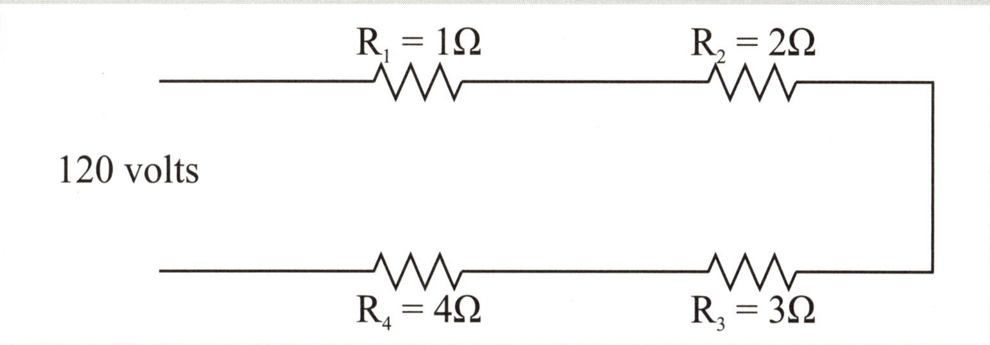

FIGURE 2-7

1. The total resistance is ______________ ohm(s).
2. The total current is ______________ ampere(s).
3. The voltage drop across R_1 is ______________ volt(s).
4. The voltage drop across R_2 is ______________ volt(s).
5. The voltage drop across R_3 is ______________ volt(s).
6. The voltage drop across R_4 is ______________ volt(s).
7. The power consumed by R_1 is ______________ watt(s).
8. The power consumed by R_2 is ______________ watt(s).
9. The power consumed by R_3 is ______________ watt(s).
10. The power consumed by R_4 is ______________ watt(s).
11. The total wattage of the circuit is ______________ watt(s).

Parallel Circuits

In a parallel circuit, part of the current flows from the voltage source through every resistor. The total current is the sum of the currents from each resistor. The same voltage from the voltage source flows through each resistor equally.

The properties of a parallel circuit are

- The total resistance is less than any one resistor.
- Current is additive.
- Voltage remains the same throughout the circuit.
- Voltage drop is the same as the voltage source in all resistors.
- Power is additive.

In a series circuit, if more resistors are added, the total resistance will increase. In a parallel circuit, if more resistors are added, the total resistance will decrease.

The formula to find the total resistance in a parallel circuit is

$$R_T = \frac{1}{\frac{1}{R_1} + \frac{1}{R_2} + \frac{1}{R_3}}$$

EXAMPLE

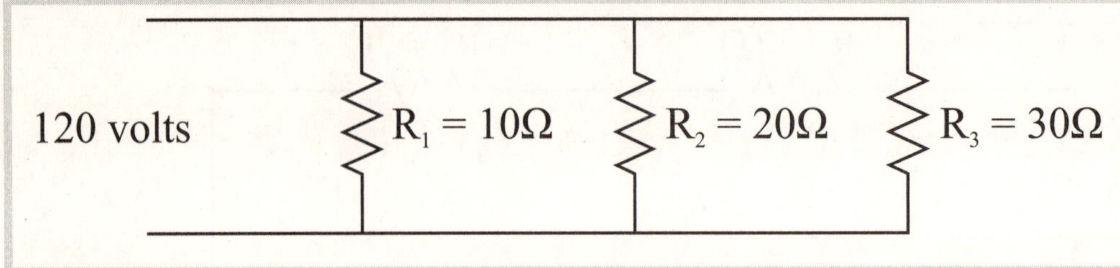

FIGURE 2-8

NOTE

In a parallel circuit, the total resistance is *always* less than the smallest resistor.

The total resistance will be less than any one resistor:

$$R_T = \frac{1}{\frac{1}{R_1}+\frac{1}{R_2}+\frac{1}{R_3}} = \frac{1}{0.1 + 0.05 + 0.033333} = \frac{1}{0.183333}$$

$$R_T = \frac{1}{0.183333} = 5.45456$$

$R_T = 5.45456$ ohms

Current is additive:

Current through R_1 is $I = 120 \div 10 = 12$ amperes
Current through R_2 is $I = 120 \div 20 = 6$ amperes
Current through R_3 is $I = 120 \div 30 = 4$ amperes
$I_{Total} = I_1 + I_2 + I_3$
$I_{Total} = 12 + 6 + 4$
$I_{Total} = 22$ amperes

Voltage remains the same through each resistor:

Voltage drop = voltage source
Voltage drop across R_1 is $E_{Drop} = 120$ volts
Voltage drop across R_2 is $E_{Drop} = 120$ volts
Voltage drop across R_3 is $E_{Drop} = 120$ volts

Power is additive:

$W = E \times I$
Power consumed by R_1 is $W = 120 \times 12 = 1440$ watts
Power consumed by R_2 is $W = 120 \times 6 = 720$ watts
Power consumed by R_3 is $W = 120 \times 4 = 480$ watts
$W_{Total} = R_1$ watts $+ R_2$ watts $+ R_3$ watts
$W_{Total} = 1440 + 720 + 480$
$W_{Total} = 2640$ watts

When all resistors in a parallel circuit are of equal value, the resistance of one resistor can be divided by the number of resistors to find the total resistance of the circuit.

EXAMPLE

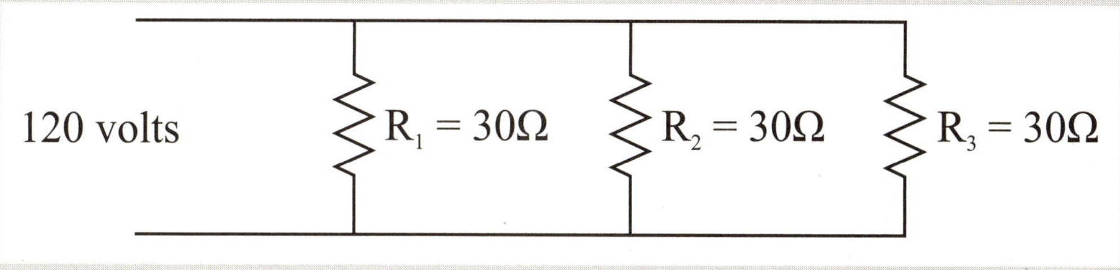

FIGURE 2-9

Divide the resistance of one resistor by the number of resistors.

$R_T = 30 \div 3$

$R_T = 10$ ohms

PROBLEMS

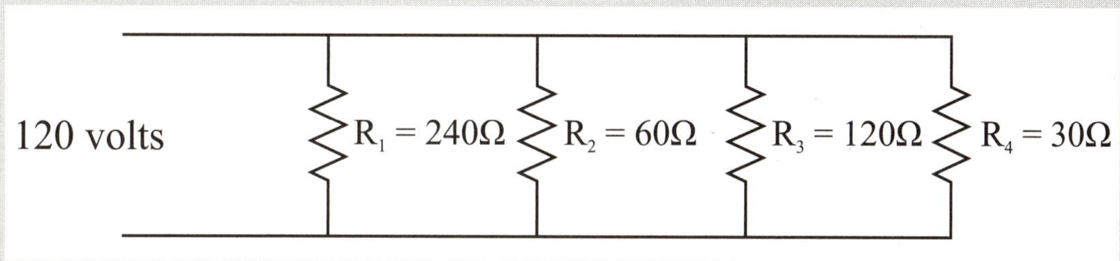

FIGURE 2-10

1. The total resistance is ______________ ohm(s).
2. The current draw of R_1 is ______________ ampere(s).
3. The current draw of R_2 is ______________ ampere(s).
4. The current draw of R_3 is ______________ ampere(s).
5. The current draw of R_4 is ______________ ampere(s).
6. The total current draw of this circuit is ______________ ampere(s).
7. The power consumed by R_1 is ______________ watt(s).
8. The power consumed by R_2 is ______________ watt(s).
9. The power consumed by R_3 is ______________ watt(s).
10. The power consumed by R_4 is ______________ watt(s).
11. The total wattage of the circuit is ______________ watt(s).

Alternative Parallel Resistance Method

There is an alternative method for finding the total resistance in a parallel circuit. Instead of calculating the total resistance first, calculate the total current in the circuit. Then, use Ohm's law to find the total resistance.

EXAMPLE

What is the total resistance in the drawing below?

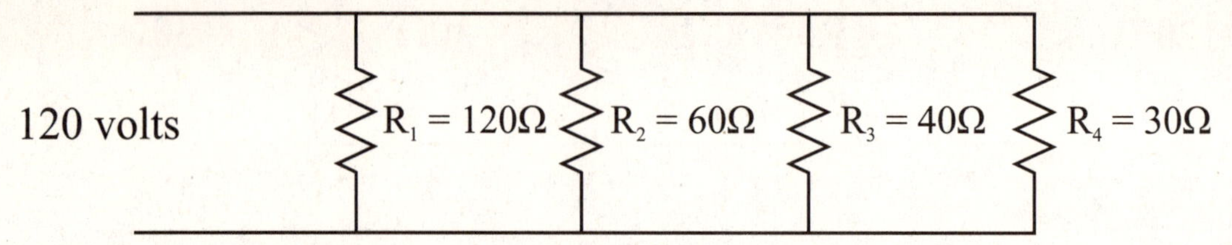

FIGURE 2-11

Step 1. To find the total resistance, start by finding the total current.
What is the current through R_1?

$$\text{Amps} = \frac{\text{Volts}}{\text{Ohms}}$$

$$I = \frac{E}{R}$$

$$I = \frac{120}{120} = 1 \text{ amp}$$

What is the current through R_2?

$$\text{Amps} = \frac{\text{Volts}}{\text{Ohms}}$$

$$I = \frac{E}{R}$$

$$I = \frac{120}{60} = 2 \text{ amps}$$

What is the current through R_3?

$$\text{Amps} = \frac{\text{Volts}}{\text{Ohms}}$$

$$I = \frac{E}{R}$$

$$I = \frac{120}{40} = 3 \text{ amps}$$

What is the current through R_4?

$$\text{Amps} = \frac{\text{Volts}}{\text{Ohms}}$$

$$I = \frac{E}{R}$$

$$I = \frac{120}{30} = 4 \text{ amps}$$

$I_T = I_1 + I_2 + I_3 + I_4 = 1 + 2 + 3 + 4 = 10$ amps. The total current is 10 amperes.

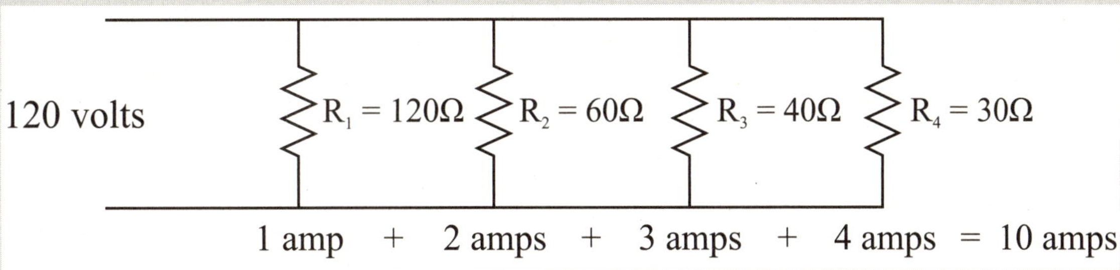

FIGURE 2-12

Step 2. Now that current and voltage are known, the total resistance can be calculated.

$$\text{Ohms} = \frac{\text{Volts}}{\text{Amps}}$$

$$R = \frac{E}{I}$$

$$R = \frac{120}{10} = 12 \text{ ohms}$$

Series-Parallel Circuits

A series-parallel circuit is a combination of at least one series circuit and at least one parallel circuit. Diagrams containing series-parallel circuits may look very complicated and impossible to solve, but they are simply a combination of small series and small parallel circuits. To find the total resistance of a series-parallel circuit, simply convert the circuit to a series circuit.

EXAMPLE

Find the total resistance of the circuit below.

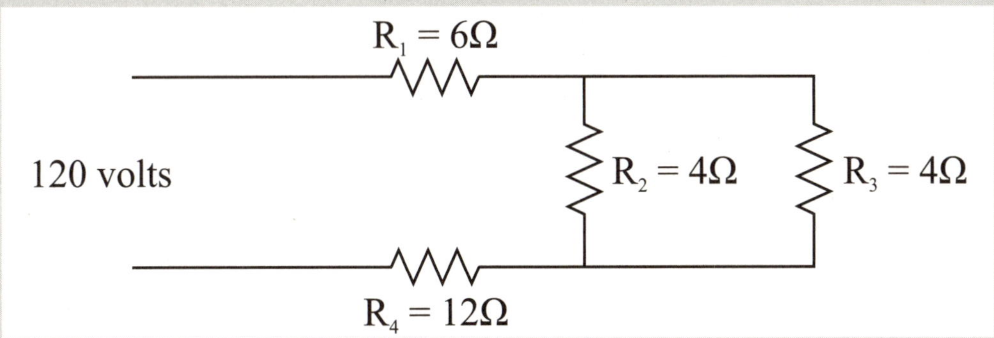

FIGURE 2-13

Step 1. R_2 and R_3 are in parallel with each other. The first step is to reduce a series-parallel circuit to a simple series circuit. Find the total resistance of R_2 and R_3.

$R_{2 \text{ and } 3} = 4 \div 2$ (both resistors are of equal value)

$R_{2 \text{ and } 3} = 2$ ohms

continues

continued

Step 2. This circuit can be redrawn to show a simple series circuit.

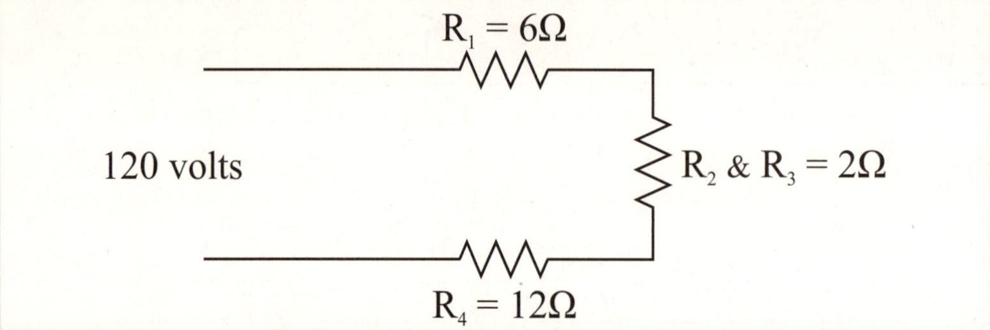

FIGURE 2-14

Step 3. The total resistance of this series circuit can be found by adding the three resistors.

$R_T = 6 + 2 + 12 = 20$

$R_T = 20$ ohms

Step 4. Now that the total resistance is known, the total current can be found.

$I = E \div R$

$I = 120 \div 20 = 6$

I = 6 amperes

Step 5. With any two of the three factors, the total wattage can be found.

$W = E \times I$

$W = 120 \times 6$

W = 720 watts

PROBLEMS

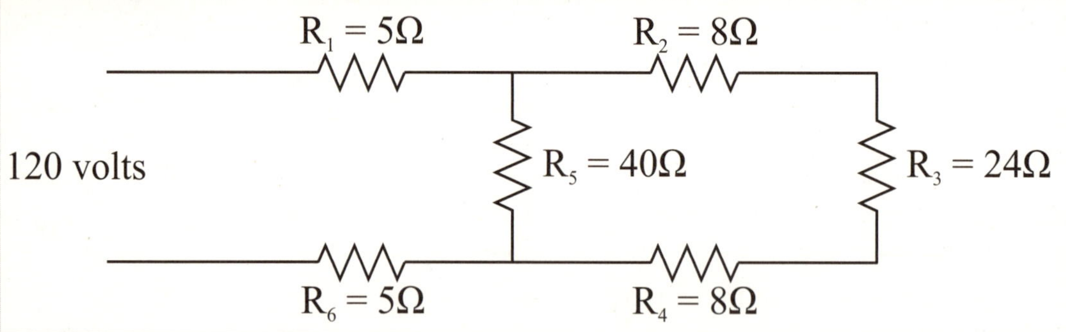

FIGURE 2-15

1. The total resistance is ______________ ohm(s).
2. The total current draw of the circuit is ______________ ampere(s).
3. The voltage drop across R_1 is ______________ volt(s).
4. The current draw across R_6 is ______________ ampere(s).
5. The voltage drop across R_6 is ______________ volt(s).
6. The total wattage of the circuit is ______________ watt(s).

END-OF-CHAPTER PROBLEMS

1. What is the voltage in the drawing below? The 20-ohm resistor has a current reading of 5 amperes.
   a) 120 volts
   b) 280 volts
   c) 480 volts
   d) None of these

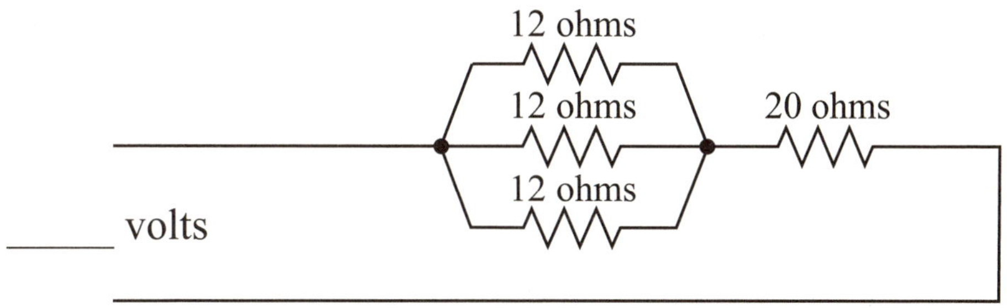

FIGURE 2-16

2. Which of the meters below are hooked up correctly?
   a) 1 only
   b) 2 only
   c) 3 only
   d) 1, 2, and 3

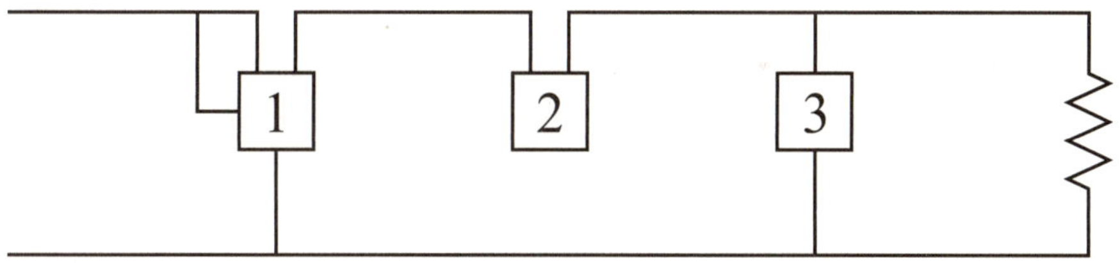

FIGURE 2-17

3. How much current is being drawn by the circuit below?
   a) 3 amperes
   b) 6 amperes
   c) 12 amperes
   d) 24 amperes

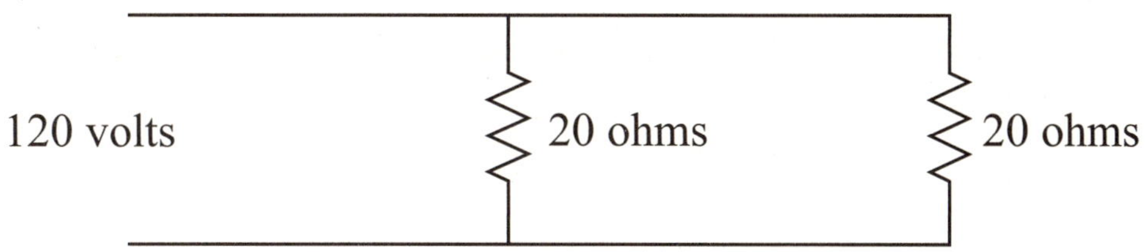

FIGURE 2-18

4. How much power is being consumed by the circuit below?

a) 103 watts

b) 206 watts

c) 240 watts

d) 420 watts

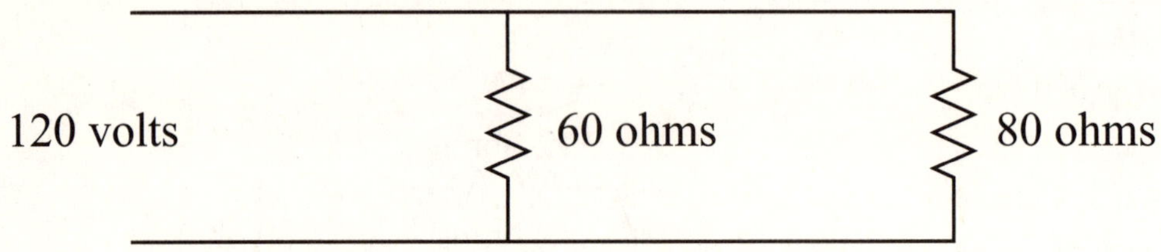

FIGURE 2-19

The drawing below is to be used with Questions 5 through 8.

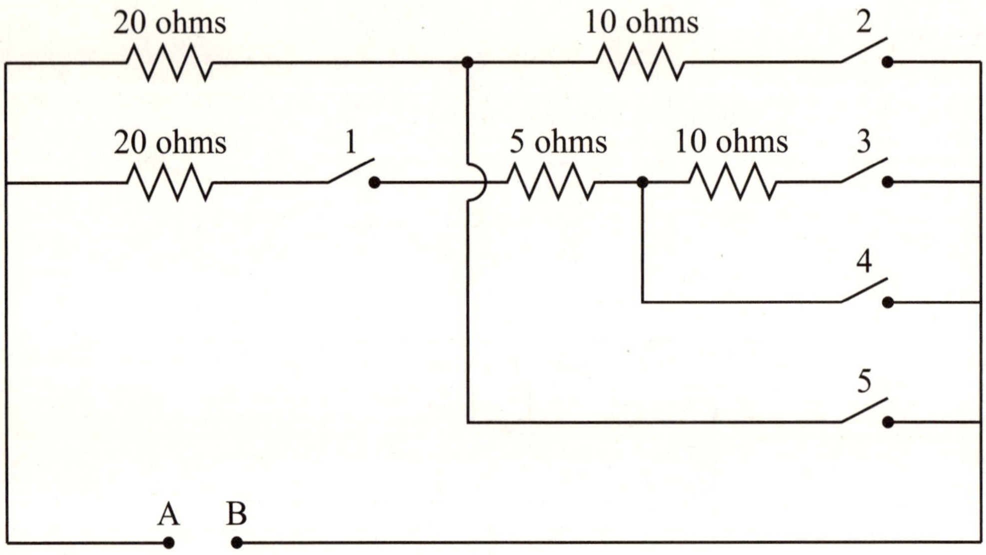

FIGURE 2-20

5. What is the voltage between A and B when switches 1 and 3 are closed? The 5-ohm resistor is drawing 4 amperes.

a) 60 volts

b) 100 volts

c) 120 volts

d) 140 volts

6. What is the resistance when switches 1 and 2 are closed?

a) 10 ohms

b) 20 ohms

c) 30 ohms

d) None of these

7. The resistance between A and B is lowest when switches ______ are closed.
   a) 3, 4, and 5
   b) 2, 3, and 4
   c) 2, 3, 4, and 5
   d) 1, 2, 3, 4, and 5
8. The resistance between A and B is lowest when switches ______ are open.
   a) 2 and 3
   b) 2 and 4
   c) 2, 3, 4, and 5
   d) 1, 2, 3, 4, and 5
9. The batteries below are 1½ volts each. What is the total voltage of the batteries?
   a) 1½ volts
   b) 6 volts
   c) 9 volts
   d) 12 volts

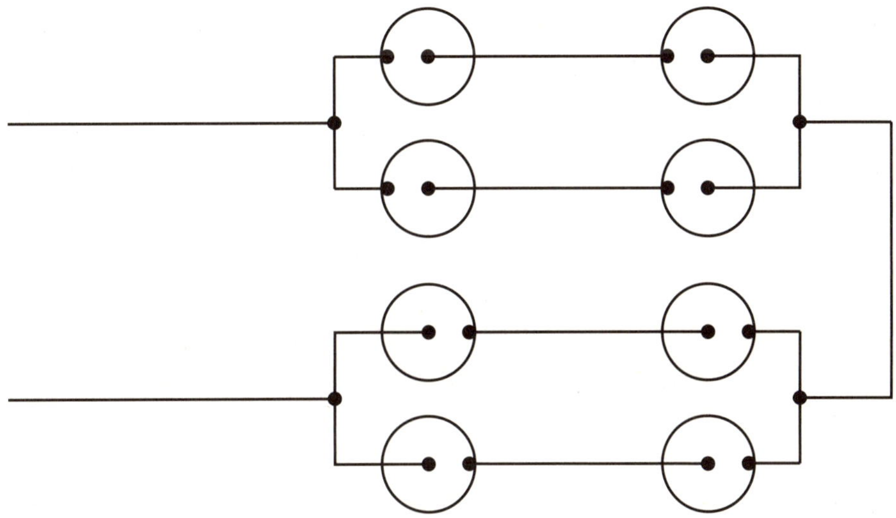

FIGURE 2-21

10. What is the resistance of two 500-kcmil uncoated copper conductors that are running in parallel? The total length of each conductor is 1000 feet.
    a) 0.0129 ohms
    b) 0.0258 ohms
    c) 0.0516 ohms
    d) None of these

11. What is the current of the circuit in the drawing below?
   a) 90 amperes
   b) 60 amperes
   c) 40 amperes
   d) 20 amperes

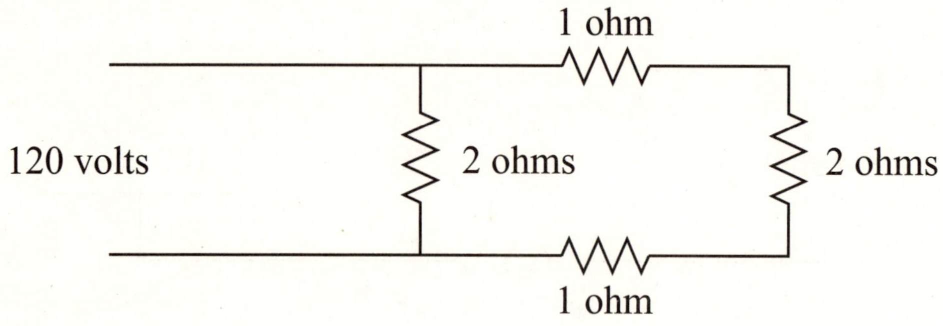

FIGURE 2-22

12. What is the resistance of a 6500-watt load at 115 volts?
   a) 0.02 ohms
   b) 2.03 ohms
   c) 56.52 ohms
   d) 113.04 ohms
13. What type of meter is represented by the drawing below?
   a) Watt meter
   b) Ammeter
   c) Voltmeter
   d) Ohmmeter

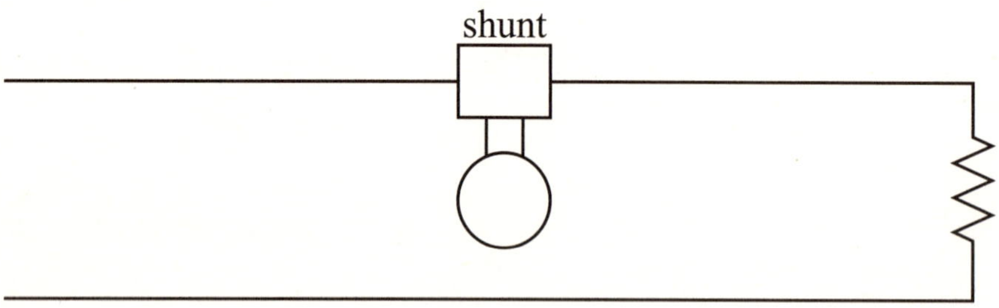

FIGURE 2-23

14. What amount of current is being drawn by four 150-watt, 120-volt heaters that are connected in series to a source voltage of 600 volts?
   a) 1 ampere
   b) 1.56 amperes
   c) 5 amperes
   d) 6.24 amperes

15. What is the voltage at the load?

a) 115 volts

b) 116.4 volts

c) 117.5 volts

d) 120 volts

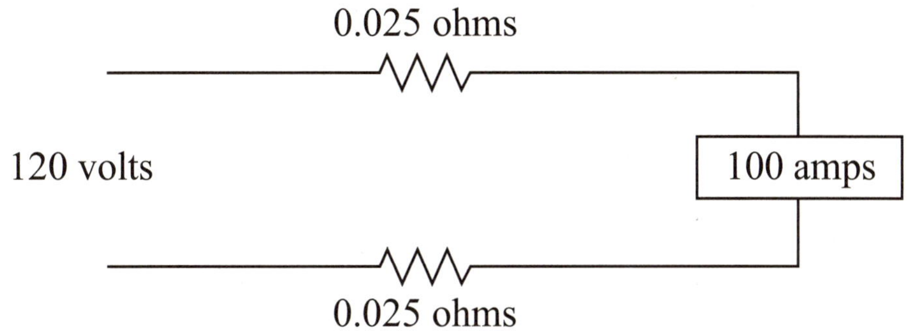

FIGURE 2-24

16. What is the amperage through the 2-ohm resistor?

a) 4 amperes

b) 3 amperes

c) 1 ampere

d) 0.4 ampere

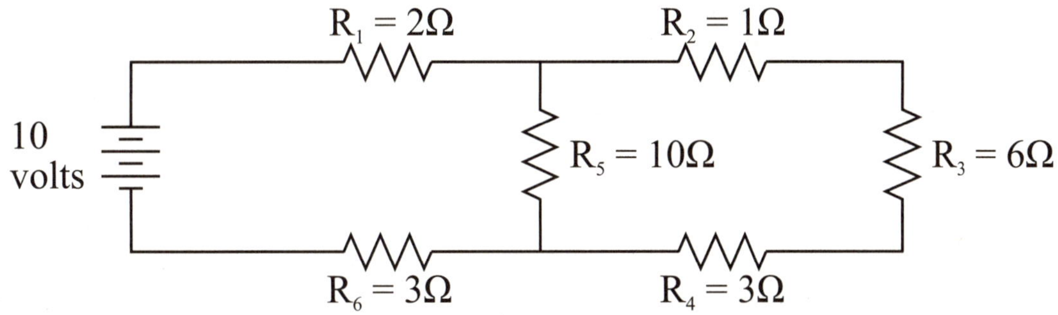

FIGURE 2-25

17. The circular mil area of a 12-AWG THHN, cu, uncoated, stranded conductor is

a) 6530

b) 1.98

c) 0.092

d) 0.030

18. If the load in the following drawing draws 120 amperes, the voltage across the load will be ______.

a) 114 volts

b) 115 volts

c) 117.5 volts

d) 120 volts

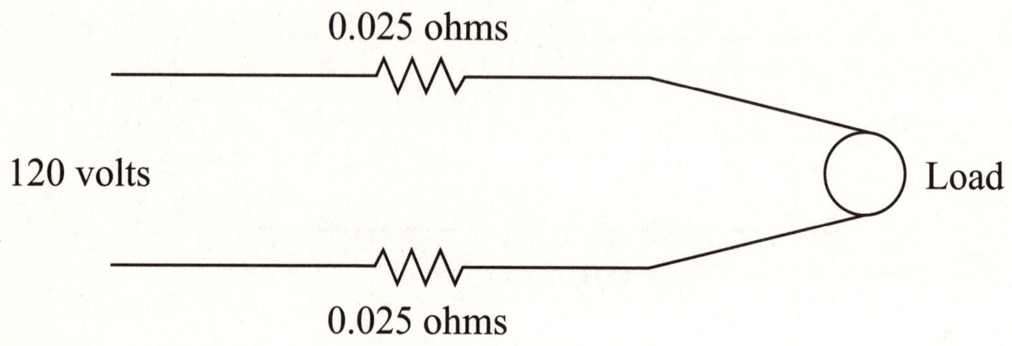

FIGURE 2-26

19. What is the resistance in the drawing below?

a) 12 ohms

b) 20 ohms

c) 22 ohms

d) 25 ohms

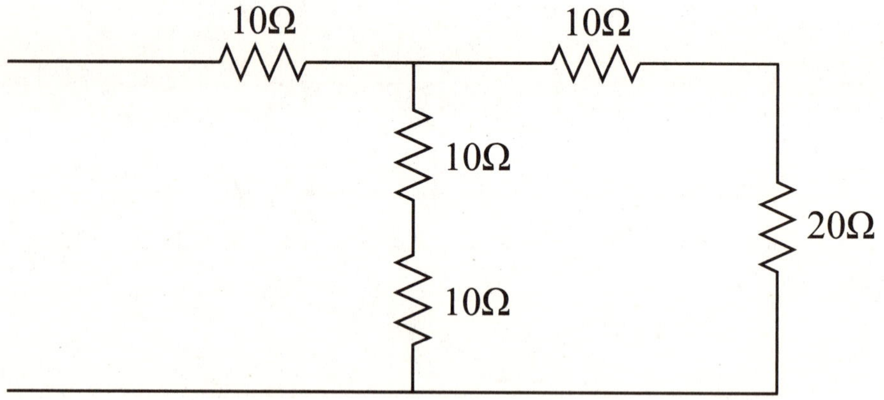

FIGURE 2-27

20. The voltage at the source is 120 volts. The voltage at the load is 115 volts. What is the current in the circuit? The load is 6500 watts.

a) 9.03 amperes

b) 18.06 amperes

c) 54.17 amperes

d) 56.52 amperes

21. Each of eight devices consumes 200 watts at 120 volts. What is the current draw of the connected load?
   a) 1.7 amperes
   b) 4.8 amperes
   c) 9 amperes
   d) 13.3 amperes
22. What is the resistance in the drawing below?
   a) 13.3 ohms
   b) 15 ohms
   c) 23.3 ohms
   d) 25 ohms

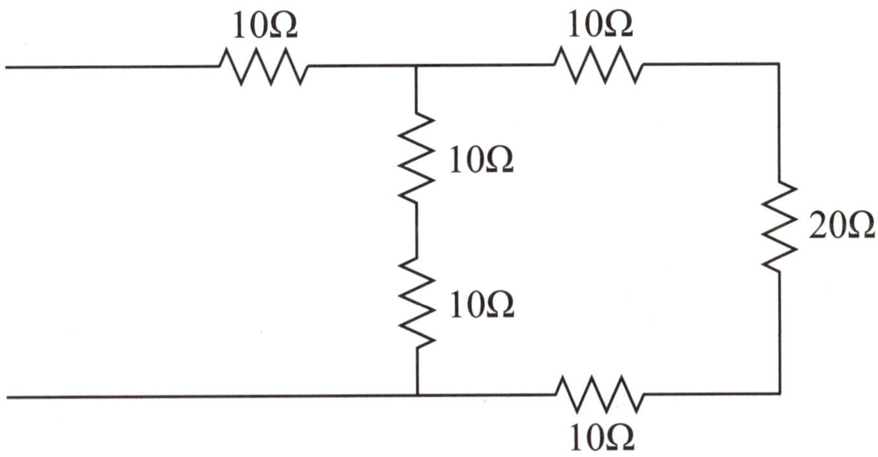

FIGURE 2-28

23. What is the power consumed by a load having a resistance of 40 ohms at 120 volts?
   a) 3 watts
   b) 360 watts
   c) 3600 watts
   d) 4800 watts

Neutral Conductors

CHAPTER
3

Neutral and Grounded Conductors Are Not the Same

The definition of a word, as found in the *NEC,* may be slightly different from the one the electrician uses in everyday conversation. An example of this is the word *neutral.* For most electricians, a neutral conductor is thought of as any white wire. The *NEC* has different meanings for a *neutral* conductor and a *grounded* conductor. All white conductors are not neutrals. A white or gray conductor has to be a grounded conductor but does not have to be a neutral conductor. A conductor with three continuous white stripes on other than green insulation is also permitted as a grounded circuit conductor. [200.7(A)]

Although neutral is not defined in the *NEC,* an insight can be gained from 310.15(B)(4). Certain neutral conductors can be omitted when applying the adjustment factors for more than three current-carrying conductors in a raceway or cable. Although neutral and grounded conductors are marked the same, grounded conductors must be counted as current-carrying conductors. A neutral conductor carries only the unbalanced current from other conductors in a circuit. [310.15(B)(4)(a)] A neutral can be in a single-phase 3-wire system or a three-phase 4-wire system.

An example of a neutral conductor is shown in the drawing below. There are two circuits that share a common grounded conductor. The voltage between the ungrounded conductors is 240 volts. The voltage between each ungrounded conductor and the common grounded conductor is 120 volts.

EXAMPLE

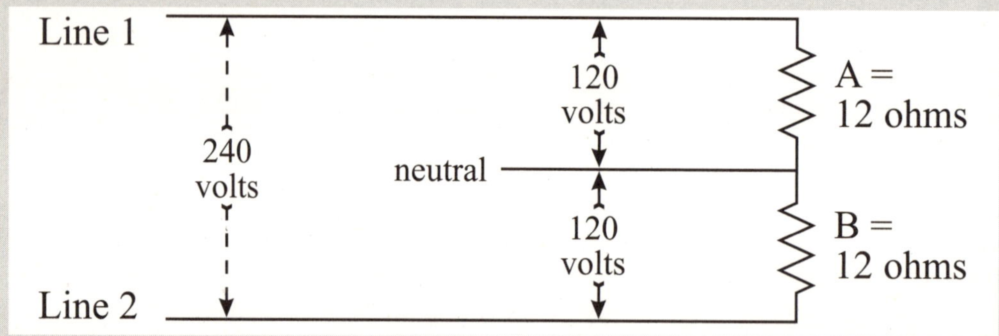

FIGURE 3-1

A white conductor in a circuit having only two conductors is *not* a neutral conductor. In a 2-wire circuit the grounded conductor is carrying the same amount of current as the ungrounded conductor. The definition explains that the neutral conductor carries only the unbalanced current. One property of a series circuit says that the current remains the same throughout the circuit. The grounded conductor of a 2-wire circuit carries the balanced current of the ungrounded conductor. To calculate the current in **Figure 3-1**, the circuit can be redrawn to show two separate series circuits.

EXAMPLES

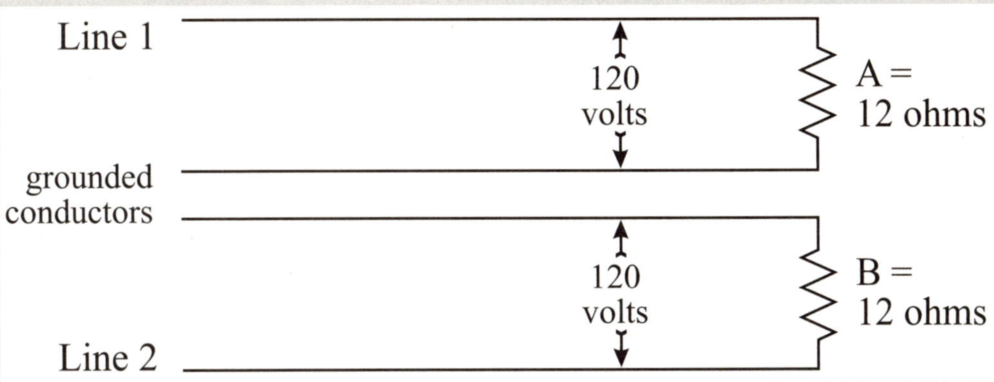

FIGURE 3-2

The current through resistor A is 10 amperes. I = E ÷ R = 120 ÷ 12 = 10 amperes.

The current through resistor B is 10 amperes. I = E ÷ R = 120 ÷ 12 = 10 amperes.

As also stated in 310.15(B)(4), there must be three or more conductors. In the example below, line 1 and line 2 are connected to a voltage potential of 240 volts. Line 1 to neutral has a voltage potential of 120 volts, and line 2 to neutral also has a potential difference of 120 volts.

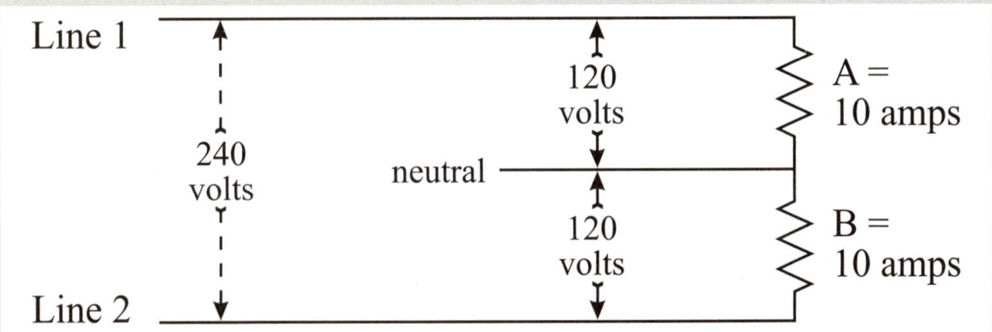

FIGURE 3-3

When current is flowing through both resistors (from two separate voltage sources of 120 volts each) and each resistor is drawing the same amount of current, there will not be any current on the neutral (except with nonlinear loads.) Because of ac voltage, the current on the neutral will be canceled. The unbalanced current on the neutral will be zero amperes. This is known as a balanced system.

Find the maximum neutral current by deenergizing the circuit that draws the least amount of current. The maximum neutral current in **Figure 3-3** is 10 amperes.

In the circuit in **Figure 3-4**, the resistance is not balanced.

The current on the neutral (with both circuits energized) will be the difference between the two resistors. This is also known as the unbalanced current. Because there is amperage on the neutral (5 amperes), this is known as an unbalanced system. The maximum neutral current in Figure 3-4 is 15 amperes.

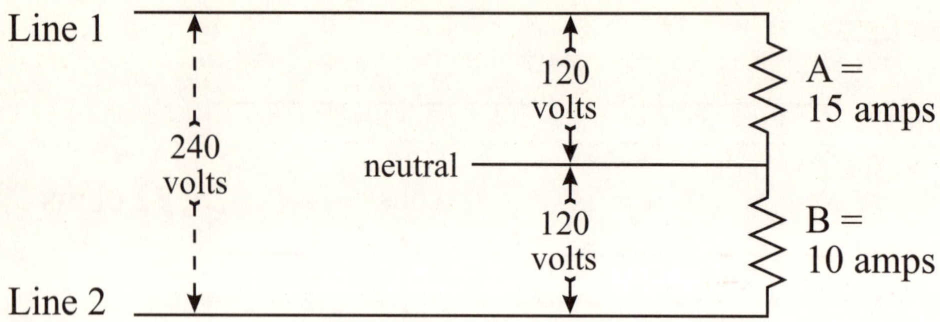

FIGURE 3-4

One potential problem with a multiwire circuit that shares a common grounded conductor is that the ungrounded conductors could end up on the same phase. Below is an example of a circuit that was installed correctly, but somebody has rearranged or rewired the panel. Lines 1 and 2 were originally on different phases, but now they are on the same phase.

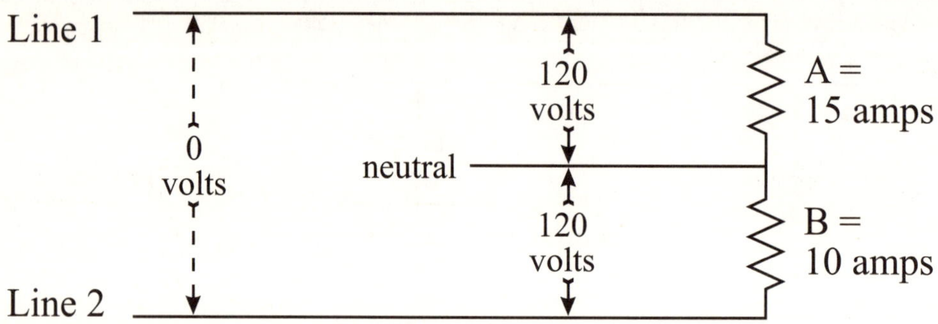

FIGURE 3-5

The ungrounded (hot) conductors will carry the same amount of current as they did before the conductors were moved. An amp meter clamped around line 1 (in the panel) reads 15 amperes. Clamped around line 2, the meter reads 10 amperes. The neutral no longer carries the unbalanced current; it now carries all of the current from lines 1 and 2. The neutral conductor is now carrying a load of 25 amperes. If undetected, this installation could be a problem. If the conductor is rated for a maximum of 20 amperes, a fire hazard exists.

Multiwire Branch Circuits

The definition of a multiwire branch circuit is in Article 100—Definitions: "A branch circuit that consists of two or more ungrounded conductors that have a voltage between them, and a grounded conductor that has equal voltage between it and each ungrounded conductor of the circuit and that is connected to the neutral or grounded conductor of the system."

In accordance with the definition, there must be a potential difference between each of the lines feeding the branch circuit.

EXAMPLES

Single-phase multiwire branch circuit

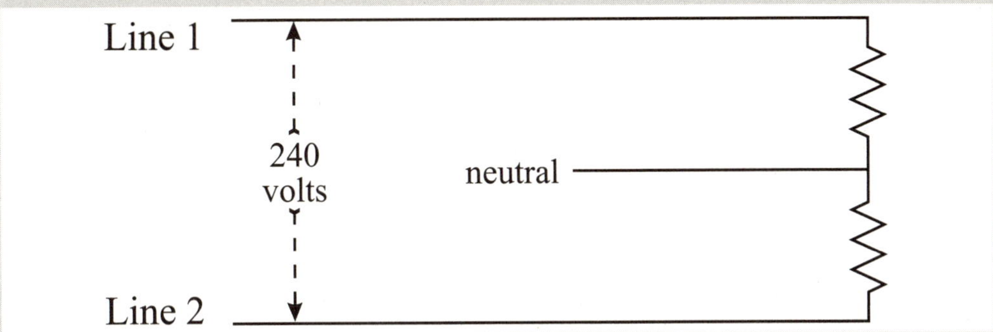

FIGURE 3-6

There is a potential difference of 240 volts between line 1 and line 2.

Three-phase multiwire branch circuit

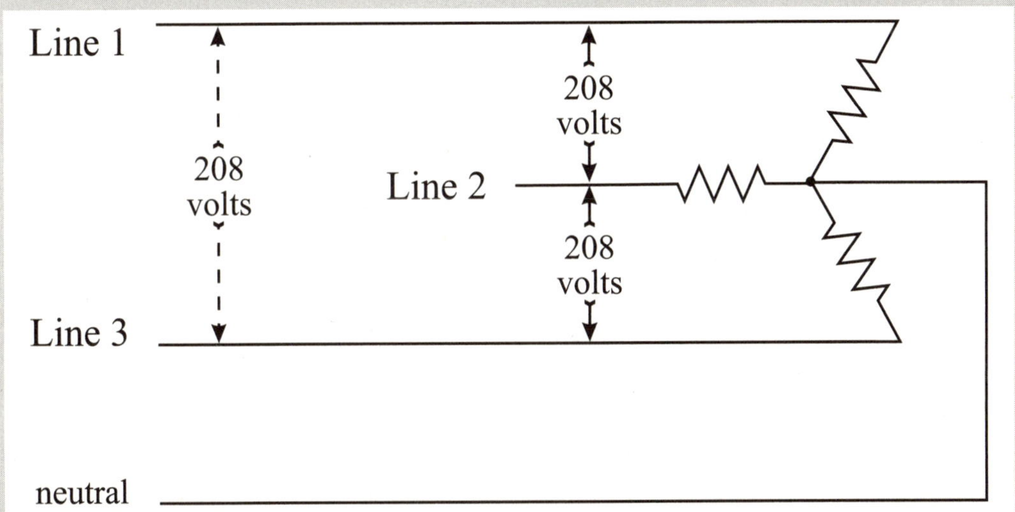

FIGURE 3-7

There is a potential difference of 208 volts between any two lines.

The definition also states that there is one neutral conductor with an equal potential difference between it and each ungrounded conductor.

EXAMPLES

Single-phase multiwire branch circuit

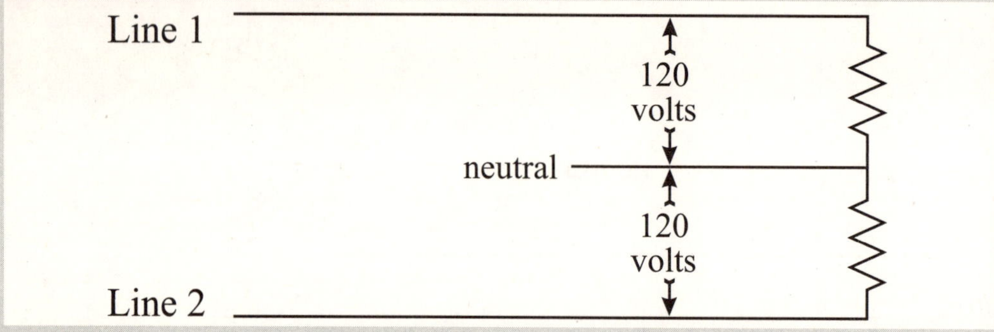

FIGURE 3-8

There is a potential difference of 120 volts between line 1 and the neutral. There is an equal potential difference of 120 volts between line 2 and the neutral.

Three-phase multiwire branch circuit

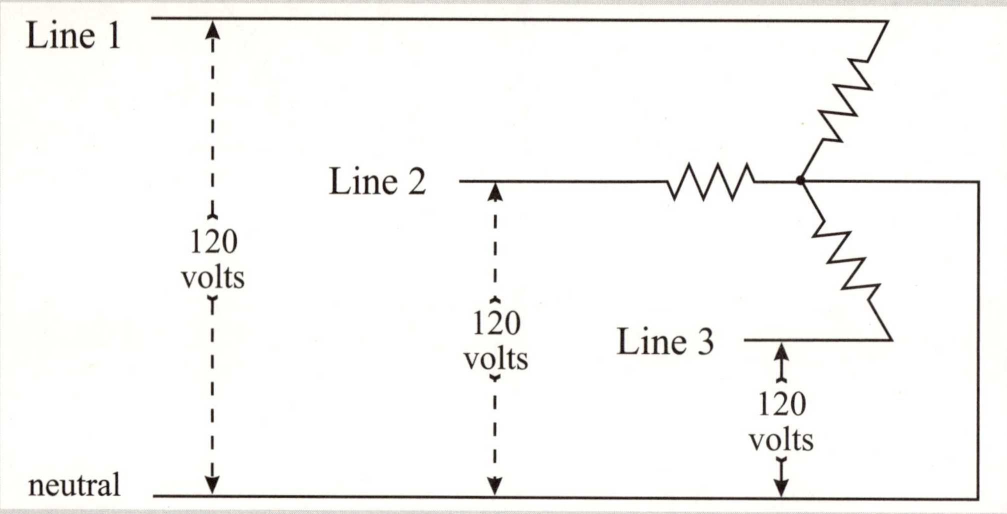

FIGURE 3-9

There is an equal potential difference of 120 volts between any line and the neutral.

A multiwire branch circuit shall not depend on a device to keep the grounded conductor continuous. [300.13(B)] When installing a receptacle in a multiwire branch circuit, the grounded conductors must be tied together. Install a jumper wire from the grounded conductors to the receptacle. A proper installation will help to ensure that there will be a proper return path to ground in the circuit.

A multiwire branch circuit incorrectly installed could be a hazard to persons and property. A neutral that has become open could induce a higher voltage on equipment than the rating allows.

Following is a diagram of a multiwire circuit that has been installed in a restaurant. Two commercial appliances are connected to two receptacle outlets that are part of a multiwire circuit. The letter A represents a bread warmer and the letter B represents a dough mixer. Each appliance is rated for 120 volts and is being supplied with 120 volts.

PROBLEMS

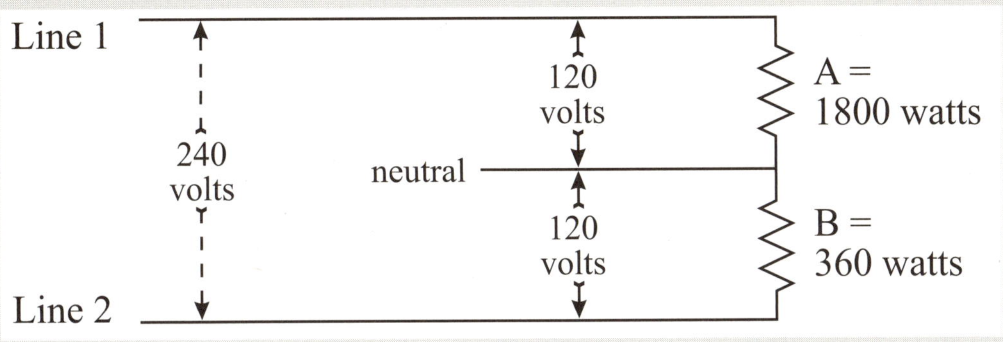

FIGURE 3-10

A = bread warmer
B = dough mixer

1. The current draw of line 1 is ______________ ampere(s).
2. The current draw of line 2 is ______________ ampere(s).
3. The resistance of A is ______________ ohm(s).
4. The resistance of B is ______________ ohm(s).
5. The voltage drop across resistor A is ______________ volt(s).
6. The voltage drop in resistor B is ______________ volt(s).

In the diagram below, the neutral conductor has lost its continuity as a return path to ground. The two circuits that were each originally 120 volts have now become one circuit. There are two resistors in this series circuit with a source voltage of 240 volts. The properties of a series circuit will now apply to the drawing below.

PROBLEMS

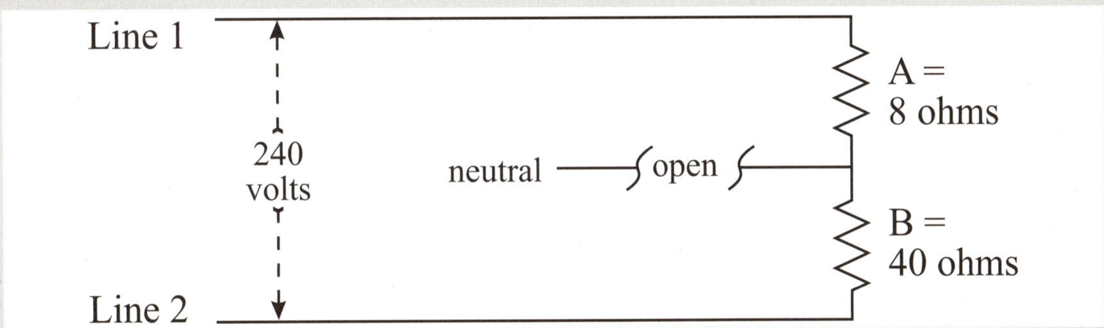

FIGURE 3-11

A = bread warmer
B = dough mixer

1. The total resistance of this series circuit is ______________ ohm(s).
2. The current draw through resistor A is ______________ ampere(s).
3. The current draw through resistor B is ______________ ampere(s).
4. The voltage drop across resistor A is ______________ volt(s).
5. The voltage drop across resistor B is ______________ volt(s).

END-OF-CHAPTER PROBLEMS

1. The feeder neutral load shall be the _____ of the load determined by Article 220.
   a) maximum unbalance
   b) average demand
   c) maximum connected
   d) average unbalance
2. An insulated grounded conductor of 6 AWG or smaller shall be identified by _____ along its entire length.
   a) a continuous white outer finish
   b) a continuous gray outer finish
   c) three continuous white stripes on other than green insulation
   d) any of these
3. A green wire with yellow stripes, used in branch circuits, may be used as a(n) _____.
   I. grounded conductor
   II. equipment grounding conductor
   III. ungrounded conductor
   a) I only
   b) II only
   c) I or II
   d) I, II, or III
4. For devices with screw shells, the terminal for the _____ shall be the one connected to the screw shell.
   a) ungrounded conductor
   b) grounding conductor
   c) grounded conductor
   d) bonding conductor
5. Where connected to a 120-volt circuit, the conductor in a flexible cord consisting of one or more ridges or grooves located on the exterior of the cord must be the _____ conductor.
   a) grounded
   b) grounding
   c) ungrounded
   d) circuit
6. When portable lamps are used with Edison-base lampholders, the grounded conductor shall be identified and attached to the _____ of the attachment plug.
   a) screw shell
   b) identified blade
   c) both A and B
   d) neither A nor B

7. The neutral conductor _____.
   a) carries the balanced current
   b) carries the unbalanced current
   c) is always the white grounded conductor
   d) is always the green conductor
8. The continuity of the grounded conductor must not be broken at receptacles _____.
   a) not having a grounded conductor
   b) in general-purpose branch circuits
   c) in multiwire branch circuits
   d) in 2-wire branch circuits
9. *Identified* as used in the *NEC,* in reference to a conductor or its terminals, means that such conductor or terminal is to be recognized as _____.
   a) grounded
   b) bonded
   c) colored
   d) marked
10. A _____ is a system or circuit conductor that is intentionally grounded.
    a) neutral conductor
    b) grounding conductor
    c) grounding electrode conductor
    d) grounded conductor
11. A(n) _____ is a branch circuit consisting of two or more ungrounded conductors having a voltage between them and a grounded conductor having equal voltage between it and each ungrounded conductor of the circuit and that is connected to the neutral or grounded conductor of the system.
    a) multipurpose branch circuit
    b) general purpose branch circuit
    c) multiwire branch circuit
    d) individual branch circuit
12. Receptacles shall have the terminal intended for connection to the grounded conductor identified _____.
    I. by a metal or metal coating that is substantially white in color
    II. by the word *white*
    III. by the letter W located adjacent to the identified terminal
    a) I only
    b) I or II
    c) II or III
    d) I, II, or III

CHAPTER 4

Cooking Equipment and Clothes Dryers

Cooking Equipment Calculations

The *National Electrical Code* gives demand factors for residential and commercial cooking equipment. These demand factors include feeders and branch circuits. Section 220.55 covers household electric ranges, wall-mounted ovens, counter-mounted cooking units, and other household cooking appliances over 1¾ kW (1750 watts). Table 220.55 is used to calculate demand factors for household cooking equipment. Notes 1 through 5 following the table are to be used with it.

EXAMPLE

Calculate the demand loads for the following ranges individually:

One range at 9 kW = _____

One range at 10 kW = _____

One range at 11 kW = _____

One range at 12 kW = _____

Step 1. Looking at Table 220.55, find the number of appliances.

Step 2. Choose Column A, Column B, or Column C.

All of these answers are found in Column C. The correct answer for all four questions is 8 kW.

In the problem below, the demand factors can be found in both Columns B and C.

EXAMPLE

One range at 8 kW = _____

Step 1. Column B because the range is between 3½ kW and 8¾ kW.

Column C because the range is not over 12 kW.

Step 2. Find the demand from both columns and select the lower.

Column B = 8 × 80% = 6.4 kW

Column C = 8 kW

The correct answer is 6.4 kW.

When there is more than one piece of equipment, simply find the number of appliances, lay a ruler (or straight edge) under that number, and follow it across to Column A, B, or C.

EXAMPLES

What is the demand load for thirty 12-kW ranges?

Column C = 15 kW plus 1 kW for each range

15 + 30 = 45 kW

What is the demand load for fifty 9-kW ranges?

Column C = 25 kW plus ¾ kW for each range

25 + (0.75 × 50) =

25 + 37.5 = 62.5 kW

Note 3 to Table 220.55 states the following:

"Over 1¾ kW through 8¾ kW. In lieu of the method provided in Column C, it shall be permissible to add the nameplate ratings of all household cooking appliances rated more than 1¾ kW but not more than 8¾ kW and multiply the sum by the demand factors specified in Column A or B for the given number of appliances. Where the rating of cooking appliances falls under both Column A and Column B, the demand factors for each column shall be applied to the appliances for that column, and the results added together."

EXAMPLES

What is the demand load for six 8-kW ranges?

Column B = 6 (ranges) × 8 (kW) × 43% = 20.64 kW

Column C = 21 kW

The answer is 20.64 kW.

What is the demand load for seven 8-kW ranges?

Column B = 7 × 8 × 40% = 22.4 kW

Column C = 22 kW

The answer is 22 kW.

If the kW of the ranges (when calculating demand factors) can be found in Columns A or B, always compare it to Column C and select the lowest kW.

EXAMPLE

What is the demand load for the following?

- Four 5-kW ovens
- Seven 8.5-kW ranges
- Nine 3.5-kW cooktops

Step 1. Find the sum (total kW) of the appliances and then find the total number of appliances.

4 × 5 = 20.0 kW

7 × 8.5 = 59.5 kW

9 × 3.5 = 31.5 kW

4 + 7 + 9 = 20 units

Total number of appliances = 20

20.0 + 59.5 + 31.5 = 111.0 kW

Total rating = 111 kW

Step 2. As all this cooking equipment falls under Column B, find the demand factor percent for the number of appliances and multiply the total kW by that percentage.

111 × 28% = 31.08 kW

Column C = 35 kW

As Column C is a larger number, Column B (31.08 kW) is the correct answer.

NOTE

Remember: Always check Column C.

Ranges with the Same Rating

For ranges rated between 12.5 kW and 27 kW, the maximum demand in Column C "shall be increased 5 percent for each additional kilowatt of rating or major fraction thereof by which the rating of individual ranges exceeds 12 kW." [Table 220.55, Note 1]

EXAMPLE

What is the demand load for a 13-kW range?

Step 1. Find the demand for one 12-kW range.
Column C = 8 kW

Step 2. Because 13 kW is one more than 12 kW, increase that rating by 5 percent.
8 × 5% = 0.4 + 8 = 8.4 kW
The demand load for a 13-kW range is 8.4 kW.
The demand load for 13.4 kW = 8 × 5% = 0.4 + 8 = 8.4 kW.
The demand load for 13.5 kW = 8 × 10% = 0.8 + 8 = 8.8 kW.

NOTE

Remember: A major fraction is 0.5 or larger.

Ranges with Different Ratings

When there are one or more ranges over 12½ kW, all the ratings of the ranges will be added together and divided by the number of appliances (ranges) to find an average kW. After an average kW is known, find the demand factor across from the total number of appliances. Then multiply the demand factor by the correct percentage for each kW (or major fraction thereof) over 12½ kW. [Table 220.55, Note 2]

EXAMPLE

What is the demand load for the following ranges?

- Six 14-kW ranges
- Ten 15-kW ranges
- Four 18-kW ranges

Step 1. Find the total number of units and the total kW load.
6 × 14 = 84
10 × 15 = 150
4 × 18 = 72
20 units and 306 kW

Step 2. Find the average kW load by dividing the total kW by the number of units.
306 ÷ 20 = 15.3

Step 3. Drop the .3 because it is not a major fraction.
15.3 = 15 kW

Step 4. Calculate the percentage of increase.
15 - 12 = 3 × 5 = 15%

Step 5. Calculate the minimum demand.
The load demand for 20 ranges from Column C is 35 kW.
35 × 15% = 5.25 + 35 = 40.25 kW
The total demand load for all of these ranges is 40.25 kW.

NOTE

When finding an average of the range ratings, any range rated less than 12 kW must be changed to 12 kW before calculating the average rating.

EXAMPLE

What is the demand load for the following ranges?

- Six 15-kW ranges
- Four 10-kW ranges
- Ten 8-kW ranges

Step 1. Change any range rated less than 12 kW to 12 kW. Then find the total number of units and the total kW load.

6 × 15 = 90

4 × ~~10~~ 12 = 48

10 × ~~8~~ 12 = 120

Total number of units is 20

Total kW load is 258

Step 2. Find the average kW load by dividing the total kW by the number of units.

258 ÷ 20 = 12.9

Step 3. Because .9 is a major fraction, change 12.9 to 13.

Step 4. Calculate the percentage of increase.

13 - 12 = 1 × 5 = 5%

Step 5. Calculate the minimum demand.

The load demand for 20 ranges from Column C = 35 kW.

35 × 5% = 1.75 + 35 = 36.75 kW

The total demand load for all of these ranges is 36.75 kW.

NOTE

Highlight "of same rating" in Note 1.

Highlight "of unequal ratings" in Note 2.

Highlight "using 12 kW for any range rated less than 12 kW" in Note 2.

Cooking Equipment Used in Instructional Programs

The demand load calculations for household cooking appliances over 1¾ kW that are used in instructional programs are found in Table 220.55 as expressed in Note 5.

EXAMPLES

What is the demand load for a school home economics class that has five 12-kW ranges?

Although these ranges are used in a commercial type of occupancy, they are used in an instructional program. Therefore, the demand can be calculated from Table 220.55.

The demand for five 12-kW ranges from Column C is 20 kW.

What is the demand load for a school that has five 12-kW ranges?

The demand factors for these ranges are not found in 220.55. The question does not stipulate that the ranges are used in an instructional program. The ranges are probably located in the school cafeteria and therefore must be calculated as commercial cooking equipment.

The answer for this question is found in 220.56 and Table 220.56.

12 × 5 × 70% = 42 kW

NOTE

Highlight "instructional programs" in Note 5.

Branch Circuit Demand Loads for Cooking Equipment

All of the demand loads calculated to this point have been for feeders and services. The fourth note under Table 220.55 provides instructions for branch circuits and can be split into three sections.

1. To find the branch circuit demand factor for one range, use Table 220.55.

EXAMPLES

What is the branch circuit demand load for an 8-kW range?

Column B = 80%

8 × 80% = 6.4 kW

The branch circuit must be sized to carry a 6.4-kW load.

What is the branch circuit demand load for a 15-kW range?

Column C = 8 kW

15 - 12 = 3 × 5 = 15% increase

Increase Column C = 8 × 0.15 = 1.2 + 8 = 9.2 kW.

The branch circuit must be sized to carry a 9.2-kW load.

2. Where supplying one wall-mounted oven, the branch circuit must be rated no less than the oven's nameplate rating. Likewise, where supplying one counter-mounted cooking unit (cooktop), the branch circuit must have a rating at least that of the nameplate rating of the cooktop.

EXAMPLE

What is the branch circuit demand load, in kW, for a 6-kW cooktop?

This question is asking for the minimum branch-circuit size in kilowatts. No calculation is necessary because the branch circuit must be rated at least 6 kW. The branch circuit must be sized to carry a 6-kW load.

3. When a single branch circuit is feeding a counter-mounted cooking unit and one or two wall-mounted ovens (in the same room), add the nameplate ratings of the appliances. Treat the total kW rating as one range and find the demand load in Table 220.55. This applies to two specific installations:
   - If the branch circuit is feeding one counter-mounted cooking unit and one wall-mounted oven (in the same room)
   - If the branch circuit is feeding one counter-mounted cooking unit and two wall-mounted ovens (in the same room)

Do not apply this calculation method if there are two wall-mounted ovens and no counter-mounted cooking units.

EXAMPLES

What is the branch circuit demand load for a 5-kW cooktop and two 3½-kW ovens? All of the appliances are in the same room fed by one branch circuit.

5 + 3.5 + 3.5 = 12 kW

The demand load from Column C for one range at 12 kW is 8 kW.

The branch circuit must be sized to carry an 8-kW load.

NOTE

Highlight "Branch-Circuit Load" in Note 4 of Table 220.55.

What is the branch circuit load for a 5-kW cooktop and a 3-kW oven? All of the appliances are in the same room fed by one branch circuit.

5 + 3 = 8 kW

The demand factor percent from Column B is 80%.

8 × 80% = 6.4 kW

The branch circuit must be sized to carry a 6.4-kW load.

Commercial Kitchen Equipment

The feeder demand factors for commercial electric cooking equipment, dishwasher booster heaters, water heaters, and other kitchen equipment are found in Table 220.56. Add the nameplate ratings of the kitchen equipment and multiply by the demand factor percent.

EXAMPLE

What is the demand load for the following commercial equipment?

- Two 14-kW ranges
- One 12-kW grill
- One 10-kW deep fryer
- One 5-kW dishwasher

Step 1. Find the total number of units and the total kW.

2 × 14 = 28

1 × 12 = 12

1 × 10 = 10

1 × 5 = 5

Total number of units is 5.

Total kW is 55.

Step 2. Find the demand factor percent for 5 units in Table 220.56.

The demand factor percent for five units of equipment is 70%.

Step 3. To calculate the feeder demand, multiply the total kW by the demand factor percent.

55 × 70% = 38.5 kW

The feeder demand for this kitchen equipment is 38.5 kW.

NOTE

The feeder demand load can never be smaller than the total kW rating of the two largest pieces of equipment.

EXAMPLE

What is the demand load for the following commercial equipment?

- Five 3-kW ovens
- One 5-kW cooktop
- Two 20-kW ranges

Step 1. Find the total number of units and the total kW.

5 × 3 = 15

1 × 5 = 5

2 × 20 = 40

Total number of units is 8.

Total kW is 60.

Step 2. Find the demand factor percent for 8 units in Table 220.56.

The demand factor percent for eight units of equipment is 65%.

Step 3. To calculate the demand load, multiply the total kW by the demand factor percent.

60 × 65% = 39 kW

Step 4. Compare the calculated load with the sum of the two largest pieces of equipment.

Although the calculated load is 39 kW, it must not be less than the minimum demand of the two largest pieces of equipment. The sum of the two largest equipment loads in this example is 40 kW (20-kW range + 20-kW range). Therefore, 39 kW is an incorrect answer. The demand load for this kitchen is 40 kW.

NOTE

Highlight the following sentence in 220.56: "However, in no case shall the feeder or service calculated load be less than the sum of the largest two kitchen equipment loads."

Feeder or Service Neutral Loads for Ranges

The neutral demand for a feeder or service supplying household electric ranges, wall-mounted ovens, and counter-mounted cooking units can have an additional demand factor of 70 percent applied to the calculation after the demand load has been determined in accordance with Table 220.55 for ranges. [220.61(B)(1)]

EXAMPLE

What is the feeder neutral demand load for five 12-kW ranges?

Column C = 20 kW

20 kW × 70% = 14 kW

The feeder neutral demand load is 14 kW.

What is the amperage of the neutral conductor if the service is 120/240 volts?

14,000 ÷ 240 = 58 amperes

A further demand factor of 70 percent is permitted for that portion of the unbalanced (neutral) load in excess of 200 amperes. [220.61(B)(2)]

When finding the feeder neutral load, check the amperage to see if it is more than 200 amperes. If it is, multiply the amount in excess of 200 by 70%.

NOTE

If this were a test question one answer would be 70 kW. This is the wrong answer!

EXAMPLE

What is the feeder neutral demand load for one hundred 10-kW ranges on a 120/240 volt service?

Column C = 25 + (100 × 0.75) = 100 kW

100 kW × 70% = 70 kW

Step 1. Find the amperage of the neutral demand load.

70,000 ÷ 240 = 291.6667 amperes

Step 2. Multiply the excess of 200 amperes by 70%, and then add it back to the 200 amperes.

291.6667 - 200 = 91.6667 (excess of 200 amperes)

91.6667 × 70% = 64.1667

200 + 64.1667 = 264.1667 amperes

Step 3. To change the amperage back to kW, simply multiply the amperes by the voltage and divide by 1000.

264.1667 × 240 = 63,400 ÷ 1000 = 63.4 kW

The feeder neutral demand load for one hundred 10-kW ranges on a 120/240 volt service is 63.4 kW.

NOTE

According to *NEC* 210.19(A)(3), "for ranges of 8¾ kW or more rating, the minimum branch-circuit rating shall be 40 amperes," and the neutral conductor for a range shall not be smaller than 10 AWG.

Clothes Dryer Calculations (Standard Method)

Clothes dryer demand load calculation procedures are located in 220.54. The load for household electric clothes dryers in a dwelling unit must be 5000 watts (volt-amperes) or the nameplate rating, whichever is larger, for each dryer served.

EXAMPLE

What is the single-family load contribution for a 4.5-kW clothes dryer?

a) 4.5 watts b) 5.0 watts c) 4500 watts d) 5000 watts

A 4500-watt (4.5 × 1000 = 4500) clothes dryer is less than the required 5000-watt minimum as specified in 220.54. The single-family load contribution for a 4.5-kW clothes dryer (using the standard method) is 5000 watts. The correct answer is d).

NOTE

Use the standard method unless the question specifically stipulates to use the optional method.

Feeder or service calculations with five (or more) dryers can be derated with the demand factors (percents) in Table 220.54. Simply multiply the total dryer load by the demand factor (percent) for the given number of dryers. Remember to use 5000 watts or the nameplate rating, whichever is larger.

EXAMPLE

What is the dryer demand load for a 10-unit multifamily dwelling? Each unit has a 4-kW clothes dryer and the service is 120/240 volt single-phase.

a) 20 kW b) 25 kW c) 40 kW d) 50 kW

Step 1. Change the 4-kW clothes dryers to 5 kW.

Step 2. Find the demand factor percent across from 10 units in Table 220.54 (50%).

Step 3. Solve the problem.

10 (units) × 5 (kW) × 50% = 25 kW

The correct answer is b), 25 kW.

Where the number of dryers is greater than 11, a calculation may be necessary to find the demand factor percent. When the number of dryers is more than 11 but less than 23, calculate the mathematical equation inside the parentheses first. Then subtract the number inside the parentheses from 47. The result of the equation will be the demand factor (percent) for 12 to 22 clothes dryers.

EXAMPLE

What is the dryer demand load for twenty 5-kW household electric clothes dryers? The service is 120/240 volt single-phase.

a) 35 kW b) 38 kW c) 47 kW d) 100 kW

The demand factor (percent) formula is % = 47 - (number of dryers - 11).

Step 1. Perform the math inside the parentheses.

(number of dryers - 11) = 20 - 11 = 9

Step 2. Subtract 9 from 47.

47 - 9 = 38%

Step 3. Solve the problem.

20 (units) × 5 (kW) × 38% = 38 kW

The correct answer is b), 38 kW.

NOTE

Where the number of dryers is 12 to 22, the demand starts at 46% (for 12 dryers) and drops 1% with each additional dryer. For example, 13 = 45%, 14 = 44%, 15 = 43%, 16 = 42%, etc.

When there are 24 to 42 dryers, calculate the mathematical equation inside the parentheses first. Then calculate the mathematical equation inside the brackets. Next, subtract the number inside the brackets from 35. The result of this equation will be the demand factor (percent) for 24 to 42 clothes dryers.

NOTE

Where the number of dryers is 24 to 42, the demand starts at 34.5% (for 24 dryers) and drops 0.5% with each additional dryer. For example, 25 = 34%, 26 = 33.5%, 27 = 33%, 28 = 32.5%, 29 = 32%, etc.

EXAMPLE

What is the dryer demand load for forty 4-kW household electric clothes dryers? The service is 120/240 volt single-phase.

a) 42.4 kW b) 52 kW c) 53 kW d) 54 kW

The demand factor (percent) formula is % = 35 - [0.5 × (number of dryers - 23)].

Step 1. Perform the math inside the parentheses.

(number of dryers - 23) = 40 - 23 = 17

Step 2. Find the product of the numbers in the brackets.

0.5 × 17 = 8.5

Step 3. Subtract 8.5 from 35.

35 - 8.5 = 26.5%

Step 4. Because the clothes dryers are less than 5 kW, change 4 kW to 5 kW.

Step 5. Solve the problem.

40 (units) × 5 (kW) × 26.5% = 53 kW

The correct answer is c), 53 kW.

Feeder or Service Neutral Loads for Clothes Dryers

The neutral demand for a feeder or service supplying electric dryers can have an additional demand factor of 70 percent applied to the calculation after the demand load has been determined in accordance with Table 220.54 for dryers. [220.61(B)(1)]

EXAMPLE

In a residential dwelling, the main feeder neutral load for a 4.5-kW clothes dryer would be _____.

a) 5.0 kW b) 4.3 kW c) 3.5 kW d) 3.0 kW

Step 1. Change the 4.5-kW clothes dryer to 5 kW.

Step 2. Multiply 5 kW by 70% for the answer.

5 × 70% = 3.5 kVA

The correct answer is c).

Clothes Dryer Calculations (Optional Method)

Some exam questions ask for clothes dryer demand loads calculated by the optional method. This will require the use of Table 220.84. Although 220.84 provides optional calculation procedures for three or more multifamily dwelling units, Table 220.84 is sometimes used on exams to find clothes dryer demand loads only. Use the nameplate rating of the clothes dryer, even if it is less than 5000 watts. [220.84(C)(3)]

EXAMPLE

What is the service demand load for a 30-unit multifamily dwelling where each unit contains a 4.5-kW clothes dryer? The service is 120/240 volt, single-phase. (Use the optional method.)

a) 40.5 kW
b) 44.55 kW
c) 45 kW
d) 135,000 watts

Step 1. Find the demand factor (percent) for 30 units from Table 220.84 (33%).

Step 2. Solve the problem.

30 (units) × 4.5 (nameplate kW) × 33% = 44.55

The correct answer is b).

NOTE

Standard method = number of units × 5000 watts (minimum) × Table 220.54. Optional method = number of units × nameplate × Table 220.84.

END-OF-CHAPTER PROBLEMS

1. What is the feeder demand load for two 10-kW ranges?
   a) 20 kW
   b) 10 kW
   c) 11 kW
   d) 13 kW
2. What is the feeder demand load for twenty-six 6.75-kW ranges?
   a) 42.12 kW
   b) 26 kW
   c) 52.7 kW
   d) 41 kW
3. What is the feeder demand load for an apartment complex that has eight 13.5-kW ranges?
   a) 23 kW
   b) 24.15 kW
   c) 25.3 kW
   d) 108 kW
4. What is the demand load, for the service, for two 3½-kW ovens and one 5-kW cooktop? All of the appliances are supplied from a single branch circuit and are located in the same room of a house.
   a) 12 kW
   b) 8 kW
   c) 9.25 kW
   d) 6.6 kW

5. What is the service demand load for twenty 12½-kW commercial ranges that are being used in an instructional training program for cooks?
   a) 35 kW
   b) 36.75 kW
   c) 162.5 kW
   d) 175 kW
6. What is the service demand load for an apartment complex with the following ranges? The service is 120/240-volt single-phase.
   - Six 12-kW ranges
   - Four 8-kW ranges
   - Ten 15-kW ranges

   a) 35 kW
   b) 36.75 kW
   c) 38.5 kW
   d) 40.25 kW
7. What is the service demand load for thirty-eight 14-kW countertop ranges?
   a) 53 kW
   b) 58.3 kW
   c) 117 kW
   d) 532 kW
8. What is the minimum demand load for ten 8-kW household ranges?
   a) 25 kW
   b) 25.6 kW
   c) 27.2 kW
   d) 80 kW
9. What is the service demand load for an apartment complex with the following ranges? The service is 120/240-volt single-phase.
   - Six 8-kW ranges
   - Four 12-kW ranges
   - Eight 15-kW ranges

   a) 13.3 kW
   b) 33 kW
   c) 34.65 kW
   d) 38 kW
10. What is the minimum demand load for twenty-three 8-kW household ranges?
   a) 37 kW
   b) 38 kW
   c) 47.84 kW
   d) 184 kW

11. What is the demand load for a 200-unit apartment with a 4-kW clothes dryer in each unit? The service is 120/240-volt single-phase.
   a) 1000 kW
   b) 800 kW
   c) 250 kW
   d) 200 kW
12. A 40-unit apartment building has a 12-kW range in each unit. What is the approximate load, in amperes, on the neutral? The transformer is 240/120-volt single-phase.
   a) 270 amperes
   b) 229 amperes
   c) 220 amperes
   d) 160 amperes
13. What size neutral conductor is required for a 13.55-kW, 240-volt range in a house?
   a) 4-AWG THHN CU
   b) 6-AWG THHN CU
   c) 8-AWG THHN CU
   d) 10-AWG THHN CU
14. A school home economics class has twenty 8-kW ranges. What is the demand load?
   a) 35 kW
   b) 44.8 kW
   c) 56 kW
   d) 160 kW
15. A building has twelve apartments. Each apartment has an 8-kW electric range on 230/115 volts. What is the load on the neutral?
   a) 82 amperes
   b) 94 amperes
   c) 117 amperes
   d) 134 amperes
16. What is the feeder demand load for a restaurant with two 14-kW stoves, a 6-kW oven, a 4-kW dishwasher, and a 3-kW booster heater?
   a) 28 kW
   b) 28.7 kW
   c) 30.75 kW
   d) 41 kW
17. A restaurant being remodeled will have the following equipment. What is the minimum feeder demand for this group of kitchen equipment (include intermittent use as equipment)?
   - One 12-kW stove
   - One 6-kW oven
   - One 8-kW oven
   - One 3-kW dishwasher
   - One 3-kW booster heater

   a) 22.4 kW
   b) 20 kW
   c) 25.6 kW
   d) 32 kW

18. What is the demand load in amperes for the following 480-volt, three-phase, 3-wire commercial kitchen equipment?
- Two 5000-watt water heaters
- Four 3000-watt fryers
- Two 6000-watt ovens

a) 13
b) 18
c) 21
d) 27

19. What is the minimum service demand load for a custom home with two 4-kW wall ovens that are supplied by one branch circuit?

a) 5.2 kW
b) 6.4 kW
c) 8 kW
d) 11 kW

20. In a 5000 ft^2 house, the main feeder neutral load for a 5-kVA clothes washer/dryer would be ______.

a) 5.0 kVA
b) 4.3 kVA
c) 3.5 kVA
d) 3.0 kVA

21. A motel has a laundromat for customer use with six 1500-watt washers and three 4500-watt dryers. The load to be added to the service for the washers and dryers is ______.

a) 24,000 watts
b) 22,500 watts
c) 18,600 watts
d) 16,875 watts

22. What is the branch-circuit demand load for one 8-kW oven, one 7-kW oven, and one 8-kW cooktop? All of the appliances are supplied from a single branch circuit and are located in the same room of a house.

a) 4.4 kW
b) 12 kW
c) 15 kW
d) 23 kW

23. What is the demand load for a 12-unit apartment with a 5-kW clothes dryer in each unit?

a) 27.6 watts
b) 27,600 watts
c) 28,200 watts
d) 60,000 watts

24. What is the branch-circuit demand load for one 10-kW oven, one 9-kW oven, and one 12-kW cooktop? All of the appliances are supplied from a single branch circuit and are located in the same room of a house.
    a) 8 kW
    b) 10 kW
    c) 15.6 kW
    d) 31 kW
25. What is the minimum demand load for 32 apartments with an 8-kW range in each apartment?
    a) 32 kW
    b) 47 kW
    c) 56.32 kW
    d) 256 kW
26. A 60-unit apartment building has a 4-kW dryer in each apartment. By using the optional method, what is the service demand load?
    a) 57.6 kW
    b) 60 kW
    c) 75 kW
    d) 240 kW
27. A motel has a laundromat for customer use with five 1500-watt washers and four 4500-watt dryers. The load to be added to the service for the washers and dryers is _____.
    a) 22.5 kW
    b) 23.5 kW
    c) 25.5 kW
    d) 27.5 kW

Voltage Drop

CHAPTER 5

NOTE

Always remember: A conductor is a resistor.

Voltage Drop Through Conductors

Up to this point, voltage drop has been calculated only through loads and resistors. No consideration has been given to any voltage drop through the conductors. As all conductors have resistance, in practical applications, there is always voltage drop through the conductors.

The resistance of a conductor is inversely proportional to its circular mil area. In other words, the larger the circular mil area, the smaller the resistance.

The resistance of a conductor is directly proportional to its length. Therefore, the longer the conductor, the larger the resistance.

EXAMPLE

A swimming pool pump is 100 feet from the main panel. The rating of the pump is 15 amps at 120 volts.

Step 1. Draw the problem.

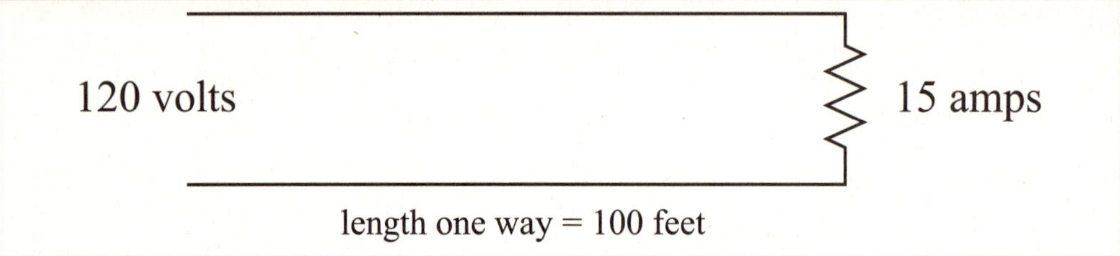

FIGURE 5-1

In the Ohm's law calculations previously discussed, conductor resistance was not introduced into any of the theory problems. In practical applications, however, conductor resistance must be considered.

Step 2. Because every conductor is a resistor, the problem needs to be redrawn.

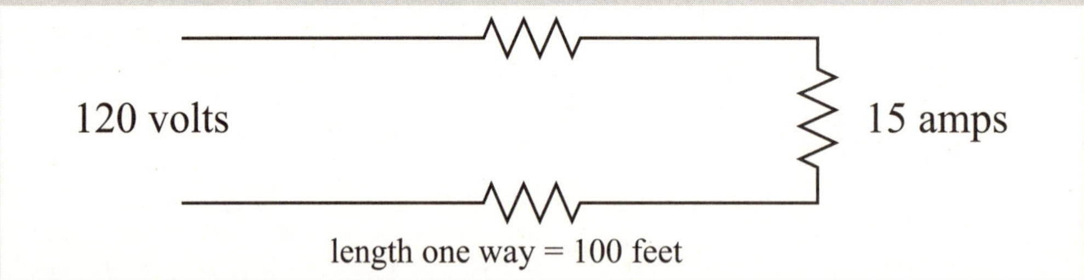

FIGURE 5-2

The properties of a series circuit now apply to this problem. With some resistance in the conductors, part of the voltage will drop through the conductors.

The higher the resistance through a conductor, the higher the voltage drop through that conductor. Such a voltage drop in the conductors could result in a shortage of voltage feeding the load.

The voltage that is dropped through the conductors is a percentage of the source voltage.

The *National Electrical Code* gives a recommended percentage of voltage drop: "Conductors for branch circuits as defined in Article 100, sized to prevent a voltage drop exceeding 3 percent at the farthest outlet of power, heating, and lighting loads, or combinations of such loads, and where the maximum total voltage drop on both feeders and branch circuits to the farthest outlet does not exceed 5 percent, provide reasonable efficiency of operation." [210.19(A)(1) FPN No. 4]

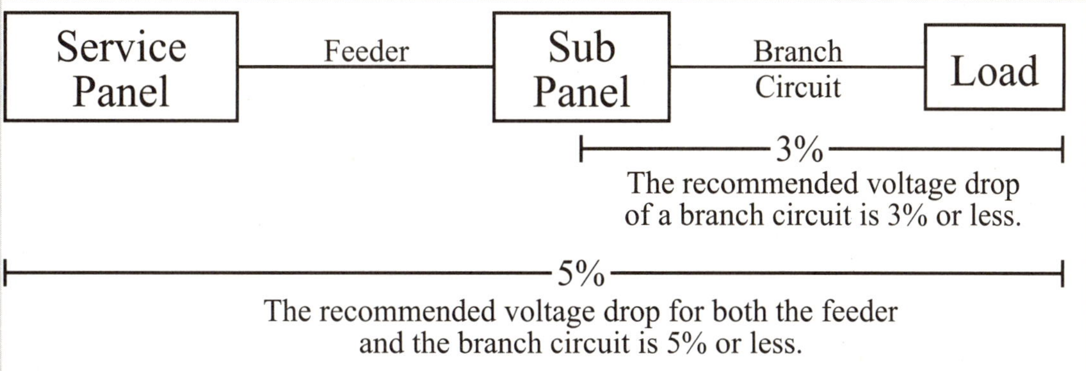

FIGURE 5-3

The recommended voltage drop is a percentage of the source voltage. To calculate the allowable voltage drop for a branch circuit, multiply the source voltage by 3%.

EXAMPLES

What is the recommended voltage drop for a 120-volt branch circuit?

120 × 3% = 3.6 volts

The maximum allowable voltage drop for a 120-volt circuit is 3.6 volts.

What is the recommended voltage drop for a 208-volt branch circuit?

208 × 3% = 6.24 volts

The maximum allowable voltage drop for a 208-volt circuit is 6.24 volts.

Feeders

The wording pertaining to the recommended percentage of voltage drop is exactly the same for feeders as it is for branch circuits: "Conductors for feeders as defined in Article 100, sized to prevent a voltage drop exceeding 3 percent at the farthest outlet of power, heating, and lighting loads, or combinations of such loads, and where the maximum total voltage drop on both feeders and branch circuits to the farthest outlet does not exceed 5 percent, will provide reasonable efficiency of operation." [215.2(A)(3) FPN No. 2]

The total voltage drop from the main to the last load cannot exceed 5%.

NOTE

The recommendation for feeders is not necessarily 2%.

EXAMPLE

If the voltage drop in the branch circuit is only 2%, then voltage drop in the feeder can be 3%.

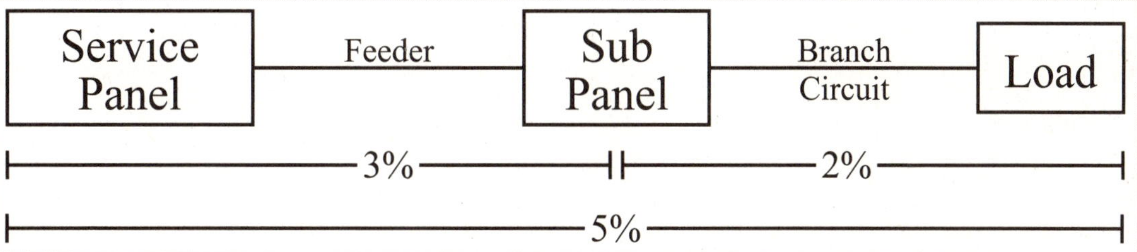

FIGURE 5-4

The feeder voltage drop can be any percentage, as long as the feeder and branch circuit added together do not exceed 5%.

- If the branch circuit is 1%, the feeder can be 4%.
- If the branch circuit is 2.5%, the feeder can be 2.5%.
- If the branch circuit is 1.5%, the feeder can be 3.5%.

EXAMPLE

What is the recommended voltage drop for a 240-volt feeder? The branch circuit has a voltage drop of 5.5 volts.

240 × 5% = 12 volts (total allowed for both the branch circuit and the feeder)

12 - 5.5 = 6.5 volts

The maximum allowable voltage drop for this 240-volt feeder is 6.5 volts.

Resistance of Conductors

Resistance of conductors is found in *NEC* Chapter 9, Tables 8 and 9. These tables give the resistance of copper and aluminum conductors. The resistance for each conductor is given in ohms/kFT (ohms per 1000 feet). These tables are used often in electrical examinations, from finding the circular-mil area of conductors to finding the resistance of conductors.

Sometimes it may be necessary to find the resistance for 1000 feet.

EXAMPLES

What is the dc resistance for 1000 feet of 12-AWG uncoated stranded copper?

Look at Table 8 to find the resistance for 1000 feet of 12-AWG stranded copper. There are two different numbers given for a 12-AWG conductor. AWG sizes 18 through 8 give a different resistance for solid and stranded conductors. Look at the fourth column, Quantity, under Conductors and Stranding. This column gives the number of strands for the conductor listed. A conductor with only one strand is a solid wire. A conductor with more than one strand is a stranded wire.

The resistance for 1000 feet of 12-AWG stranded copper is 1.98 ohm.

What is the dc resistance for 1000 feet of 6-AWG copper?

Notice that the above example does not specify coated or uncoated copper. A coated conductor is not a copper-clad conductor. It is a copper wire that has been coated or tinned so that the insulation will adhere to the copper. Coated conductors are rarely used today because of the synthetic types of insulation, like TW, THW, and THHN. Assume uncoated copper unless coated copper is specified.

The resistance for 1000 feet of 6-AWG copper is 0.491 ohm.

Sometimes it may be necessary to calculate the resistance for a length other than 1000 feet.

EXAMPLE

What is the dc resistance for 300 feet of 6-AWG copper?

Step 1. Find the resistance for 1 foot of the conductor.

0.491 ÷ 1000 = 0.000491 ohms per foot

Step 2. Multiply the resistance per foot by the length of the conductor.

0.000491 × 300 = 0.1473

The resistance for 300 feet of 6-AWG copper is 0.1473 ohm.

The calculation can be shortened by taking the length that is given and moving the decimal point three places to the left, then multiplying that number by the resistance.

EXAMPLE

What is the dc resistance for 525 feet of 500-kcmil aluminum?

Step 1. Find the resistance for 1000 feet of 500-kcmil aluminum.

0.0424

Step 2. Move the decimal point three places to the left (0.525), in the length that is given, and multiply it by the resistance.

$0.0424 \times 0.525 = 0.02226$

The resistance for 525 feet of 500-kcmil aluminum is 0.02226 ohm.

Ohm's law formulas, together with Tables 8 and 9, can be used to calculate the answer to a voltage drop question. The length can be found if the type of conductor and resistance are known.

EXAMPLES

A branch circuit has a source voltage of 120 volts. The voltage at the load is 116 volts. The branch circuit conductors are 12-AWG uncoated stranded copper conductors. The amperage reading at the load is 10 amps. What is the length (one way) from the source to the load?

Step 1. Draw the problem.

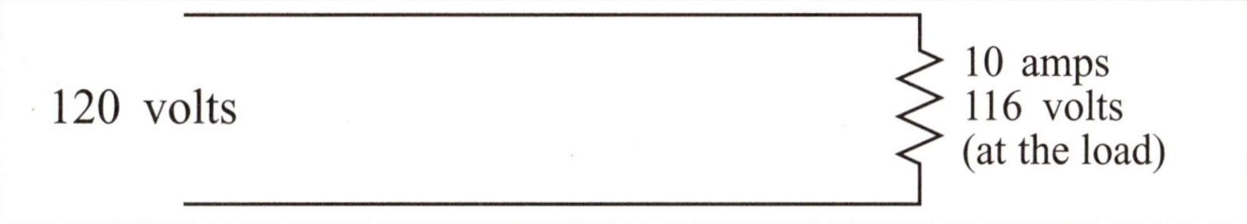

FIGURE 5-5

Because of resistance, the other 4 volts are in the conductors.

Step 2. In a 2-wire, single-phase, branch circuit the voltage drop is divided evenly between the two conductors. In this example, two volts are dropped in each conductor. Because each conductor is a resistor, the problem can be redrawn to show a total of three resistors.

The properties of a series circuit will apply to this problem.

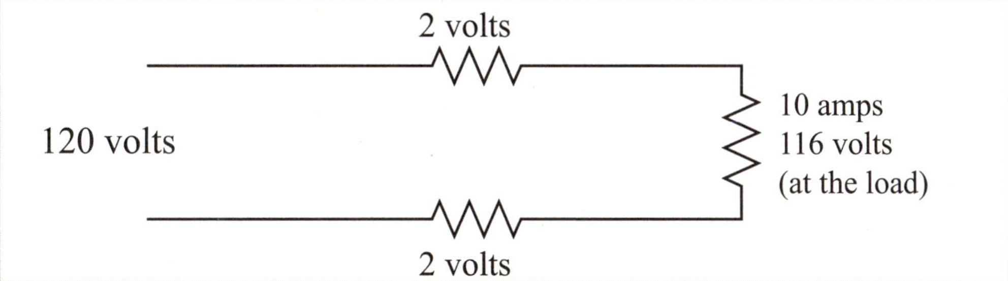

FIGURE 5-6

Step 3. One property of a series circuit is that the current remains the same through the circuit. The amperage through each resistor is 10 amperes.

Step 4. As there are two known factors (volts and amperes), the resistance of each conductor can be calculated.

$R = E \div I = 2 \div 10 = 0.2$ ohm

continues

continued

Step 5. The resistance for 1000 feet of 12-AWG uncoated stranded copper is 1.98 ohms.

The resistance for 1 foot is 0.00198 ohm.

Step 6. Divide the resistance of one conductor by the resistance for 1 foot.

0.2 ÷ 0.00198 = 101.0101 = 101

The length for each conductor is approximately 101 feet.

What is the approximate length of a 6-AWG THW copper conductor that has a resistance of 0.055 ohm?

Step 1. The resistance of the conductor is given: 0.055 ohm.

The resistance for 1000 feet of 6-AWG uncoated copper is 0.491 ohm.

The resistance for 1 foot is 0.000491 ohm.

Step 2. Divide the resistance given for the conductor by the resistance for 1 foot.

0.055 ÷ 0.000491 = 112.01629 = 112

The answer to this problem is 112 feet.

Ohm's Law and Voltage Drop

There are different methods for calculating voltage drop. One method uses the Ohm's law formulas. When amperage and resistance are known, voltage (or in this case voltage drop) can be found.

EXAMPLE

What is the voltage drop of a branch circuit with a voltage source of 240 volts? The circuit is drawing 20 amps and is being fed with a 10-AWG solid copper conductor. The length from the source to the load is 145 feet.

Step 1. Draw the problem.

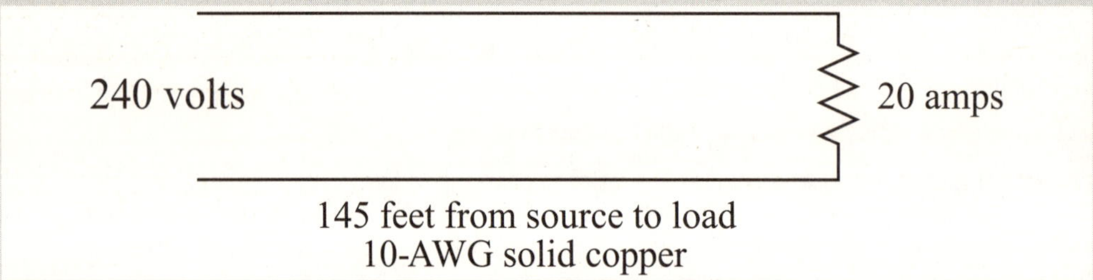

FIGURE 5-7

Step 2. Find the total resistance for one conductor.

The resistance for 1000 feet of 10-AWG solid copper conductor is 1.21 ohms.

Step 3. Divide the resistance by 1000 to find the resistance per foot.

1.21 ÷ 1000 = 0.00121

Step 4. Multiply the length by the resistance per foot.

145 × 0.00121 = 0.17545 ohm

Step 5. Calculate the voltage drop in one conductor.

E = I × R

E = 20 × 0.17545 = 3.509

The voltage drop in one conductor is 3.509 volts.

Step 6. Multiply the voltage drop of one conductor by 2.

3.509 × 2 = 7.018 volts

The total voltage drop of this circuit is approximately 7 volts. This is within the recommended voltage drop, because 3% of 240 volts is 7.2 volts.

Voltage Drop Formulas

Single-phase:

$$V_D = \frac{2\ K\ L\ I}{Cmil}$$

Three-phase:

$$V_D = \frac{1.732\ K\ L\ I}{Cmil}$$

Table 5-1 Formula Definitions

V_D	Volts dropped in a circuit
2	Multiplying factor for single-phase (2 represents the length of the conductors in a single-phase circuit)
1.732	Multiplying factor for three-phase (the square root of 3 represents the length of the conductors in a three-phase circuit)
K	The factor obtained by multiplying the conductor's circular mil area by one foot of the conductor's resistance Approximate K factor for copper = 12.9 Approximate K factor for aluminum = 21.2
L	Length or distance from the voltage source to the load (one way only)
I	Actual current used by the load NOTE: Do not use 125% for motors and continuous loads.
Cmil	Circular-mil area of the conductor (Chapter 9, Table 8)

Table 5-2 Formula Variations

	Single-phase	Three-phase
To find the volts dropped	$V_D = \frac{2\ K\ L\ I}{Cmil}$	$V_D = \frac{1.732\ K\ L\ I}{Cmil}$
To find the conductor size	$Cmil = \frac{2\ K\ L\ I}{V_D}$	$Cmil = \frac{1.732\ K\ L\ I}{V_D}$
To find the distance (one way)	$L = \frac{Cmil\ V_D}{2\ K\ I}$	$L = \frac{Cmil\ V_D}{1.732\ K\ I}$
To find the current in amperes	$I = \frac{Cmil\ V_D}{2\ K\ L}$	$I = \frac{Cmil\ V_D}{1.732\ K\ L}$

EXAMPLE

What size conductor is needed for a load of 10 amps that has a voltage source of 120 volts? The load is 100 feet from the source.

Step 1. Draw the problem.

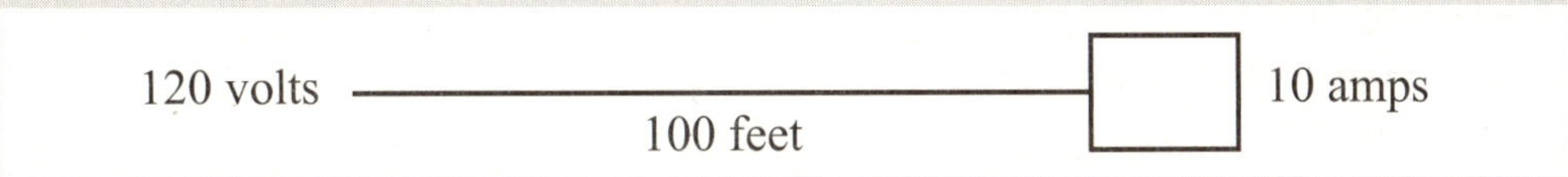

FIGURE 5-8

continues

continued

Step 2. Find the correct voltage drop formula.

The question is asking for the wire size or circular mil size. The voltage is single-phase.

$$\text{Cmil} = \frac{2\ K\ L\ I}{V_D}$$

Step 3. Find all of the known factors and insert them into the formula.

2 = 2 (This is a single-phase circuit.)

K = 12.9 (When copper or aluminum is not specified, use copper. When wire size is not given, use 12.9 for copper and 21.2 for aluminum.)

L = 100 (This is the length (one way).)

I = 10 (This is the current.)

V_D = 3.6 (3% × 120 volts)

Step 4. Multiply letters that are side by side.

$$\text{Cmil} = \frac{2\ K\ L\ I}{V_D} = \frac{2 \times 12.9 \times 100 \times 10}{3.6} = \frac{25{,}800}{3.6} = 7166.6667 = 7167$$

Cmil = 7167

Step 5. Now, look at the Area Circular Mils column in Table 8 of Chapter 9. The circular-mil area for a 12-AWG conductor is 6530. A conductor size of at least 7167 is required to stay within the recommended 3% voltage drop. The next conductor size is 10 AWG with a circular mil size of 10,380.

The size of conductor needed for a load of 10 amperes at 120 volts, 100 feet from the source, is 10-AWG copper.

Always look to see if the type of conductor has been specified, because different conductors have different K values.

EXAMPLE

What size aluminum conductor is required for a load of 30 amperes at 240 volts? The load is 120 feet from the source.

Step 1. Draw the problem.

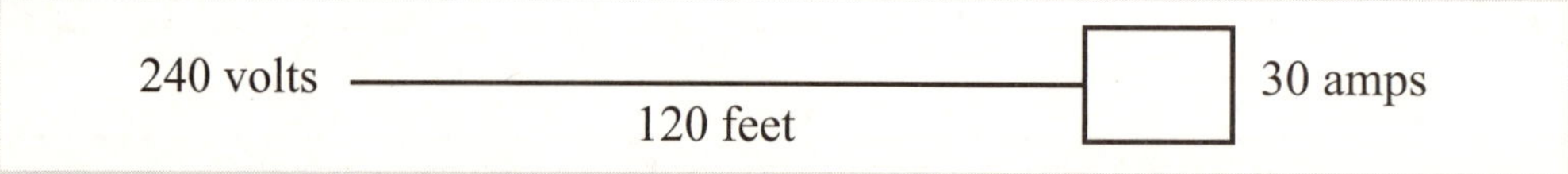

FIGURE 5-9

Step 2. Find the correct voltage drop formula.

The question is asking for the wire size or circular-mil size.

The voltage is single-phase.

$$\text{Cmil} = \frac{2\ K\ L\ I}{V_D}$$

Step 3. Find all of the known factors and insert them into the formula.

2 = 2 (This is a single-phase circuit.)

K = 21.2 (An aluminum conductor has been specified. The approximate K factor for aluminum is 21.2.)

L = 120 (This is the length (one way).)

I = 30 (This is the current.)

V_D = 7.2 (3% × 240 volts)

Step 4. Multiply the numbers on the top line and divide by the numbers on the bottom line.

$$\text{Cmil} = \frac{2\ K\ L\ I}{V_D} = \frac{2 \times 21.2 \times 120 \times 30}{7.2} = \frac{152{,}640}{7.2} = 21{,}200$$

Cmil = 21,200

Step 5. Now look at the Area Circular Mils column in Table 8. The next wire size larger than 21,200 circular mils is 6 AWG with a circular mil area of 26,240.

The size conductor needed for this calculation is 6-AWG aluminum.

Finding Exact K Values

As previously discussed, when the conductor size is not given, an approximate K value may be used (12.9 for copper and 21.2 for aluminum). Each conductor has a different K value. If the conductor size is known, the exact K can be calculated and inserted into the formula.

To find the exact K value, multiply the circular-mil area by the ohm/kFT and divide by 1000.

$$K = \frac{\text{Cmil} \times R}{1000}$$

EXAMPLE

What is the exact K for a 6-AWG copper conductor?

Step 1. Find the circular-mil area and the resistance of the conductor.

Cmil = 26,240

R = 0.491

Step 2. Multiply Cmil by R and divide by 1000.

$$K = \frac{26{,}240 \times 0.491}{1000} = 12.88384$$

The K value for a 6-AWG copper conductor is 12.88384.

NOTE

If the conductor size is known, the exact K should replace the approximate K value.

EXAMPLE

What is the voltage drop of a 208-volt, single-phase circuit that has a load of 20 amperes and is 200 feet away? The circuit is being fed with 8-AWG stranded THW conductors.

Step 1. Draw the problem.

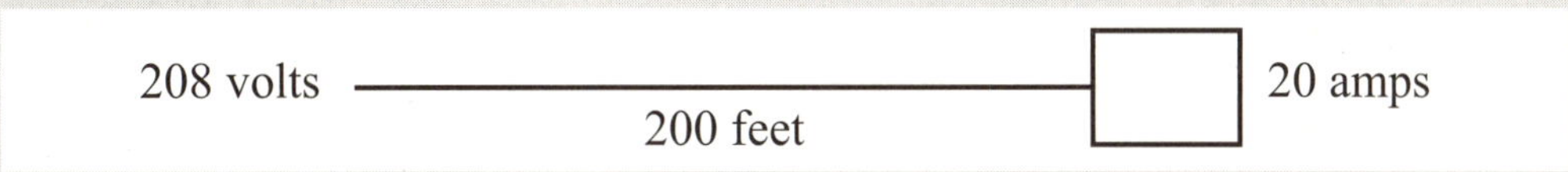

FIGURE 5-10

Step 2. Find the correct voltage drop formula.

$$V_D = \frac{2\ K\ L\ I}{\text{Cmil}}$$

continues

continued

Step 3. Find all of the known factors and insert them into the formula.

$$K = \frac{16{,}510 \times 0.778}{1000} = 12.84478$$

L = 200

I = 20

Cmil = 16,510

$$V_D = \frac{2 \times 12.84478 \times 200 \times 20}{16{,}510} = \frac{102{,}758.24}{16{,}510} = 6.224 \text{ volts}$$

The voltage drop in the circuit above is 6.224 volts.

The recommended voltage drop for 208 volts is 6.24 volts. The voltage drop in this example is within the *NEC* recommendation.

To find the maximum load (current), find the voltage drop formula that begins with I =. Remember that if the conductor size is known, the exact K should replace the approximate K value.

EXAMPLE

What is the maximum load permitted on an 8-AWG THW copper conductor that is supplying a 240-volt branch circuit 140 feet from the source?

Step 1. Draw the problem.

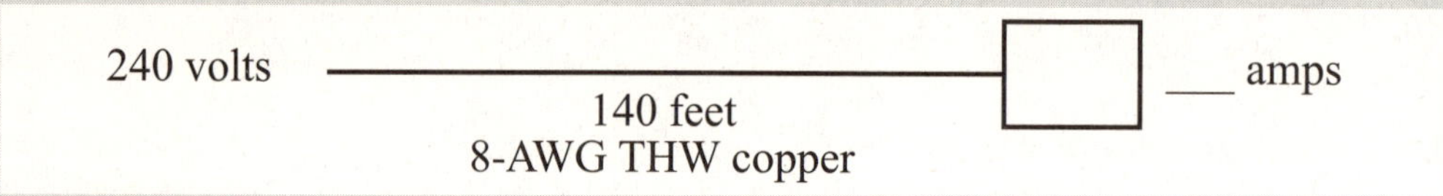

FIGURE 5-11

Step 2. Find the correct voltage drop formula.

$$I = \frac{\text{Cmil } V_D}{2\ K\ L}$$

Step 3. Find all of the known factors and insert them into the formula.

$$K = \frac{16{,}510 \times 0.778}{1000} = 12.84478$$

L = 140

Cmil = 16,510

V_D = 7.2

$$I = \frac{16{,}510 \times 7.2}{2 \times 12.84478 \times 140} = \frac{118{,}872}{3{,}596.5384} = 33 \text{ amps}$$

The maximum load allowed, without exceeding the recommendation, is 33 amperes.

The next example is asking for maximum distance. Note that the voltage is three-phase.

EXAMPLE

What is the maximum distance allowed for 6-AWG THW copper conductors that have a load of 56 amperes? The voltage is 208-volt, three-phase.

Step 1. Draw the problem.

FIGURE 5-12

Step 2. Find the correct voltage drop formula.

$$L = \frac{\text{Cmil } V_D}{1.732 \text{ K I}}$$

Step 3. Find all of the known factors and insert them into the formula.

$$K = \frac{26{,}240 \times 0.491}{1000} = 12.88384$$

I = 56

Cmil = 26,240

$V_D = 6.24$

$$L = \frac{26{,}240 \times 6.24}{1.732 \times 12.88384 \times 56} = \frac{163{,}737.6}{1{,}249.629409} = 131 \text{ feet}$$

The maximum length allowed, without exceeding the recommendation, is 131 feet.

Up to this point, all conductors have been 75° C (167° F). The resistance numbers in Tables 8 and 9 are based on a temperature of 75° C. For 60° C (140° F) and 90° C (194° F) conductors, the K values will be different. For a conductor with a higher temperature rating, the resistance will be higher. For a conductor with a lower temperature rating, the resistance will be lower. As the K values are derived from resistance, they too will change.

The formula for temperature change is at the bottom of Table 8:

$$R_2 = R_1 \left[1 + \alpha (T_2 - 75)\right], \text{ where } \alpha_{cu} = 0.00323,\ \alpha_{AL} = 0.00330 \text{ at } 75° C$$

Table 5-3 Definitions of Formulas

R_1	Resistance (or K value) of the original temperature
R_2	Resistance (or K value) of the new temperature
α	Coefficient for copper or aluminum
T_2	Temperature of the new conductor (60° C or 90° C)

EXAMPLE

What is the exact K for a 12-AWG THHN solid copper conductor?

$$R_1 = \frac{\text{Cmil} \times R}{1000} = \frac{6530 \times 1.93}{1000} = 12.6029$$

$\alpha = 0.00323$

$T_2 = 90$

continues

continued

$R_2 = R_1 [1 + \alpha (T_2 - 75)]$
$R_2 = 12.6029 [1 + 0.00323 (90 - 75)]$
$R_2 = 12.6029 [1 + 0.00323 (15)]$
$R_2 = 12.6029 [1 + 0.00323 \times 15]$
$R_2 = 12.6029 [1 + 0.04845]$
$R_2 = 12.6029 \times 1.04845$
$R_2 = 13.2135105$
The K value for 12-AWG THHN solid copper conductor is 13.2.

However, there is a much easier method for finding the resistance (and K value) of 60° C or 90° C conductors.

- 90° C wire is approximately 5% above 75° C wire.
- 60° C wire is approximately 5% below 75° C wire.
- Resistance for 90° C conductors = Resistance in Table 8 × 1.05
- Resistance for 60° C conductors = Resistance in Table 8 ÷ 1.05

EXAMPLES

What is the resistance for 1000 feet of 10-AWG THHN solid copper wire?
$1.21 \times 1.05 = 1.2705$

What is the resistance for 500 feet of 1-AWG TW copper wire?
$0.154 \times 0.500 = 0.077 \div 1.05 = 0.0733$

Because the K values are derived from circular mils and resistance, when there is a change in resistance, there will be a change in the K value.

- Exact K for 90° C = K × 1.05
- Exact K for 60° C = K ÷ 1.05

EXAMPLES

What is the exact K for a 3-AWG THHN copper conductor?

$$K = \frac{\text{Cmil} \times R}{1000} = \frac{52{,}620 \times 0.245}{1000} = 12.8919 \times 1.05 = 13.536495$$

What is the exact K for a 4-AWG TW aluminum conductor?

$$K = \frac{\text{Cmil} \times R}{1000} = \frac{41{,}740 \times 0.508}{1000} = 21.20392 \div 1.05 = 20.19420952$$

Table 9, Alternating-Current Resistance

Skin effect and induction change the resistance in ac circuits but do not affect dc circuits. Because there are several factors that alter the resistance, there is a table that gives the ohms per 1000 feet for ac circuits.

Table 9 gives the resistance of conductors in PVC, aluminum, and steel raceways. Like Table 8, the resistance is given in ohms per 1000 feet.

NOTE

In Table 9, draw a wider line across the page between 1 AWG and 1/0 AWG. Highlight below the line in the PVC, Aluminum, and Steel Conduit columns for both copper and aluminum conductors (Columns 4-9).

EXAMPLE

What is the ac resistance of 1000 feet of 500-kcmil aluminum in a steel conduit?

R = 0.045

The resistance of a conductor is slightly higher in an ac circuit. Factors that change the resistance in an ac circuit, such as skin effect, are negligible in smaller conductors. When finding the resistance or calculating the K value, any wire smaller than 1/0 AWG can be found in Table 8. The resistance does not increase substantially until the conductor size reaches 1/0 and above.

The resistance and K values in Table 9 are found in the same manner as in Table 8.

EXAMPLES

What is the resistance of 450 feet of 250-kcmil copper in a steel conduit?

$0.054 \times 0.450 = 0.0243$

What is the K value for a 400-kcmil THW copper conductor in a PVC conduit?

$$K = \frac{\text{Cmil} \times R}{1000} = \frac{400{,}000 \times 0.033}{1000} = 13.2$$

When doing a voltage drop calculation for ac conductor sizes 1/0 AWG and larger, use Table 9. If a conduit type is not specified, assume steel.

EXAMPLE

What is the voltage drop of a 208-volt, three-phase circuit that has a load of 300 amps and is 400 feet away? The circuit is being fed with 500-kcmil THW copper conductors.

Step 1. Draw the problem.

FIGURE 5-13

Step 2. Find the correct voltage drop formula.

$$V_D = \frac{1.732 \text{ K L I}}{\text{Cmil}}$$

Step 3. Find all of the known factors and insert them into the formula.

$$K = \frac{500{,}000 \times 0.029}{1000} = 14.5$$

L = 400

I = 300

Cmil = 500,000

$$V_D = \frac{1.732 \times 14.5 \times 400 \times 300}{500{,}000} = \frac{3{,}013{,}680}{500{,}000} = 6.02736 \text{ volts}$$

The voltage drop in the circuit above is approximately 6 volts.

Voltage Drop with a Junction Box in the Circuit

When a junction box is installed in a branch circuit, the recommended voltage drop is still only 3%. Conductors of different materials, different types of insulation, and even different sizes can be joined in a junction box. Two different calculations must be performed if any part of the conductor is different.

NOTE

- Use copper unless aluminum is specified. [110.5]
- Use single-phase unless three-phase is stated.
- Use 75° C conductors unless a different insulation is given.
- Assume ac unless dc is specified.
- If the type of conduit is not specified, assume steel.
- Do not make the mistake of thinking a junction box is a sub feed panel.

EXAMPLE

What size aluminum conductors are required between the junction box and the load in the drawing below? (Do not exceed *NEC* recommendations.)

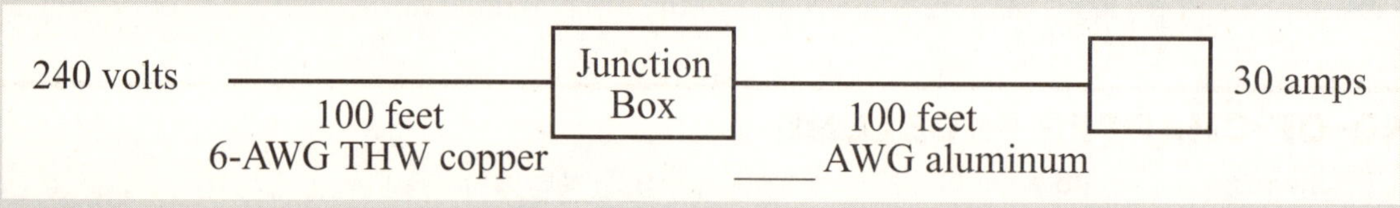

FIGURE 5-14

Step 1. Find the amount of voltage drop in the conductor that is given.

$$V_D = \frac{2\ K\ L\ I}{Cmil} = \frac{2 \times 12.88384 \times 100 \times 30}{26{,}240} = 2.946 \text{ volts}$$

The 6-AWG conductors in this problem have a voltage drop of 2.946 volts.

Step 2. Find the amount of allowable voltage drop for the rest of the circuit. The recommended voltage drop for 240 volts is 7.2 volts in a branch circuit.

7.2 - 2.946 = 4.254 volts

The remaining voltage drop is 4.254 volts.

Step 3. Calculate the size conductor needed for the rest of the circuit. Remember to insert only the remaining voltage drop into the formula.

$$Cmil = \frac{2\ K\ L\ I}{V_D} = \frac{2 \times 21.2 \times 100 \times 30}{4.254} = 29{,}901 \text{ Cmils}$$

The size aluminum conductors needed to comply with *NEC* recommendations are 4-AWG conductors.

Alternative Voltage Drop Formulas Utilizing Ohm's Law

These formulas use the conductor's resistance (R) instead of the conductor's resistivity factor (K). The conductor's resistance per thousand feet (ohm/1000 ft) must replace the R. These resistance values are located in Tables 8 and 9 of *NEC* Chapter 9.

Table 5-4 Alternative Voltage Drop Formulas

To find voltage drop	$V_D = \frac{2\ R\ L\ I}{1000}$
To find the length	$L = \frac{V_D \times 1000}{2\ R\ I}$
To find the current	$I = \frac{V_D \times 1000}{2\ R\ L}$

EXAMPLE

What is the voltage drop of a 208-volt, single-phase circuit that has a load of 20 amperes and is 200 feet away? The circuit is being fed with 8-AWG stranded THW conductors.

Step 1. Find the resistance of the 8-AWG conductors from Table 8.

Uncoated copper (ohm/1000 ft) = 0.778

Step 2. Select the correct formula and replace the letters with the known factors.

$$V_D = \frac{2 \times 0.778 \times 200 \times 20}{1000} = 6.224 \text{ volts}$$

NOTE

The example to the left is also on page 59, where it was calculated with the voltage drop formula.

END-OF-CHAPTER PROBLEMS

1. What is the dc resistance of two 1000-foot 500-kcmil THW conductors that are in parallel?
   a) 0.0145 ohm
   b) 0.0258 ohm
   c) 0.0129 ohm
   d) 0.0516 ohm
2. What is the voltage across the load in the drawing below?
   a) 100 volts
   b) 105.5 volts
   c) 115 volts
   d) 120 volts

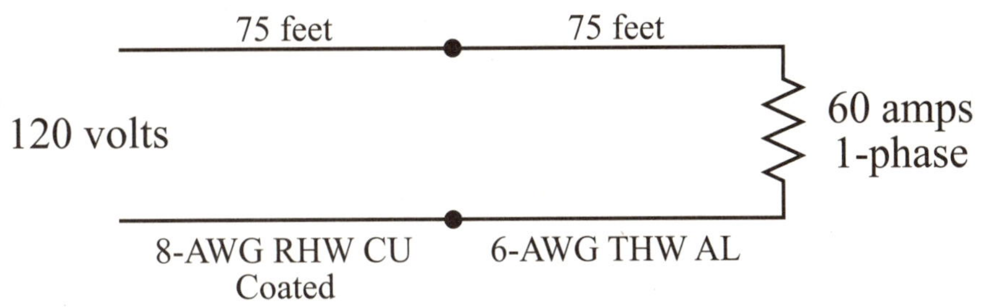

FIGURE 5-15

3. When measured with an ohmmeter, the conductor resistance is 0.021 ohm and the conductor is 1000-kcmil THWN aluminum in a steel raceway. The length of the conductor would be approximately _____ feet.
   a) 875
   b) 991
   c) 1010
   d) 1143

4. What is the approximate length of a conductor that is connecting the service to the load? The conductor is a 6-AWG THW bare uncoated copper conductor with a resistance of 0.035 ohm.
   a) 0.071′
   b) 0.491′
   c) 71′
   d) 1296′
5. What is the resistance of a conductor that is carrying 400 amps and has a voltage drop of 6.4 volts? The source voltage is 208-volt, 3-phase.
   a) 0.016 ohm
   b) 0.0214 ohm
   c) 0.061 ohm
   d) 62.5 ohms
6. What is the resistance of 500 feet of 1000-kcmil AL in a steel conduit?
   a) 0.045 ohm
   b) 0.025 ohm
   c) 0.023 ohm
   d) 0.0125 ohm
7. What size THW conductors are required to carry a load of 120 kVA for a distance of 100 feet? The voltage is 208-volt, three-phase.
   a) 2/0 AWG THW
   b) 3/0 AWG THW
   c) 300 kcmil THW
   d) 400 kcmil THW
8. What size conductors are required to feed a 5-hp, 230-volt, three-phase motor that is 150 feet from the source? Do not exceed *NEC* recommendations.
   a) 12 AWG THW
   b) 12 AWG THHN
   c) 10 AWG THW
   d) 8 AWG THW
9. What is the resistance for 1000 feet of 4/0-AWG THW AL in an aluminum conduit? The circuit is ac.
   a) 0.067 ohm
   b) 0.10 ohm
   c) 0.11 ohm
   d) 0.0608 ohm

10. What 75° C conductors are required to carry 900 amperes if the conductors are paralleled?
    a) 700 kcmil TW
    b) 600 kcmil THW
    c) 500 kcmil THHN
    d) 700 kcmil THW
11. A 4-AWG TW conductor is good for how many kW at 120 volts?
    a) 8.4
    b) 17
    c) 70
    d) 8400
12. What is the dc resistance for two sets of 500-kcmil THWN aluminum conductors that are run in parallel? The total length of each conductor is 1000 feet.
    a) 0.0129 ohm
    b) 0.0212 ohm
    c) 0.0424 ohm
    d) 0.0848 ohm
13. What is the ac resistance of 450 feet of 1/0-AWG RHW AL in an aluminum conduit?
    a) 0.20 ohm
    b) 0.21 ohm
    c) 2.00 ohms
    d) 0.0945 ohm
14. What is the resistance of a conductor that is carrying 400 amps and has a voltage drop of 7.2 volts? The source voltage is 208 volts, three-phase.
    a) 0.0108 ohm
    b) 0.2880 ohm
    c) 0.0180 ohm
    d) 0.009 ohm
15. What size conductors are required to feed a 5-hp, 115-volt, single-phase motor that is 150 feet from the source? Do not exceed *NEC* recommendations.
    a) 1 AWG THW
    b) 2 AWG THW
    c) 3 AWG THW
    d) 4 AWG THW

The drawing below refers to Question 16.

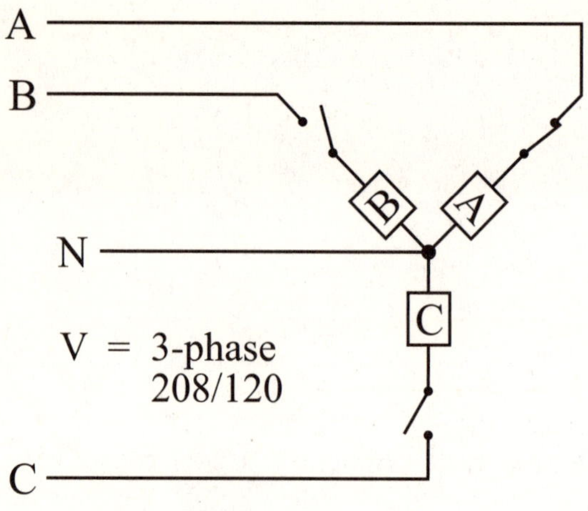

FIGURE 5-16

16. If load A is 7.2 kVA and the conductor size is 6-AWG THW copper, how far from the source can load A be located and not exceed 3% voltage drop?

a) 30.5′

b) 61′

c) 102′

d) 106′

17. Three 4-AWG THW conductors are connecting a 3-hp, 208-volt, three-phase motor to a panel. The voltage drop is 3%. What is the resistance of each conductor in ohms?

a) 0.34

b) 1104

c) 0.308

d) 1.104

18. What is the voltage drop on two 12-AWG stranded uncoated THW conductors 150 feet long, connecting a ½-hp, 115-volt, single-phase motor? The source voltage is 115 volts.

a) 5.8212 volts

b) 5.6742 volts

c) 3.4500 volts

d) None of these

19. What is the ac resistance of 545 feet of 500-kcmil AL THW wire in an aluminum conduit?

a) 0.023 ohm

b) 0.026 ohm

c) 0.0424 ohm

d) 0.048 ohm

20. What is the resistance of 500 feet of 500-kcmil THW aluminum in a magnetic conduit?

a) 0.0212 ohm

b) 0.0215 ohm

c) 0.0225 ohm

d) 0.024 ohm

21. What size aluminum conductors are required to carry a load of 22 kVA for a distance of 215 feet? The voltage is 208-volt, three-phase. Do not exceed *NEC* recommendations.
  a) 3 AWG THW
  b) 2 AWG THW
  c) 1 AWG THW
  d) 1/0 AWG THW

The drawing below refers to Question 22.

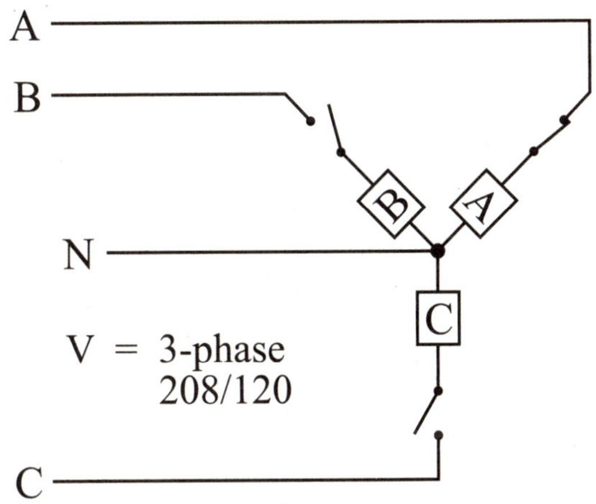

FIGURE 5-17

22. What would load A be in amps if located 90′ from the source and fed by 3-AWG THW conductors and limited to 3% voltage drop?
  a) 72 amps
  b) 82 amps
  c) 92 amps
  d) 102 amps

23. What is the ac resistance for 1000 feet of 4/0-AWG THW copper in a steel raceway?
  a) 0.0608 ohm
  b) 0.062 ohm
  c) 0.063 ohm
  d) 0.067 ohm

24. Where bare conductors are permitted by other sections of the *NEC*, the dimensions for bare conductors in _____ shall be permitted.
  a) Table 310.13
  b) Table 310.16
  c) Chapter 8, Table 9
  d) Chapter 9, Table 8

25. What is the voltage at the load in the drawing below?
   a) 105.5 volts
   b) 108 volts
   c) 112 volts
   d) 116 volts

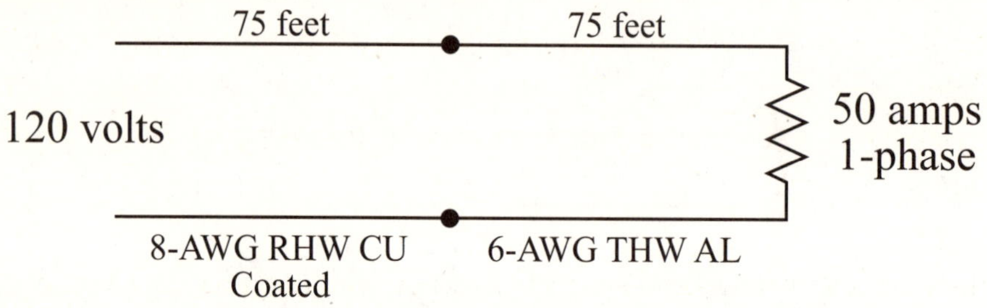

FIGURE 5-18

The drawing below refers to Question 26.

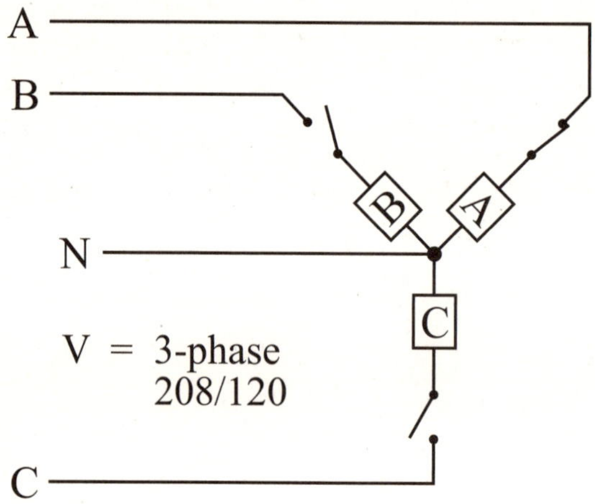

FIGURE 5-19

26. If load A is 12 kVA and the conductor size is a 1-AWG THW copper conductor, how far from the source can load A be located and not exceed 3% voltage drop?
   a) 116′
   b) 134′
   c) 404′
   d) 701′

27. What size conductors are required to feed a 240-volt, single-phase motor with a full load current of 25 amperes? The length is 225 feet. Do not exceed *NEC* recommendations.

a) 2 AWG

b) 4 AWG

c) 6 AWG

d) 8 AWG

28. What is the dc resistance for 1500 feet of 8-AWG uncoated stranded copper conductor?

a) 0.778 ohm

b) 1.167 ohms

c) 1.179 ohms

d) 1.214 ohms

29. What is the voltage drop on 6-AWG uncoated copper conductors having a load of 6150 volt-amps at 240-volts, single-phase? The circuit is an ac, 2-wire circuit and the length is 450 feet.

a) 5.66 volts

b) 9.8 volts

c) 11.3 volts

d) 11.8 volts

30. What is the approximate length of a 6-AWG THW uncoated copper conductor that has a resistance of 0.04 ohm?

a) 10′

b) 82′

c) 164′

d) 410′

CHAPTER 6

Conductor Ampacity Correction Factors

The maximum conductor ampacities are listed in Tables 310.16 through 310.20 of the *National Electrical Code*. A definition of ampacity is in Chapter 1, Article 100: "**Ampacity.** The current, in amperes, that a conductor can carry continuously under the conditions of use without exceeding its temperature rating."

The maximum ampacities given in Tables 310.16 through 310.20 are based on a maximum of three current carrying conductors and a maximum ambient temperature of 30° C (86° F). When there are more than three current carrying conductors or the ambient temperature is more than 30° C, the ampacities listed must be reduced or derated.

When the ambient temperature exceeds 30° C, the table ampacity must be multiplied by the correction factors. The ambient temperature correction factors are on the pages where the ampacities are listed. When there are more than three current carrying conductors, the table ampacity must be multiplied by the correction factors in Table 310.15(B)(2)(a).

There are four questions relating to conductor ampacities.

1. **What is the table ampacity?**

   This is the ampacity taken straight from the tables in the *NEC* before any correction factors have been applied.

2. **What is the adjusted table ampacity?**

   This is the maximum ampacity after the correction factors have been applied. It is also the maximum ampacity if the load is rated as noncontinuous.

3. **What is the maximum overcurrent protective device?**

   After the adjusted table ampacity is known, the overcurrent protection can be determined. Where the adjusted ampacity does not correspond to a standard size fuse or circuit breaker, the next higher standard size shall be permitted only if it does not exceed 800 amperes. [240.4(B)] See 240.6(A) for standard ampere ratings of fuses and breakers. Certain conductors in Table 310.16 are marked with an asterisk (*), referencing 240.4(D), which limits the maximum overcurrent protection for 14-AWG, 12-AWG, and 10-AWG conductors. [240.4(D)]

4. **What is the maximum continuous load?**

   A load where the maximum current is expected to continue for three hours or more is a continuous load. For branch-circuit and feeder conductors supplying continuous loads, the ampacity of the load must be multiplied by 125 percent before the application of any adjustment or correction factors. [210.19(A) and 215.2(A)] Where a branch circuit or feeder supplies a continuous load, the overcurrent protective device must not be less than 125 percent of the continuous load. [210.20(A) and 215.3] The full load current on conductors supplying a single motor, used in a continuous duty application, shall also be multiplied by 125 percent. [430.22(A)]

Ambient Temperature Correction Factors

Ambient temperature is the temperature of the environment surrounding a conductor. It is the temperature of the environment through which the conductor or raceway is installed. This environment may include air, water, earth, or a combination thereof. The ambient temperature chosen for the correction factor has to be the highest possible temperature from the source to the load.

Ampacity correction factors for ambient temperatures are found at the bottom of Tables 310.16 through 310.20. These tables are used to correct the ampacity of conductors by multiplying the conductor's ampacity by the appropriate factor.

EXAMPLES

What is the ampacity of a 6-AWG THW conductor in a boiler room that reaches temperatures of 135° F?

Step 1. Look at Table 310.16 to find the conductor's ampacity.

6-AWG THW = 65 amperes (the table ampacity)

Step 2. Find the ambient temperature correction factor at the bottom of Table 310.16. Follow down the same column of the conductor and find the factor that intersects with the given temperature.

135° F = 0.58

This number is found under the 75° C (167° F) column (where THW is located) across from 132-140° F.

Step 3. Multiply the two factors. The result is the conductor's adjusted table ampacity.

65 × 0.58 = 37.7 = 38 amperes (the adjusted table ampacity)

The maximum ampacity of a 6-AWG conductor in this temperature is only 38 amperes. Because of the high temperature surrounding the conductors, the heat generated through the conductor would not dissipate if more than 38 amperes were applied.

What is the maximum overcurrent protective device in the previous example? The adjusted table ampacity in the last example was 38 amperes. This ampacity does not correspond with a standard size fuse or circuit breaker. The next higher standard size is 40 amperes.

Because the conductor is in an ambient temperature of 135° F, the maximum size fuse or breaker permitted to protect this 6-AWG THW copper conductor is 40 amperes.

125 Percent Versus 80 Percent

Before the 2002 edition of the *National Electrical Code,* the article covering panelboards contained a requirement pertaining to overcurrent protection for continuous loads. The requirement stipulated that if the load was continuous, the total load on any overcurrent device located in a panelboard could not exceed 80 percent of its rating. This requirement was removed because it was stated in other articles. Although the wording was different, the meaning was the same.

In accordance with 210.20(A), the rating of the branch-circuit overcurrent device must not be less than 125 percent of the continuous load. The old requirement for panelboards stated that if the load was continuous, the load on the overcurrent device could not exceed 80 percent of its rating. Both statements are saying the same thing; it depends on the point of view. The reciprocal of 125 percent is 80 percent. If the conductor ampacity is known and the load will be continuous, multiply the conductor's ampacity by 80 percent. If the fuse or circuit breaker size is known and the load will be continuous, multiply the rating of the fuse or circuit breaker by 80 percent. If the load is known, and the load will be continuous, multiply by 125 percent to size the conductor or overcurrent device.

EXAMPLES

What is the maximum continuous load permitted on a 20-ampere breaker?

20 × 80% = 16 amperes

What size circuit breaker is required for a 16-ampere continuous load?

16 × 125% = 20-ampere breaker

The asterisk (*) in Table 310.16 plays a very important role when calculating overcurrent protection for 14-AWG, 12-AWG, and 10-AWG conductors. Unless specifically permitted in 240.4(E) or 240.4(G), the overcurrent protection shall not exceed 15 amperes for 14-AWG, 20 amperes for 12-AWG, and 30 amperes for 10-AWG copper. Aluminum and copper-clad conductors also have maximum sizes: 15 amperes for 12 AWG and 25 amperes for 10 AWG. [240.4(D)]

EXAMPLES

What is the maximum overcurrent protection for a 10-AWG THHN conductor in an ambient temperature of 40° C?

The table ampacity is 40 amperes. The adjusted table ampacity is 36.4 amperes.

$40 \times 0.91 = 36.4 = 36$ amperes

The maximum overcurrent protective device for this conductor is 30 amperes. Normally when the adjusted ampacity of a conductor does not match a standard size fuse or breaker, the next size up is permissible. Because this is a 10-AWG conductor, the overcurrent protection is limited to a maximum ampacity of 30 amperes.

What is the maximum ampacity for 12-AWG THW conductors that are feeding a continuous load? The ambient temperature is 90° F.

The table ampacity is 25 amperes. The adjusted table ampacity is 24 amperes.

$25 \times 0.94 = 23.5 = 24$ amperes

In accordance with 240.4(D), the maximum overcurrent protective device is 20 amperes. Because of the overcurrent device and because the load is continuous, the maximum load permitted is 16 amperes.

$20 \times 80\% = 16$ amperes

To check this answer, multiply 16 amperes by 125 percent. The result will be the maximum overcurrent protective device permitted on a 20-ampere fuse or breaker feeding a continuous load.

$16 \times 125\% = 20$ amperes

The correction factors can also be used to find a conductor size for a certain load at a certain temperature. To find the adjusted table ampacity, simply divide the load by the temperature correction factor.

EXAMPLE

What size THW conductors are required for a 70-ampere load on a roof that reaches temperatures of 120° F?

a) 6 AWG b) 4 AWG c) 3 AWG d) 2 AWG

$$\text{Adjusted Table Ampacity (ATA)} = \frac{\text{Load}}{\text{Temperature Correction Factor}}$$

$$\text{ATA} = \frac{70}{0.75} = 93.3 = 93 \text{ amperes}$$

Look at Table 310.16 and find a THW conductor that has an ampacity rating of at least 93 amperes. The size THW conductors required for a 70-ampere load, in a temperature of 120° F, are 3-AWG THW copper conductors.

It may be easier to use an alternative method during an electrical exam. Start by derating each conductor's ampacity in the answers. The first conductor with enough ampacity for the required load (after derating) is the correct answer. The required ampacity after derating is 70 amperes.

6 AWG = $65 \times 0.75 = 49$

This conductor is not large enough before derating.

4 AWG = $85 \times 0.75 = 64$

This conductor is not large enough after derating.

3 AWG = $100 \times 0.75 = 75$

After derating, this conductor can carry the load.

Table 310.15(B)(2)(a) Correction Factors

Table 310.15(B)(2)(a) is another ampacity correction factor. This table is for raceway or cable fill. Tables 310.16 through 310.20 are rated for three conductors or fewer. Table 310.15(B)(2)(a) gives the multiplication factors for more than three conductors.

The following shall not be counted as current carrying conductors:

- Conductors of different systems [310.15(B)(2)(a) Ex. No.1]
- Conductors in cable trays [310.15(B)(2)(a) Ex. No. 2]
- Conductors in nipples of 24 inches or less [310.15(B)(2)(a) Ex. No. 3]
- Neutral conductors of normally balanced circuits containing three or more conductors [310.15(B)(4)(a)]
- A grounding or bonding conductor [310.15(B)(5)]

The following shall be counted as current carrying conductors:

- All ungrounded (hot) conductors
- The neutral conductor in a 3-wire circuit consisting of two phase wires and the neutral that is fed from a 4-wire, 3-phase wye-connected system [310.15(B)(4)(b)]
- The neutral conductor of a 4-wire, 3-phase wye circuit where the major portion of the load consists of nonlinear loads, because there are harmonic currents present in the neutral conductor. Examples include fluorescent lights, lights with ballast, and computer equipment. [310.15(B)(4)(c)]
- The grounded conductor of any 2-wire circuit

EXAMPLES

What is the ampacity of a 6-AWG THW conductor in a conduit that has a total of five conductors?

Because nothing is specified, count all five conductors as current carrying conductors.

The ampacity of 6-AWG THW from Table 310.16 = 65 amperes

Table 310.15(B)(2)(a) correction factor for 5 conductors = 80%

65 × 80% = 52 amperes

The maximum current a 6-AWG THW copper conductor can carry in a conduit with four other conductors is 52 amperes.

What is the maximum overcurrent protection for 10-AWG THW conductors in an 18″ conduit? The conduit contains 12 conductors.

The table ampacity is 35 amperes. The adjusted table ampacity is 35 amperes.

The conduit is a nipple. Conductors in nipples (24 inches or less) are not derated.

The maximum overcurrent protective device is 30 amperes.

Although a 35-ampere overcurrent protective device is a standard rating (240.6), it is not permissible. Because of 240.4(D), a 10-AWG conductor is limited to a maximum ampacity of 30 amperes.

The correction factors for Table 310.15(B)(2)(a) can also be used to find a conductor size for a certain load when the load and the number of conductors are known. To find the adjusted table ampacity, simply divide the load by the correction factor in Table 310.15(B)(2)(a).

EXAMPLE

What size THW conductors are required for a load of 70 amperes? The conductors are in a conduit with a total of six conductors.

a) 6 AWG b) 4 AWG c) 3 AWG d) 2 AWG

$$\text{Adjusted Table Ampacity (ATA)} = \frac{\text{Load}}{\text{Table 310.15(B)(2)(a) Correction Factor}}$$

$$\text{ATA} = \frac{70}{0.80} = 87.5 = 88 \text{ amperes}$$

Now, look at Table 310.16 and find a THW conductor that has an ampacity rating of at least 88 amperes. The size conductors required for this 70-ampere load are 3-AWG THW.

It may be easier to use an alternative method during an electrical exam. Start by derating each conductor's ampacity in the answers. The first conductor with enough ampacity for the required load (after derating) is the correct answer. The required ampacity, after derating, is 70 amperes.

6 AWG = 65 × 80% = 52

This conductor is not large enough before derating.

4 AWG = 85 × 80% = 68

This conductor is not large enough after derating.

3 AWG = 100 × 80% = 80

After derating, this conductor can carry the load.

Combination Ambient Temperature and Table 310.15(B)(2)(a) Correction Factors

Until now, ambient temperature and Table 310.15(B)(2)(a) correction factors have been separate. There are times when both correction factors will apply.

EXAMPLE

What is the ampacity of a 6-AWG THW conductor in a boiler room that reaches temperatures of 115° F? There are a total of five conductors in the conduit.

The ampacity of 6 AWG THW from Table 310.16 = 65 amperes

The multiplication factor for 115° F = 0.75

The Table 310.15(B)(2)(a) correction factor for five conductors = 80%

65 × 0.75 × 80% = 39 amperes

The maximum current that a 6-AWG THW copper wire can carry in a conduit with four other conductors in a boiler room that reaches temperatures of 115° F is 39 amperes.

The correct conductor size can be found by dividing the load by both the ambient temperature correction factor and the Table 310.15(B)(2)(a) correction factor.

EXAMPLE

What size THW conductors are required for a load of 70 amperes on a roof that is run in a conduit with a total of six conductors? The roof will reach temperatures of 120° F.

a) 3 AWG b) 2 AWG c) 1 AWG d) 1/0 AWG

$$\text{ATA} = \frac{\text{Load}}{\text{Temperature Correction} \times \text{Table 310.15(B)(2)(a)}}$$

$$\text{ATA} = \frac{70}{0.75 \times 0.80} = 116.7 = 117 \text{ amperes}$$

Now look at Table 310.16 and find a THW conductor that has an ampacity rating of at least 117 amperes. 1-AWG THW copper conductors are required for this 70-ampere load.

Alternative Method (The required ampacity, after derating, is 70 amperes.):
3 AWG = 100 × 0.75 × 80% = 60
This conductor is not large enough after derating.
2 AWG = 115 × 0.75 × 80% = 69
This conductor is not large enough after derating.
1 AWG = 130 × 0.75 × 80% = 78
After derating, this conductor can carry the load.

Ambient Temperature and Table 310.15(B)(2)(a) Correction Factors for a Single Motor

Branch-circuit conductors supplying a single motor used in a continuous duty application must have an ampacity of at least 125 percent of the motor full-load current rating. [430.22(A)] Do not use the nameplate rating of the motor. Use the full-load currents found in Tables 430.247 through 430.250.

The ampacity of conductors that feed the motor must be at least 125 percent of the full-load current of the motor. If ambient temperature and/or Table 310.15(B)(2)(a) applies to a single motor, multiply the full-load current by 125 percent, then continue with the calculation.

EXAMPLES

What size THW conductors are required for a 2-hp, 240-volt motor in a room where the ambient temperature is 120° F? The conduit feeding the motor contains six conductors.

a) 14 AWG b) 12 AWG c) 10 AWG d) 8 AWG

Step 1. Find the full-load current of the motor.
12 amperes [Table 430.248]

Step 2. Insert 12 amperes into the formula as the load and multiply by 125 percent.

$$\text{ATA} = \frac{\text{Motor FLC} \times 125\%}{\text{Temperature Correction} \times \text{Table 310.15(B)(2)(a)}}$$

$$\text{ATA} = \frac{12 \times 125\%}{0.75 \times 80\%} = 25 \text{ amperes}$$

The size conductors required are 12-AWG THW copper conductors.

Alternative Method:
If the conductor ampacity is known, and the load is a single motor, multiply the conductor's ampacity by 80 percent. Because the alternative method derates the conductor's ampacity, use the reciprocal of 125 percent. This is the first 80 percent in the formula below. The temperature correction factor is 0.75 and the Table 310.15(B)(2)(a) correction factor is 80%. The required ampacity, after derating, is 12 amperes.

14 AWG = 20 × 80% × 0.75 × 80% = 9.6

This conductor is not large enough.

12 AWG = 25 × 80% × 0.75 × 80% = 12

This conductor can carry the load.

When a conductor size is known and the maximum full-load current of a single motor is needed, multiply the conductor's ampacity by the ambient temperature correction factor, Table 310.15(B)(2)(a) correction factor, and by 80%.

EXAMPLE

What is the maximum ampacity a 6-AWG THHN conductor can carry to a motor when the ambient temperature is 103° F and there are seven conductors in the conduit?

6-AWG THHN = 75 amperes

Temperature correction at 103° F = 0.91

Seven conductors in the conduit = 70%

Conductor derating factor for a single motor = 80% (reciprocal of 125%)

$75 \times 0.91 \times 70\% \times 80\% = 38.22 = 38$ amperes

The maximum full-load current for a motor in the above problem is 38 amperes.

Ambient Temperature and Table 310.15(B)(2)(a) Correction Factors for Continuous Loads

For branch-circuit and feeder conductors supplying continuous and noncontinuous loads, the minimum conductor size (before the application of any adjustment or correction factors) must have an allowable ampacity not less than the noncontinuous load plus 125 percent of the continuous load. [210.19(A) and 215.2(A)] If an exam question specifies the load is continuous and does not mention that any of it is noncontinuous, calculate the total load as continuous. If ambient temperature is not a factor and the number of current-carrying conductors is less than four, multiply the load by 125 percent and select the conductor.

EXAMPLE

What size THW conductors are required for a 95-ampere continuous load?

a) 3 AWG b) 2 AWG c) 1 AWG d) 1/0 AWG

Step 1. Because the load is continuous, multiply the 95-ampere load by 125 percent.

$95 \times 125\% = 118.8 = 119$ amperes

Step 2. Select a THW conductor from Table 310.16 that can carry at least 119 amperes.

1-AWG THW copper

The minimum size THW conductors required for a 95-ampere continuous load are 1-AWG THW copper conductors.

When continuous loads and conductor correction factors are in the same problem, several factors must be considered. The ampacity of conductors feeding continuous loads must not be less than 125 percent of the continuous load. The overcurrent device must protect the conductors after all correction factors have been applied.

EXAMPLE

What size THW conductors are required for a 95-ampere continuous load in an ambient temperature of 31° C?

a) 3 AWG b) 2 AWG c) 1 AWG d) 1/0 AWG

It is not necessary to multiply the load by 125 percent before applying the ambient temperature. The correction factor for a THW conductor in an ambient temperature of 31° C is 0.94.

$$\text{Adjusted Table Ampacity (ATA)} = \frac{95}{0.94} = 101 \text{ amperes}$$

In accordance with Table 310.16, a 2-AWG copper conductor can carry 101 amperes.

Although a 2-AWG THW copper conductor can carry the load in an ambient temperature of 31° C, another calculation is necessary. Because the load is continuous, the conductors must have an allowable ampacity of at least 125 percent of the continuous load.

95 × 125% = 118.8 = 119 amperes = 1 AWG

Selecting the larger of the two conductors will satisfy both calculations. The minimum size THW conductors required for this example are 1-AWG THW copper conductors.

Overcurrent Protection

If the load is continuous, overcurrent protection must be considered. The overcurrent protective device must provide overcurrent protection for the conductors in accordance with their ampacity. [240.4] After calculating and selecting the conductor size, calculate the overcurrent protective device for the continuous load. The rating of the branch-circuit or feeder overcurrent device must not be less than 125 percent of the continuous load. (If the exam question includes noncontinuous loads, include the noncontinuous loads at 100 percent.) Next, multiply the selected conductor's ampacity by the correction factors. Where the adjusted ampacity does not correspond to a standard size fuse or circuit breaker, the next higher standard size shall be permitted. This provision does not apply when the rating exceeds 800 amperes. (See 240.4(B).) If the rating of the next higher size fuse or breaker is less than was calculated for the continuous load, the size of the conductor is not large enough.

EXAMPLE

What size THW conductors are required for a 95-ampere continuous load? The conduit contains six conductors.

a) 3 AWG b) 2 AWG c) 1 AWG d) 1/0 AWG

Step 1. Apply the conductor ampacity correction factor to the load in the question. The correction factor for six current-carrying conductors in a raceway is 80 percent.

$$\text{ATA} = \frac{95}{0.80} = 118.8 = 119 \text{ amperes} = 1 \text{ AWG}$$

Step 2. Select a THW conductor from Table 310.16 that can carry at least 119 amperes.

1-AWG THW copper

Although 1-AWG THW copper conductors can carry the 95 amperes after derating, this calculation is not finished.

Step 3. Find the minimum overcurrent protective device for this continuous load. Multiply the continuous load by 125 percent and select the overcurrent protective device.

95 × 125% = 118.8 = 119 amperes

Because 119 amperes is not a standard size fuse of circuit breaker, the next higher standard size is permitted. The minimum size overcurrent protection for a 95-ampere continuous load is 125 amperes.

Step 4. Determine if a 125-ampere fuse or circuit breaker will protect the conductor after the conductor's ampacity has been derated.

Multiply the conductor's ampacity (from Table 310.16) by the correction factor in Table 310.15(B)(2)(a).

130 × 80% = 104 amperes

The ampacity of a 1-AWG THW copper conductor is 130 amperes.

The next higher standard size fuse or circuit breaker above 104 is 110 amperes. [240.6(A)] The maximum size fuse or circuit breaker permitted to provide overcurrent protection for a 1-AWG THW copper conductor (with six current-carrying conductors in a raceway) is 110 amperes.

continues

NOTE

If the conductor insulation in the previous question were THHN, 1 AWG would be the correct answer. A 1-AWG THHN copper conductor has a table ampacity of 150 amperes.

continued

Because of the continuous load, this circuit must be protected by a 125-ampere fuse or circuit breaker. Therefore, 1-AWG THW copper conductors are not large enough. Try the calculation again using the next larger conductor. The ampacity of a 1/0-AWG THW copper conductor is 150 amperes.

$150 \times 80\% = 120$ amperes

The next standard size fuse or circuit breaker above 120 amperes is 125.

The minimum size conductors permitted for a 95-ampere continuous load (with six current-carrying conductors in a conduit) are 1/0-AWG THW copper conductors.

It is not necessary to multiply the load by 125 percent before applying ambient temperature correction factors and Table 310.15(B)(2)(a) correction factors. (See Example D3(a) in Annex D of the *National Electrical Code.*)

EXAMPLES

What size THW conductors are required for a 95-ampere continuous load in an ambient temperature of 31° C? The conduit contains six conductors.

a) 2 AWG b) 1 AWG c) 1/0 AWG d) 2/0 AWG

Step 1. Apply the conductor ampacity correction factor to the load in the question. The correction factor for a THW conductor in an ambient temperature of 31° C is 0.94. The correction factor for six current-carrying conductors in a raceway is 80 percent.

$$\text{ATA} = \frac{95}{0.94 \times 0.80} = 126 \text{ amperes}$$

Step 2. Select a THW conductor from Table 310.16 that can carry at least 126 amperes.

1-AWG THW copper

Step 3. Find the minimum overcurrent protective device for this continuous load. Multiply the continuous load by 125 percent and select the overcurrent protective device.

$95 \times 125\% = 118.8 = 119$ amperes

The next higher standard size fuse of circuit breaker has a rating of 125 amperes.

Step 4. Apply the correction factors to the 1-AWG THW copper conductors.

$130 \times 0.94 \times 80\% = 98$ amperes

The next higher standard size fuse or circuit breaker above 98 amperes is 100. This circuit must be protected by a 125-ampere fuse or circuit breaker. Therefore, this conductor is not large enough.

Try the calculation again using the next larger conductor. The ampacity of a 1/0-AWG THW copper conductor is 150 amperes.

$150 \times 0.94 \times 80\% = 113$ amperes

The next standard size fuse or circuit breaker above 113 amperes is 125.

The minimum size conductors permitted for a 95-ampere continuous load (in an ambient temperature of 31° C and with six current-carrying conductors in a conduit) are 1/0-AWG THW copper conductors.

What size THW conductors are required for a lighting load of 20 amperes in a commercial building where the ambient temperature is 47° C? The lighting load is continuous and the conduit feeding the load contains eight conductors.

a) 12 AWG b) 10 AWG c) 8 AWG d) 6 AWG

The correction factor for a THW conductor in an ambient temperature of 47° C is 0.75. The correction factor for eight current-carrying conductors in a raceway is 70 percent.

$$\text{ATA} = \frac{20}{0.75 \times 0.70} = 38 \text{ amperes} = 8 \text{ AWG}$$

The minimum overcurrent protective device for this continuous load is 25 amperes.

$20 \times 125\% = 25$ amperes

Apply the correction factors to the 8-AWG THW copper conductors.

$50 \times 0.75 \times 70\% = 26$ amperes

The minimum size THW conductors required for this example are 8-AWG THW copper conductors. Although 8-AWG THW conductors are required, the maximum size overcurrent protection is limited to 30 amperes. After derating, the conductor's allowable ampacity is 26 amperes. The next standard size fuse or circuit breaker above 26 amperes is 30 amperes.

END-OF-CHAPTER PROBLEMS

1. Three heater loads are connected to a 208-volt, three-phase source. Each heater is connected to a phase wire and a grounded wire. How much should the ampacity be derated, if at all? (There is also a grounding conductor in the conduit.)
   a) 70%
   b) 80%
   c) 90%
   d) Not at all
2. What is the ampacity of an 8-AWG XHHW in a wet location at 26 degrees C? There are not more than three conductors in the conduit.
   a) 50 amperes
   b) 55 amperes
   c) 60 amperes
   d) 70 amperes
3. What is the maximum overcurrent protection for each of twenty 12-AWG THW conductors pulled in a 1″ conduit nipple that is 18″ long?
   a) 25 amperes
   b) 20 amperes
   c) 15 amperes
   d) None of these
4. What size THHN copper conductors are required to carry a 70-ampere load of three banks of fluorescent lights? The lights are connected three-phase wires to one neutral. The source voltage is 208/120-volt, three-phase. The three current-carrying conductors and the neutral are in the same raceway. The ambient temperature is 30° C, and the lighting load is continuous.
   a) 6 AWG
   b) 4 AWG
   c) 3 AWG
   d) 2 AWG

5. What is the minimum size branch circuit conductor for a 7½-hp, 3-phase, 208-volt motor? There will be three conductors in the conduit in an ambient temperature of 125° F. Use THW conductors.
   a) 12 AWG
   b) 10 AWG
   c) 8 AWG
   d) 6 AWG
6. What is the ampacity of a 10-AWG THW conductor in an ambient temperature of 76° F?
   a) 24 amperes
   b) 30 amperes
   c) 35 amperes
   d) 37 amperes
7. What is the overcurrent protection for a 10-AWG THHN copper conductor at 38° C?
   a) 35 amperes
   b) 36.4 amperes
   c) 40 amperes
   d) 30 amperes
8. What is the maximum overcurrent protection for a 10-AWG THW conductor in an ambient temperature of 30° C?
   a) 25 amperes
   b) 30 amperes
   c) 35 amperes
   d) 40 amperes
9. What is the maximum ampacity for one 12-AWG THW conductor that is in a raceway with five other 12-AWG THW conductors? The conduit also contains two 12-AWG grounding conductors for a total of eight conductors in the raceway.
   a) 16 amperes
   b) 17.5 amperes
   c) 20 amperes
   d) 25 amperes
10. What is the derating factor, if any, for forty-three 12-AWG THW conductors in a conduit 18″ long?
   a) 60%
   b) 50%
   c) 35%
   d) Do not derate.
11. What size THW conductors are required to carry a 3-hp, 208-volt, single-phase motor? There will be a total of eight conductors in the conduit and the ambient temperature is 42° C.
   a) 12 AWG
   b) 10 AWG
   c) 8 AWG
   d) 6 AWG

12. What is the maximum overcurrent protection for four 12-AWG THHN conductors that are feeding lighting loads? The ambient temperature is 30° C.
   a) 20 amperes
   b) 25 amperes
   c) 30 amperes
   d) None of these
13. What size THW conductors are required to feed an 8-kVA, 240-volt, 120-gallon fixed storage-type water heater?
   a) 10 AWG
   b) 8 AWG
   c) 6 AWG
   d) 4 AWG
14. What is the maximum ampacity of a 10-AWG THHN conductor in an ambient temperature of 30° C? There are six conductors in the raceway.
   a) 32 amperes
   b) 30 amperes
   c) 24 amperes
   d) None of these
15. What is the derating factor, if any, for twenty-four conductors in a 25″ conduit nipple?
   a) 70%
   b) 50%
   c) 45%
   d) Do not derate.
16. What is the maximum ampacity for a 10-AWG THHN conductor in an ambient temperature of 38° C?
   a) 30 amperes
   b) 36.4 amperes
   c) 40 amperes
   d) None of these
17. What is the ampacity of a 3-AWG THWN copper conductor in an ambient temperature of 110° F? The conduit contains no more than three conductors.
   a) 110 amperes
   b) 100 amperes
   c) 85 amperes
   d) 82 amperes
18. What is the maximum ampacity for 4-AWG RHH conductors? They feed a motor operating in a poorly ventilated area where the ambient temperature is 43° C.
   a) 86.45 amperes
   b) 82.65 amperes
   c) 75 amperes
   d) 66.12 amperes

19. What is the maximum ampacity of a 10-AWG RHW copper conductor? There are four current-carrying conductors in the conduit and the ambient temperature is 70° F.
    a) 35 amperes
    b) 30 amperes
    c) 29.4 amperes
    d) 28 amperes
20. What is the maximum load a 4-AWG TW can carry, in kW? The conductors are feeding store parking lot lighting connected to a 120-volt source.
    a) 3–4
    b) 5–7
    c) 8–9
    d) 10–15
21. A conduit contains two sets of three-phase conductors, two neutrals, and two grounding conductors. What is the reduction (if any) per conductor? The conductors are feeding electric discharge lighting.
    a) 60%
    b) 70%
    c) 80%
    d) None of these
22. Three heater loads are connected to a 208-volt, three-phase source. Each heater is connected to a phase wire and a common neutral. How much should the ampacity be derated, if at all? There is also a grounding conductor in the conduit.
    a) 70%
    b) 80%
    c) 90%
    d) Not at all
23. If a nipple holds eight 8-AWG XHHWs and the area is defined as a wet location, the ampacity of each 8-AWG XHHW would be _____ amperes.
    a) 35
    b) 40
    c) 50
    d) 55
24. The source is 3-phase, 208/120 volt. The load is three heaters fed line to neutral. The three current-carrying conductors and the three neutrals are all in the same conduit. If the load current is 40 amperes, what size THHN conductors would be required? The ambient temperature is 30° C (all copper conductors).
    a) 10 AWG
    b) 8 AWG
    c) 6 AWG
    d) 4 AWG

25. The branch-circuit protection for a 120-gallon, 7.5 kVA, fixed storage-type water heater would be _____ amperes on 240-volt, single-phase.
   a) 35
   b) 40
   c) 50
   d) 60
26. The load current rating of a copper 10-AWG RHW is _____ amperes when there are four current-carrying conductors in a conduit and the ambient temperature is 90° F.
   a) 22.56
   b) 24.64
   c) 26.32
   d) 27.43
27. What is the maximum overcurrent protection allowed for each of three 10-AWG THHN copper conductors in an ambient temperature of 45° C?
   a) 35 amperes
   b) 30 amperes
   c) 25 amperes
   d) 20 amperes
28. A single-phase load after demand factors have been calculated is 299 amperes at 240/120 volts. The conductors are paralleled in one 3″ nonmetallic conduit. What size THW conductors are required to meet *NEC* requirements?
   a) 1/0
   b) 2/0
   c) 3/0
   d) 350 kcmil
29. What is the maximum overcurrent protection permitted for 8-AWG XHHW conductors in a wet location at 34° C? There are three conductors in the conduit.
   a) 45 amperes
   b) 50 amperes
   c) 55 amperes
   d) 60 amperes
30. What is the maximum ampacity of twenty-five 14-AWG Cu THW conductors in one conduit? The ambient temperature is 45° C.
   a) 6–7 amperes
   b) 7.1–8 amperes
   c) 8.1–9 amperes
   d) 9.1–10 amperes

31. Three heaters, not thermostatically controlled, have a line current of 44 amperes. Each heater is connected to one phase and the neutral. The six conductors are required to be _____ size THW.
    a) 12-AWG
    b) 8-AWG
    c) 4-AWG
    d) 2-AWG
32. The secondary load of a three-phase, 4-wire 480-208/120-volt transformer is 100 amps per phase, with 0 amps on the neutral, and the four conductors are in one nonmetallic conduit with the grounding conductor. Table 310.15(B)(2)(a) would require the full load current to be increased by _____.
    a) 80%
    b) 1.25%
    c) 125%
    d) 0%
33. What size THHN conductors are required for a load of 30 amperes? The conductors are in a conduit with a total of 8 conductors.
    a) 12 AWG
    b) 10 AWG
    c) 8 AWG
    d) 6 AWG
34. What is the ampacity of a 12-AWG THHN that is in a conduit with five other 12-AWG conductors and two bare 12-AWG conductors?
    a) 16 amperes
    b) 20 amperes
    c) 21 amperes
    d) 24 amperes
35. What is the minimum rating for a branch circuit supplying a 20-ampere, continuous loaded appliance?
    a) 20 amperes
    b) 25 amperes
    c) 30 amperes
    d) 35 amperes
36. A conduit contains nine 10-AWG THW conductors. The conductors are three sets of 10-2 with ground nonmetallic cables. What is the ampacity of each conductor?
    a) 25 amperes
    b) 28 amperes
    c) 30 amperes
    d) 32 amperes

Raceway Fill

CHAPTER

7

Tables for Calculating Raceway Fill

To determine the maximum number of conductors permitted in raceways, look in Chapter 9 and Annex C of the *NEC*. Use Annex C when all the conductors are the same size; use Tables 4 and 5 when different size conductors are installed in the same raceway.

According to Chapter 9, Notes to Tables, "equipment grounding or bonding conductors, where installed, shall be included when calculating conduit or tubing fill. The actual dimensions of the equipment grounding or bonding conductor (insulated or bare) shall be used in the calculation" (Note 3). The approximate area can be found in Table 5 for insulated and Table 8 (in the Area in.^{2} column) for bare conductors.

Table 1 in Chapter 9 specifies the maximum percentage of cross-sectional area in determining raceway fill. A conduit may be filled to 53 percent if a single conductor is installed in the conduit. A conduit may be filled to 31 percent if exactly two conductors are installed in the conduit. A raceway longer than twenty-four inches and containing three or more conductors has a maximum fill of 40 percent.

Table 4 provides the following information: metric designator, trade size, nominal internal diameter, 100 percent, 60 percent, 53 percent, 31 percent, and 40 percent. Table 4 covers twelve types of conduit and tubing. The Total Area 100% column is also known as the raceway's cross-sectional area. Exam questions might ask for the cross-sectional area instead of the total area.

EXAMPLE

What is the cross-sectional area, in square inches, of a 1 ¼-inch EMT raceway?

a) 0.464 b) 0.598 c) 1.380 d) 1.496

The answer can be found in Table 4 (in the Electrical Metallic Tubing subtable), in the Total Area 100% in.^{2} column. The cross-sectional area of a 1 ¼-inch EMT raceway is 1.496 in.^{2}

Conduit or tubing nipples that are 24 inches or less can be filled to 60 percent of their total cross-sectional area.[Chapter 9 Table Notes, Note 4] No calculation is necessary for conduit and tubing nipples because Table 4 contains a 60 percent column.

EXAMPLE

What is the allowable fill of a 1-inch electrical metallic tubing nipple?

a) 1.049 in.^{2} b) 0.864 in.^{2} c) 0.519 in.^{2} d) 0.346 in.^{2}

Because of the nipple in this question, the raceway can be filled to 60 percent. In Table 4 (of Chapter 9), 60 percent of a 1-inch EMT raceway is 0.519 in.^{2}

Use Table 5 to find the approximate area in square inches of insulated conductors with different types of insulation. The Table 5 column that will be used most often is the last column, Approximate Area in.^{2}

Determining Conductor Area

EXAMPLES

What is the area (in square inches) of a 10-AWG TW conductor?

Look in Table 5 and find the type of conductor. Then find the size and follow it across to the Approximate Area in.^{2} column and select the number. The area of a 10-AWG TW conductor is 0.0243 in.^{2}

What is the area square inch of a 6-AWG XHHW conductor?

a) 0.0507 **b)** 0.0590 **c)** 0.0726 **d)** 0.274

The answer is in Table 5, across from 6-AWG XHHW conductor: 0.0590.

There is a footnote at the bottom of Table 5. The footnote is for conductor types listed with an asterisk (*): RHH, RHW, and RHW-2 without outer covering. There are two types of RHH, RHW, and RHW-2 conductors: conductors with an outer covering and conductors without an outer covering.

EXAMPLES

What is the area square inch for a 12-AWG RHH conductor with an outer covering?

First, find the type RHH conductor. Be careful and pick the RHH conductor without the asterisk (*). This question is asking for an RHH conductor with an outer covering. The conductors with the asterisk (*) are without the outer covering.

The area square inch for a 12-AWG RHH conductor with an outer covering is 0.0353.

What is the area square inch for an 8-AWG RHW-2 conductor with an outer covering?

The area square inch for an 8-AWG RHW-2 conductor with an outer covering is 0.0835.

Looking for the approximate area for RHH, RHW, and RHW-2 conductors, with and without an outer covering, can be confusing. Pay close attention to the terms "with outer covering" and "without outer covering." Because of the similarity between these two terms, it's easy to choose the wrong answer on an exam. Notice how "with outer covering" could easily be mistaken for "without covering."

One way to help eliminate the possibility of choosing the wrong answer is to write down the area in.^{2} for both with and without outer covering. After writing both sets of numbers, look at the question again to see if it is asking for the conductor with or without the outer covering. The conductor with the outer covering has an extra layer of covering; therefore it will be the larger of the two numbers. The conductor without the outer covering does not have an extra layer of covering; therefore it will be the smaller of the two numbers.

EXAMPLES

What is the area square inch for a 14-AWG RHW conductor without an outer covering?

Step 1. Find the area in.^{2} for both conductor types. The two sets of numbers, for RHW with and without the outer covering, are 0.0209 and 0.0293.

Step 2. Look back at the question and see that it is asking for a conductor without the outer covering.

Step 3. Because the conductor in the question does not have the extra layer of insulation, select the smaller of the two numbers.

The area square inch for a 14-AWG RHW conductor without an outer covering is 0.0209.

What is the area square inch for a 10-AWG RHH conductor with an outer covering?

a) 0.3330 b) 0.0437 c) 0.0353 d) 0.0333

The two sets of numbers, for RHH with and without the outer covering, are 0.0333 and 0.0437. Because the conductor in the question has the extra layer of insulation, select the larger of the two numbers.

The area square inch for a 10-AWG RHH conductor with an outer covering is 0.0437.

Annex C Tables

When all of the conductors are the same size and the same type of insulation, either calculate the answer from Tables 4 and 5 or simply look up the answer in Annex C. Annex C is in the back of the *Code* book. The types of conductors are listed with different types of conduit and tubing. The same types of conduit listed in Table 4 can be found in Annex C. Annex C gives the maximum number of fixture wires or conductors permitted in conduit or tubing. Tables C1 through C12(A) are based on the 40 percent fill for three or more conductors as permitted in Table 1 of Chapter 9.

EXAMPLES

How many 10-AWG THHN conductors can be installed in 1″ electrical metallic tubing?

a) 10 b) 14 c) 16 d) 17

Table C1 allows sixteen 10-AWG THHN conductors in 1″ EMT. The correct answer is c).

How many 12-AWG THWN-2 conductors can be installed in ½″ electrical nonmetallic tubing?

a) 7 b) 8 c) 9 d) 10

Pay close attention to the type of conduit listed in the question. This question is asking for electrical nonmetallic tubing (ENT). Electrical nonmetallic tubing can be found in Table C2.

The maximum number of 12-AWG THWN-2 conductors allowed in ½″ electrical nonmetallic tubing is seven. The correct answer is a).

How many 12-AWG THHN conductors can be installed in ½″ liquidtight flexible metal (LFMC) conduit?

a) 7 b) 8 c) 9 d) 10

Nine THHN conductors are allowed in liquidtight flexible metal conduit (Table C7). The correct answer is c).

Annex C can also be used when the number of conductors is known and the size conduit is unknown.

EXAMPLES

What is the minimum size rigid metal conduit required for three 4/0-THHN conductors?

a) 1¼-inch b) 1½-inch c) 2-inch d) 2½-inch

Look in Table C8 to find the conductor type and size (THHN, 4/0). Starting at the ½″ conduit, follow the numbers across the page and stop at three (or more) conductors. Follow that column up to find the minimum trade size. The correct answer to this question is c), 2″. Actually, four 4/0-THHN conductors are permitted in 2″ rigid metal conduit.

What is the minimum size schedule 80 rigid PVC conduit required for seven 10-AWG RHW conductors with outer covering?

a) ¾-inch b) 1-inch c) 1¼-inch d) 1½-inch

continues

continued

The minimum size schedule 80 rigid PVC conduit required for seven 10-AWG RHW conductors with outer covering is 1 ¼" (answer c). If the 10-AWG RHW conductors were without the outer covering, they would fit into a 1" schedule 80 rigid PVC conduit.

What is the minimum size flexible metallic conduit required for three 4-AWG THWN conductors?

a) ¾-inch b) 1-inch c) 1 ¼-inch d) 1 ½-inch

The minimum size flexible metallic conduit required for three 4-AWG THWN conductors is 1" (answer b).

Conductors in Conduit Nipples

There will be times when it is necessary to calculate conductors in conduit nipples.

EXAMPLES

How many 14-AWG THW conductors are allowed in a ½" EMT conduit that is 21" long?

a) 8 b) 11 c) 13 d) 14

A 21" conduit is a nipple, and therefore it can be filled to 60 percent instead of only 40 percent. The allowable fill of a ½" EMT nipple is 0.182 (Table 4, Chapter 9).

The area of a 14-AWG THW conductor is 0.0139 in.^{2} (Table 5, Chapter 9).

To find the maximum number of conductors permitted, divide the area of the conduit by the area of the conductor.

$0.182 \div 0.0139 = 13.09 = 13$

Thirteen 14-AWG THW conductors can be installed in a ½" EMT conduit nipple (answer c).

How many 4/0-AWG THWN conductors are allowed in a 2" intermediate metal conduit (IMC) conduit nipple?

a) 7 b) 6 c) 5 d) 4

The allowable fill of a 2" IMC nipple is 2.178.

The area of a 4/0-AWG THWN conductor is 0.3237 in.^{2}

$2.178 \div 0.3237 = 6.7 = 6$

Six 4/0-AWG THWN conductors can be installed in a 2" IMC conduit nipple (answer b).

It will sometimes be necessary to calculate what size conduit nipple is required for certain conductors.

EXAMPLES

What is the minimum size rigid metal conduit nipple allowed for eleven 10-AWG RHH conductors with outer covering?

a) ¾-inch b) 1-inch c) 1 ¼-inch d) 1 ½-inch

Step 1. Calculate the area in.^{2} necessary for the conductors.

$11 \times 0.0437 = 0.4807$

Step 2. Look in the 60% column in Table 4 and select a conduit having an area of at least 0.4807 in.^{2} The minimum size rigid metal conduit nipple for eleven 10-AWG RHH conductors with outer covering is 1".

What size rigid metal conduit nipple is required for the following eighteen conductors?

- Four 12-AWG RHW conductors without covering
- Six 12-AWG TW conductors
- Eight 10-AWG RHH conductors with outer covering

a) 3/4-inch b) 1-inch c) 1 1/4-inch d) 1 1/2-inch

Step 1. Look in the table and find the area square inch for each conductor and multiply it by the number of conductors.

$4 \times 0.026 = 0.104$

$6 \times 0.0181 = 0.1086$

$8 \times 0.0437 = 0.3496$

Step 2. Add the numbers to find the total area required for the conductors.

$0.104 + 0.1086 + 0.3496 = 0.5622$

Step 3. Look in the 60% column in Table 4 and select a conduit having at least 0.5622 area in.^{2} The minimum size rigid metal conduit nipple allowed for the conductors in this example is 1 1/4″.

If all of the conductors are the same size (total cross-sectional area including insulation) and the calculated number of conductors includes a decimal fraction of 0.8 or larger, the calculated number can be rounded up to the next whole number (Note 7 to Chapter 9: Tables).

EXAMPLES

How many 10-AWG RHH conductors with outer covering are allowed in a 1″ EMT conduit that is 12″ long?

a) 10 b) 11 c) 12 d) 13

Step 1. Find the area in.^{2} for a 1″ EMT conduit nipple.

0.519 (Table 4, Chapter 9)

Step 2. Find the area in.^{2} of a 10-AWG RHH conductor with outer covering.

0.0437 (Table 5, Chapter 9)

Step 3. Divide the area of the conduit nipple by the area of the conductor to find the number of conductors allowed.

$0.519 \div 0.0437 = 11.8764$

Step 4. Because the calculated number has a decimal of at least 0.8, it is permissible to round the eleven up to twelve.

Twelve 10-AWG RHH conductors with outer covering are allowed in a 1″ EMT nipple (answer c).

How many 10-AWG RHW conductors without outer covering are allowed in a 3/4″ rigid metal conduit nipple?

a) 7 b) 8 c) 9 d) 10

The allowable fill of a 3/4″ rigid metal conduit nipple is 0.329.

The area in.^{2} of a 10-AWG RHW conductor without outer covering is 0.0333.

$0.329 \div 0.0333 = 9.8799$

Ten 10-AWG RHW conductors without outer covering can be installed in a 3/4″ rigid metal conduit nipple because the calculated number contains .8 or larger (answer d).

Different Sizes and Types of Conductors

When different sizes and/or types of conductors are in the same conduit, a calculation will be necessary.

NOTE

Do not use Annex C as a shortcut for raceways containing different sizes of conductors. Table C3 shows that only three 250-kcmil THHN conductors are permitted in a 2″ flexible metal conduit. Annex C does not show the remaining space available for smaller conductors. Raceways containing different sizes and/or types of conductors should be calculated.

EXAMPLES

What size EMT conduit is required for eight 12-AWG THHN and five 10-AWG THHN conductors?

a) ½-inch b) ¾-inch c) 1-inch d) 1 ¼-inch

Step 1. Find the area in.^{2} for each conductor and multiply it by the number of conductors.

12 AWG = 0.0133 × 8 = 0.1064

10 AWG = 0.0211 × 5 = 0.1055

Step 2. Add the numbers to find the total area in.^{2}

0.1064 + 0.1055 = 0.2119

Step 3. Find electrical metallic tubing in Table 4. The conduit needs to have an area of at least 0.2119 in.^{2} A ¾″ conduit has an allowable fill of 0.213 square inch.

The minimum size EMT conduit required for eight 12-AWG THHN and five 10-AWG THHN conductors is ¾″ (answer b).

What size flexible metal conduit is required for three 250-kcmil THHN copper conductors and one 2-AWG THHN copper conductor?

a) 2-inch b) 2 ¼-inch c) 2 ½-inch d) 3-inch

Step 1. Find the area in.^{2} for each of the conductors.

250-kcmil THHN copper = 0.397

2-AWG THHN copper = 0.1158

Step 2. Find the total area in.^{2} for all the conductors.

0.397 + 0.397 + 0.397 + 0.1158 = 1.3068

Step 3. Find the flexible metal conduit with a minimum area of 1.3068 in.^{2}

2″ = 1.307

The minimum size flexible metal conduit required for three 250-kcmil THHN copper conductors and one 2-AWG THHN copper conductor is 2″ (answer a).

Raceways are sometimes partially filled with conductors, and more conductors need to be added.

EXAMPLES

A ¾″ EMT conduit contains four 10-AWG RHW conductors without outer covering. How many 12-AWG RHH conductors without outer covering can be added?

a) None b) 1 c) 2 d) 3

Step 1. Find the area in.^{2} of four 10-AWG RHW conductors without covering.

0.0333 × 4 = 0.1332

Step 2. Subtract 0.1332 from the area of a ¾″ conduit to determine the remaining area in.^{2} in the ¾″ conduit.

¾″ EMT = 0.213 area in.^{2}

0.213 - 0.1332 = 0.0798 remaining area in.^{2}

Step 3. Find the area in.^{2} of a 12-AWG RHH conductor without outer covering and divide this number into the area remaining in the ¾″ conduit. 12-AWG RHH conductor without outer covering = 0.026 in.^{2}

0.0798 ÷ 0.026 = 3.069

Three 12-AWG RHH conductors without outer covering can be added (answer d).

If two 10-AWG RHH conductors with outer covering are in a 1″ liquidtight flexible metal conduit, how many 10-AWG RHH conductors without outer covering could be added?

a) 6 b) 7 c) 8 d) 9

Step 1. Find the area in.^{2} of two 10-AWG RHH conductors with covering.

0.0437 × 2 = 0.0874

Step 2. Subtract 0.0874 from the area of a 1″ conduit to determine the remaining area in.^{2} in the 1″ conduit.

0.349 - 0.0874 = 0.2616

Step 3. Find the area in.^{2} of a 10-AWG RHH conductor without outer covering and divide this number into the area remaining in the 1″ conduit.

0.2616 ÷ 0.0333 = 7.86

Step 4. Don't be tricked by the decimal .86. In accordance with Note 7 in Chapter 9, rounding up is permissible only when a conduit or tubing contains conductors that are the same size. The conductors must have the same cross-sectional area, including insulation.

Only seven 10-AWG RHH conductors without outer covering are permitted in this example (answer b).

Sometimes a question will state that a conduit is filled to a certain percentage with conductors and then will ask how many other conductors of a certain type can be added.

EXAMPLES

A 3/4″ EMT conduit is filled to 20% with 12-AWG THWN conductors. How many 10-AWG THHN conductors can be added?

a) 4 b) 5 c) 6 d) 7

A raceway containing three or more conductors is limited to 40 percent fill. In the problem, 20 percent of the raceway is full. Therefore, 20 percent is remaining. One way to find the area in.^{2} for 20 percent fill is to multiply the cross-sectional area (100%) by 20%.

0.533 × 20% = 0.1066

Another way to find the area in.^{2} for 20 percent fill is to divide the 40 percent column by two.

0.213 ÷ 2 = 0.1065

Either method will achieve the correct answer. The last step is to divide the area in.^{2} of the 10-AWG THHN into the remaining area in the conduit.

0.1066 ÷ 0.0211 = 5.05 = 5

Five 10-AWG THHN conductors can be added to the raceway in this example (answer b).

A 1″ PVC conduit (Schedule 40) nipple, 24″ in length, is filled with 6-AWG XHHW conductors that total 20% of the conduit. How many more 6-AWG XHHW conductors can be added?

a) 5 b) 6 c) 7 d) 8

Twenty percent of the conduit is full, but this time the conduit is a nipple. Conduit nipples have an allowable fill of 60 percent.

60% - 20% = 40%

This problem could be calculated like the previous example, but there is an easier method. Annex C is based on 40 percent. Since this conduit started out as a nipple, but now 20 percent of the conduit is full, 40 percent is remaining for fill.

Table C10 allows five 6-AWG XHHW conductors in 40 percent fill in PVC. Five 6-AWG XHHW conductors can be added to the conduit in this example (answer a).

END-OF-CHAPTER PROBLEMS

1. How many 6-AWG XHHW conductors can be installed in 1¼″ IMC conduit nipple that is 25″ long?
   a) 8
   b) 9
   c) 10
   d) 11
2. What is the internal diameter of a ½″ liquidtight flexible nonmetallic conduit (Type LFNC-A)?
   a) 0.312″
   b) 0.495″
   c) 0.630″
   d) 0.632″
3. How many 8-AWG XHHW conductors can be installed in a 1″ rigid metal conduit nipple 18″ long?
   a) 9
   b) 11
   c) 12
   d) 13
4. What size 18″ PVC (Schedule 40) conduit nipple is required for the following conductors? (The ambient temperature is 30° C.)
   - 4–6 AWG THW
   - 6–10 AWG RHW without outer covering

   a) ¾″
   b) 1″
   c) 1¼″
   d) 1½″
5. What is the ampacity of each of the 6-AWG THW conductors in question 4?
   a) 32.5 amperes
   b) 55 amperes
   c) 65 amperes
   d) 75 amperes
6. Eight 8-AWG THW conductors occupy _____ area in.^{2} of a 1½″ EMT nipple that is 18″ long.
   a) 0.0437
   b) 0.3496
   c) 0.4448
   d) 1.221

7. What is the allowable fill, in area in.^{2}, of a 1½″ flexible metal conduit that has more than two conductors?
   a) 0.511
   b) 0.576
   c) 0.743
   d) 1.307
8. What is the area in.^{2} of a 12-AWG RHW conductor without outer covering?
   a) 0.0182
   b) 0.026
   c) 0.0333
   d) 0.0353
9. What is the internal diameter of a ¾″ rigid PVC (Schedule 80) conduit?
   a) 0.409″
   b) 0.526″
   c) 0.722″
   d) 0.836″
10. A 1″ ENT conduit is filled to 20% with existing conductors. How many 12-AWG RHW (with outer covering) conductors can be added?
    a) 4
    b) 5
    c) 6
    d) 9
11. If 20% of a 20″ long, 1½″ rigid metal conduit was already occupied by 6-AWG XHHWs, how many more 6-AWG XHHWs could be added?
    a) 7
    b) 12
    c) 13
    d) 14
12. A 3½″ Type A PVC conduit is already filled with conductors totaling 20% of the conduit. What is the allowable area remaining, in the raceway, for conductor fill?
    I. 10.694×0.2
    II. 4.278×0.5
    III. $4.278 - (10.694 \times 0.2)$
    a) I only
    b) II only
    c) I and II
    d) I, II, and III

13. How many 12-AWG RHW conductors with outer covering can be installed in a ¾″ EMT nipple that is 21″ long?
   a) 6
   b) 7
   c) 9
   d) 12
14. What size ENT conduit nipple is required for the following conductors? (The ambient temperature is 100° F.)
   - 2–10 AWG RHW with outer covering
   - 2–12 AWG THW
   - 3–14 AWG TW

   a) ¾″
   b) 1″
   c) 1¼″
   d) ½″
15. What is the ampacity of each of the 10-AWG RHW conductors in question 14?
   a) 35 amperes
   b) 33 amperes
   c) 30 amperes
   d) 22 amperes
16. How many 12-AWG RHH conductors can be installed in a 1″ liquidtight flexible metal conduit?
   a) 9
   b) 10
   c) 12
   d) 13
17. What is the area in.^{2} of 2½″ electrical metallic tubing?
   a) 5.858
   b) 4.909
   c) 2.731
   d) 2.343
18. Two conductors may occupy not more than _____ of a 1″ rigid conduit 36″ long.
   a) 27%
   b) 31%
   c) 34%
   d) 40%

19. How many 6-AWG XHHW conductors can be installed in a 2″ intermediate metal conduit nipple 18″ long?
   a) 24
   b) 35
   c) 36
   d) 37
20. How many 12-AWG RHW conductors, with outer covering, can be installed in a 1″ liquidtight flexible nonmetallic conduit (Type LFNC-B)?
   a) 9
   b) 10
   c) 13
   d) 19
21. How many 8-AWG RHH conductors, without covering, are allowed in a 1½″ EMT conduit?
   a) 14
   b) 13
   c) 10
   d) 9
22. What is the internal diameter of a 1¼″, Schedule 40, PVC conduit?
   a) 1.255″
   b) 1.36″
   c) 1.435″
   d) 1.5″
23. A 1½″ EMT conduit 30″ in length with more than two conductors may be filled to _____ area in.^{2}
   a) 0.631
   b) 0.774
   c) 0.814
   d) 1.079
24. If 20% of a 20″ long 1½″ EMT conduit already contained conductors, what is the maximum number of 8-AWG XHHWs that could be added?
   a) 18
   b) 9
   c) 8
   d) 7
25. Two conductors may occupy not more than _____ of a 1″ EMT conduit 18″ long.
   a) 60%
   b) 53%
   c) 40%
   d) 31%

26. What is the allowable fill in area in.^{2} of a 1¼-inch flexible metal conduit nipple?

a) 0.9156

b) 0.9006

c) 0.8718

d) 0.766

27. What size PVC (Schedule 40) conduit is required for three 250-kcmil XHHW-2 aluminum phase conductors and a 4-AWG XHHW-2 aluminum equipment grounding conductor?

a) 1″

b) 1½″

c) 2″

d) 2½″

28. What is the area square inch of a 750-kcmil XHHW conductor?

a) 0.9923

b) 1.0532

c) 1.1122

d) 1.4082

29. How many 12-AWG RHW conductors without outer covering are allowed in a 1″ EMT conduit nipple 20″ long?

a) 20

b) 19

c) 13

d) 8

30. What is the maximum number of 12-AWG RHWs without outer covering that can be pulled in a 1½″ rigid metal conduit?

a) 21

b) 24

c) 32

d) 40

31. On a remodeling job, the owner wishes to reuse the existing conduit. How many additional 12-AWG THHN conductors can be added to a ½″ rigid metal conduit if it contained seven 12-AWG TW conductors?

a) 4

b) 3

c) 2

d) 1

32. What is the area square inches of a 2½″ rigid metal conduit?

a) 5.858

b) 4.866

c) 2.489

d) 1.946

33. How many 12-AWG RHW conductors without outer covering are allowed in a 1″ EMT conduit?

a) 13

b) 12

c) 10

d) 9

34. What is the allowable fill, in area in.^{2}, of a 2″ liquidtight flexible nonmetallic conduit (LFNC-B) that will contain four conductors?

a) 1.298

b) 1.337

c) 2.033

d) 3.246

35. Two-inch electrical metallic tubing measuring 20″ in length with more than two conductors may be filled to _____ area in.^{2}

a) 1.342

b) 2.013

c) 2.067

d) 3.356

CHAPTER 8

Box Fill and Junction Box Sizing

Box Fill Calculations

Boxes and conduit bodies must be of sufficient size to provide free space for all enclosed conductors. Article 314 of the *NEC* contains detailed requirements for box fill and junction box sizing. While box fill requirements are in 314.16, junction box provisions are in 314.28. Boxes and conduit bodies containing conductors 18 AWG through 6 AWG must be installed in accordance with 314.16. Calculations for these boxes are based on the sizes and numbers of conductors. Boxes and conduit bodies enclosing conductors 4 AWG or larger, under 600 volts, must also comply with the provisions of 314.28. Pull and junction boxes containing conductors 4 AWG or larger, under 600 volts, are calculated from the sizes and numbers of raceways.

Box volume calculations are found in 314.16(A). Minimum volume and maximum number of conductors are listed in Table 314.16(A) for various metal boxes. The boxes are based on the same size conductors in a box containing no fittings or devices, such as fixture studs, cable clamps, hickeys, switches, or receptacles. There shall be a reduction of conductors allowed in boxes containing fittings or devices. According to the *Code,* "in no case shall the volume of the box, as calculated in 314.16(A), be less than the fill calculation as calculated in 314.16(B)."[314.16]

Boxes 100 cubic inches or less, except for boxes listed in Table 314.16(A), must be durably and legibly marked by the manufacturer with their volume. Nonmetallic boxes must also be marked. The volume of a box or wiring enclosure is the total volume of the assembled sections. Additional volume can be gained by using plaster (mud) rings, domed covers, extension rings, or other similar item. Additional volume can be calculated only if those items either are marked with a cubic inch (or cubic centimeter) capacity, or have the same dimension as a box listed in Table 314.16(A).

A standard 4″ square metal box that has a depth of 1½″ may or may not be marked with a cubic inch (or cubic centimeter) volume on the box because it is listed in Table 314.16(A). A standard plastic, nail-on wall box must have a volume marking somewhere on the box because it is not listed in the table and because it is a nonmetallic box.

The maximum numbers of conductors for standard metal boxes are listed in Table 314.16(A). The maximum number of conductors can be found easily in boxes that do not contain any fittings or devices.

EXAMPLES

How many 12-AWG conductors are permitted in a 4″ square box that has a depth of 1½″? (There will not be any fittings or devices in the box.)

a) 6 b) 7 c) 8 d) 9

Only nine 12-AWG conductors are permitted in a 4 × 1½-inch square box (answer is d).

How many 14-AWG conductors are permitted in a 3 × 2 × 2¾-inch device box? (There will not be any fittings or devices in the box.)

a) 6 b) 7 c) 8 d) 9

Seven 14-AWG conductors are permitted in a 3 × 2 × 2¾-inch device box (answer is b).

What is the maximum number of 10-AWG XHHW copper conductors permitted in a 4 × 2⅛-inch deep octagon box?

a) 6 b) 7 c) 8 d) 9

Eight 10-AWG conductors are permitted in a 4 × 2⅛-inch octagon box (answer is c). Neither the type of insulation nor the material of the conductor is a concern in this section.

Box fill calculations are found in 314.16(B). Since fittings and devices take up space, boxes containing them must have the number of conductors reduced. Volume allowances for conductors, clamps, fittings, devices, and equipment are in 314.16(B)(1) though (B)(5).

EXAMPLE

How many 14-AWG conductors are permitted in a 3 × 2 × 2 3/4-inch device box? (The box will have two internal cable clamps and a receptacle.)

a) 3 b) 4 c) 5 d) 6

Step 1. Find the maximum number of 14-AWG conductors allowed in a 3 × 2 × 2 3/4-inch device box.

Seven 14-AWG conductors are permitted in a 3 × 2 × 2 3/4-inch device box.

Step 2. Determine the number of conductors for the items in the box.

One or more internal cable clamps: 1 conductor

One receptacle: 2 conductors

Step 3. Subtract three from the maximum number of conductors permitted in the box.

7 - 3 = 4

Four 14-AWG conductors are permitted in a 3 × 2 × 2 3/4-inch device box that contains two cable clamps and a receptacle.

The following table is a quick reference for box fill calculations. The left column lists items in 314.16(B)(1) though (B)(5). The center column shows volume allowance for items in the first column. The right column shows how the items are counted.

Table 8-1 Box Fill Calculations

Items Within the Box	Determination of Conductor Volume Allowance	Number of Conductors Counted
Each conductor originating outside the box and terminating or spliced within the box	Volume allowance for that specific conductor	1
Each conductor passing through the box without splice or termination (unbroken conductor)	Volume allowance for that specific conductor	1
Each conductor passing through the box without splice or termination (two or more times the minimum length for free conductors in 300.14)	Volume allowance for that specific conductor	2
A conductor, no part of which leaves the box (such as equipment bonding jumpers and pigtails)	Not counted	0
One or more internal cable clamps	Volume allowance for largest conductor in box	1
A cable connector with its clamping mechanism outside the box	Not counted	0
One or more luminaire (fixture) studs	Volume allowance for largest conductor in box	1
One or more hickeys	Volume allowance for largest conductor in box	1
Each yoke or strap containing one or more devices or equipment	Volume allowance for largest conductor connected to the device or equipment	2

continues

Table 8-1 *Continued*

Items Within the Box	Determination of Conductor Volume Allowance	Number of Conductors Counted
One or more equipment grounding conductors	Volume allowance for largest equipment grounding conductor in box	1
One or more isolated equipment grounding conductors	Volume allowance for largest isolated equipment grounding conductor in box	1
Small fittings such as locknuts and bushings	Not counted	0

Note: "An equipment grounding conductor or conductors or not over four luminaire fixture wires smaller than 14-AWG, or both, shall be permitted to be omitted from the calculations where they enter a box from a domed luminaire (fixture) or similar canopy and terminate within that box." [314.16(B)(1) Exception]

(Source: *NFPA's Pocket Electrical References* by Charles R. Miller, Table 10.1.)

Where all the conductors are the same size, total the number of conductors counted and subtract from the maximum number permitted in the box.

EXAMPLE

How many 14-AWG THHN conductors are allowed in a 4 × 2⅛-inch square box that already contains one fixture stud, one hickey, two cable clamps, one 14/2 nonmetallic-sheathed cable with ground, and one 14/3 nonmetallic-sheathed cable with ground?

The maximum number of 14-AWG conductors in a 4 × 2⅛" box is 15.

Table 8-2 Allowed Conductors

1 fixture stud	1 conductor
1 hickey	1 conductor
2 cable clamps	1 conductor
1 14/2 wire	2 conductors
1 14/3 wire	3 conductors
2 ground wires	1 conductor
Total number of conductors	**9**

Subtract the nine conductors counted in the box from the maximum number allowed.

15 - 9 = 6

There is enough space left in this box for six additional 14-AWG conductors.

Boxes that are not listed in Table 314.16(A) also have a maximum number of conductors permitted. Table 314.16(B) provides the volume allowance for conductors 18 AWG through 6 AWG.

EXAMPLE

What is the maximum number of 14-AWG conductors permitted in a 20 in.^{3} plastic nail-on box? (There will also be a receptacle in the box.)

a) 8 b) 9 c) 10 d) 11

Step 1. Find the volume from Table 314.16(B) for a 14-AWG conductor. The volume required for each 14-AWG conductor is 2 in.^{3}

Step 2. A receptacle counts as two conductors.

$2 \times 2 = 4$ in.^{3}

Step 3. Deduct 4 in.^{3} from the 20 in.^{3} box to determine the area remaining for conductors.

$20 - 4 = 16$ in.^{3} remaining

Step 4. Divide the cubic inch capacity remaining by the cubic inches required per conductor.

16 in.^{3} $\div$ 2 in.^{3} = 8 conductors

Eight 14-AWG conductors are allowed in a 20 in.^{3} box that contains a receptacle.

Table 314.16(A) also shows the volume for the boxes listed. This number will be needed if the box contains different size conductors.

EXAMPLE

How many 12-AWG conductors are allowed in a $4 \times 2\frac{1}{8}$-inch square box? The box has a plaster ring with a 4.5 in.^{3} capacity. There are two 14-AWG conductors connected to a switch, two 10-AWG conductors running through the box, and two 10-AWG equipment grounding conductors. Two conduits are connected to the box with EMT connectors.

a) 6 b) 7 c) 8 d) 9

Step 1. Look in Table 314.16(A) to find the volume in cubic inches for a $4 \times 2\frac{1}{8}$-inch square box.

$4 \times 2\frac{1}{8}$-inch square box = 30.3 in.^{3}

Step 2. Add that number to the volume of the plaster ring.

$30.3 + 4.5 = 34.8$ in.^{3}

Step 3. Calculate the volume in cubic inches of the items in the box.

Two 14-AWG conductors = $2 \times 2 = 4$

Two 10-AWG conductors = $2 \times 2.5 = 5$

(Each conductor running through the box counts as one conductor.)

Two 10-AWG equipment grounding conductors = $1 \times 2.5 = 2.5$

(One or more grounding conductors count as one.)

One switch = $2 \times 2 = 4$

(A device counts as two times the largest conductor connected to the device.)

The EMT connectors do not count. No allowance shall be required for a cable connector with its clamping mechanism outside the box. [314.16(B)(2)]

Step 4. Add the numbers to find the total volume already occupied.

$4 + 5 + 2.5 + 4 = 15.5$ in.^{3}

Step 5. Deduct 15.5 in.^{3} from 34.8 in.^{3} to determine the area remaining for conductors.

$34.8 - 15.5 = 19.3$ in.^{3} remaining

Step 6. Divide the cubic-inch capacity remaining by the cubic inches required for each 12-AWG conductor.

$19.3 \div 2.25 = 8.5778$

Eight 12-AWG conductors can be added to the box in the question above (answer c).

Pull and Junction Box Sizing

Pull and junction boxes enclosing conductors of 4 AWG or larger, under 600 volts, are calculated from the sizes and numbers of raceways. Specifications for sizing pull and junction boxes for conductors larger than 6 AWG are in 314.28. Besides the sizes and numbers of raceways, the types of pulls must be known. Pull and junction boxes may include one or more of the following: straight pull, angle pull, and U pull. Straight pull specifications are found in 314.28(A)(1). Requirements for angle and U pulls are found in 314.28(A)(2).

Boxes containing straight pulls are sized according to the largest raceway that enters the box. The length must be at least eight times the trade size of the largest raceway. [314.28(A)(1)]

EXAMPLES

What is the minimum length of the pull box in the drawing below?

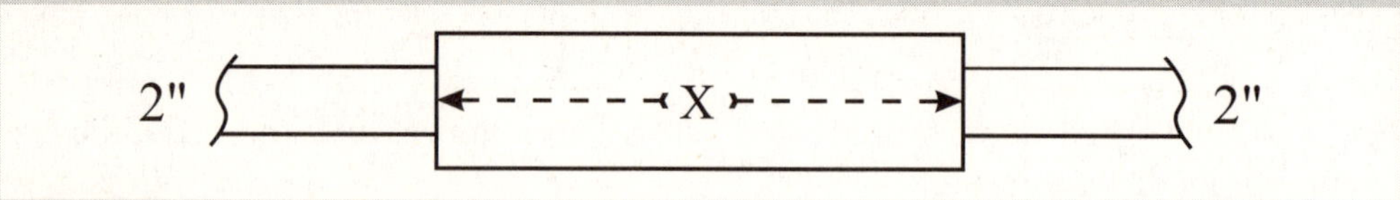

FIGURE 8-1

The minimum length is eight times the largest raceway.
8 × 2 = 16″

What is the minimum length of the pull box in the drawing below?

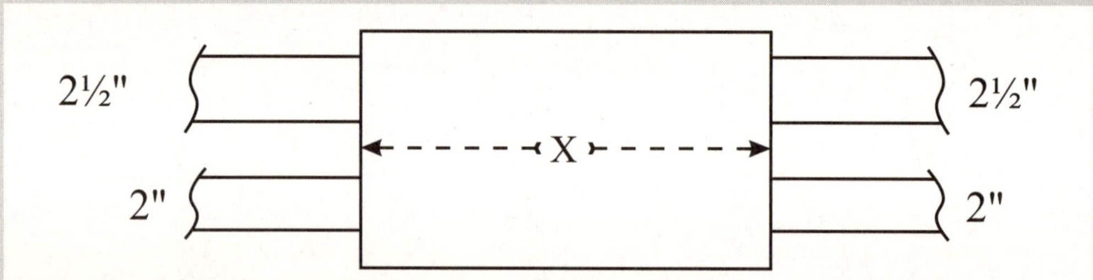

FIGURE 8-2

The minimum length is eight times the largest raceway.
8 × 2½ = 20″

What is the minimum length of the pull box in the drawing below?

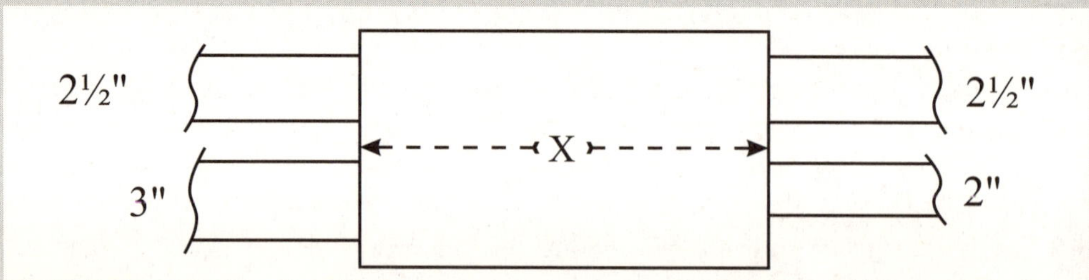

FIGURE 8-3

The minimum length is eight times the largest raceway.
8 × 3 = 24″

NOTE

The width and depth of a straight pull junction box must be large enough to provide proper installation of the raceway or cable, including the locknuts and bushings.

Pull and junction boxes may have raceways on all four walls. If all the pulls are straight pulls, calculate each dimension separately.

EXAMPLE

What is the minimum dimension of X and Y in the drawing below? All of the pulls are straight pulls.

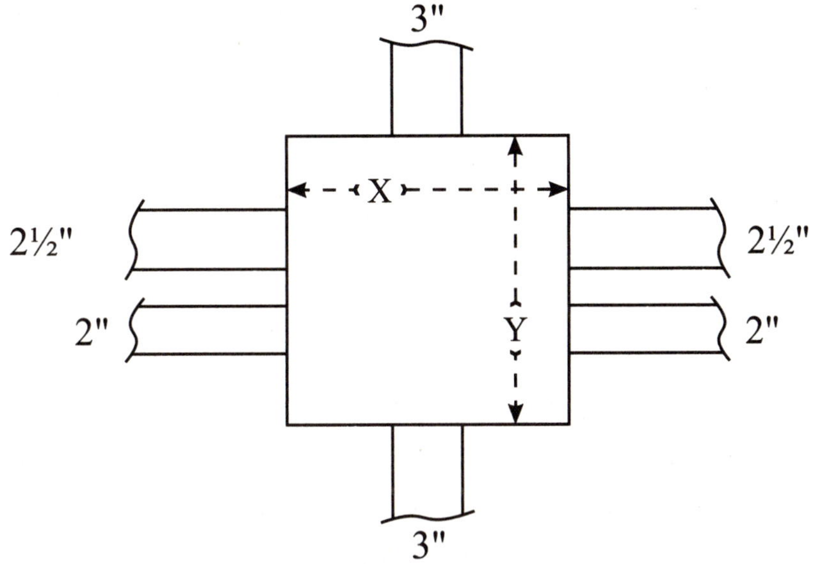

FIGURE 8-4

The minimum dimension for X is 8 × 2½ = 20″
The minimum dimension for Y is 8 × 3 = 24″

Junction boxes that have angle or U pulls are not calculated in the same way as boxes containing straight pulls. Junction boxes having conductors that are spliced must also be calculated by this method.

In accordance with 314.28(A)(2), where splices or where angle or U pulls are made, the distance from the inside of the box (where the raceway enters) to the opposite wall of the box must be at least six times the trade size of the largest raceway in a row. Where there are additional raceways entering the box, increase the distance for additional entries by the amount of the sum of the diameters of all other raceway entries in the same row on the same wall of the box. If there is more than one row on the same wall of the box, calculate each row separately. Select only the row requiring the largest dimension.

Calculate the dimension of a box with angle pulls by starting with one wall where the raceways enter the box and finding the distance to the opposite wall of the box. The calculations are based on the wall where the raceways enter to the opposite wall of the box, not where the conductors go. First, select one wall and multiply the largest raceway by six. Add to that number the trade size of all other raceways in the same row, on the same side of the box.

EXAMPLES

What is the minimum dimension of X and Y in the drawing below?

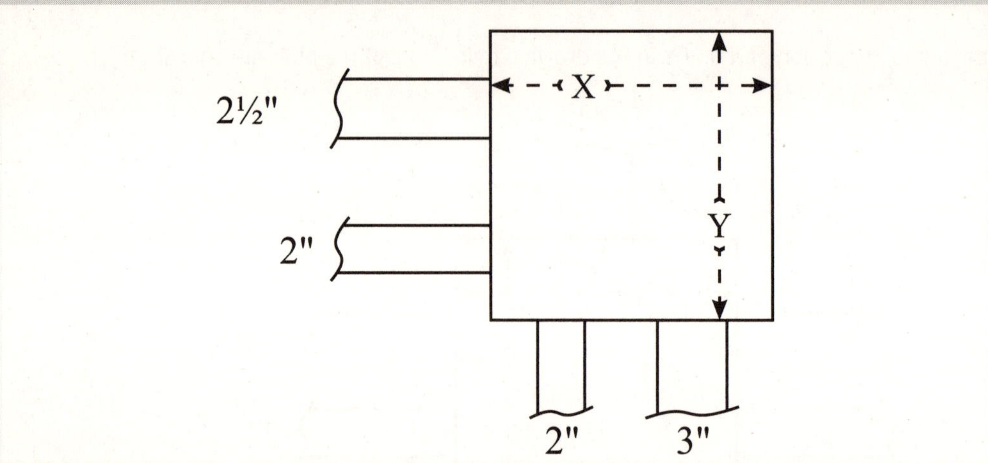

FIGURE 8-5

Step 1. As there are no raceway entries on opposite walls, there are no straight pulls. Select one wall and multiply the largest raceway by six.

Step 2. Add to that number the trade size of all other raceways on the same side of the box.

The minimum dimension for X is $6 \times 2\frac{1}{2} = 15 + 2 = 17''$

The minimum dimension for Y is $6 \times 3 = 18 + 2 = 20''$

What is the minimum dimension of X and Y in the drawing below?

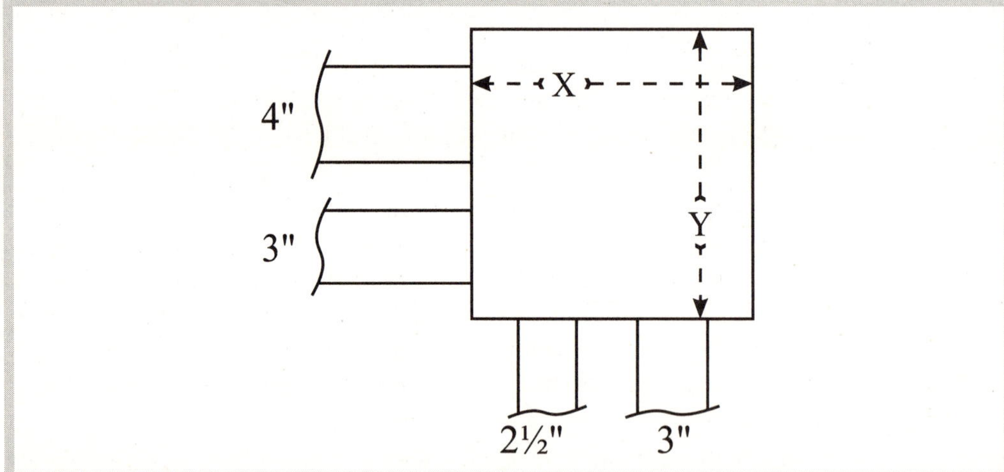

FIGURE 8-6

The minimum dimension for X is $6 \times 4 = 24 + 3 = 27''$

The minimum dimension for Y is $6 \times 3 = 18 + 2\frac{1}{2} = 20\frac{1}{2}''$

When raceways enter more than two walls of the box, calculate each wall separately. Select the largest size for the left–right (horizontal) dimension, and for the top–bottom (vertical) dimension.

EXAMPLES

What is the minimum dimension of X and Y in the drawing below? All of the pulls are angle pulls.

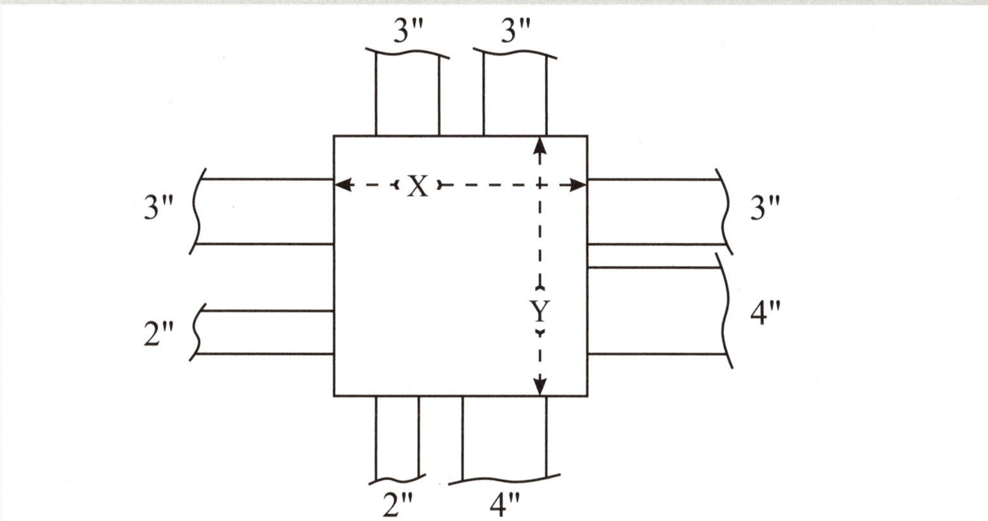

FIGURE 8-7

The minimum dimension for X is
Left side: 6 × 3 = 18 + 2 = 20″
Right side: 6 × 4 = 24 + 3 = 27″
The minimum dimension for Y is
Top: 6 × 3 = 18 + 3 = 21″
Bottom: 6 × 4 = 24 + 2 = 26″

The minimum dimension for X is 27 inches.
The minimum dimension for Y is 26 inches.

What is the minimum dimension of X and Y in the drawing below? All of the pulls are angle pulls.

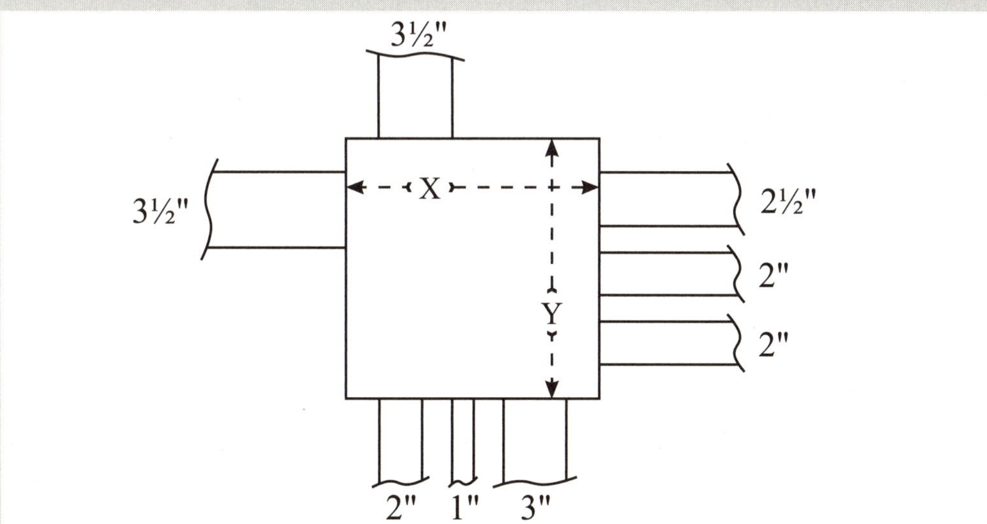

FIGURE 8-8

continues

continued

The minimum dimension for X is
Left side: $6 \times 3\frac{1}{2} = 21 + 0 = 21''$
Right side: $6 \times 2\frac{1}{2} = 15 + 2 + 2 = 19''$

The minimum dimension for Y is
Top: $6 \times 3\frac{1}{2} = 21 + 0 = 21''$
Bottom: $6 \times 3 = 18 + 2 + 1 = 21''$

The minimum dimension for X is 21 inches.
The minimum dimension for Y is 21 inches.

U pulls are calculated by the same method as angle pulls. A box with raceway entries only on one wall will have a minimum distance to the opposite wall. Multiply the trade size of the largest raceway by six and add the sum of the diameters of all other raceways entering the same wall. Raceway entries enclosing the same conductor must not be less than six times the trade diameter of the largest raceway.

EXAMPLES

What is the minimum dimension of X and Y in the drawing below?

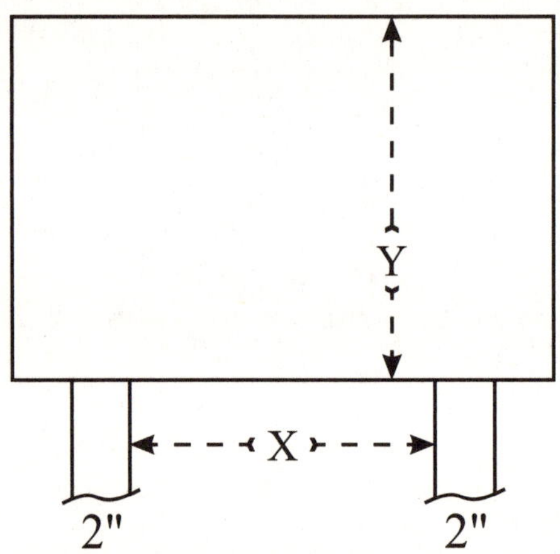

FIGURE 8-9

The minimum dimension for X is $6 \times 2 = 12''$
The minimum dimension for Y is $6 \times 2 = 12 + 2 = 14''$

What is the minimum dimension of X and Y in the drawing below?

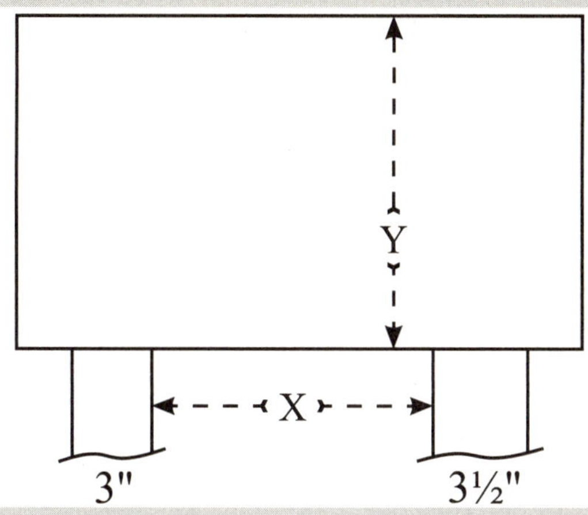

FIGURE 8-10

The minimum dimension for X is 6 × 3½ = 21″
The minimum dimension for Y is 6 × 3½ = 21 + 3 = 24″

Raceways that enclose the same conductors cannot be closer than six times the larger raceway, even if they are on different walls.

EXAMPLE

What is the minimum dimension of X, Y, and Z in the drawing below?

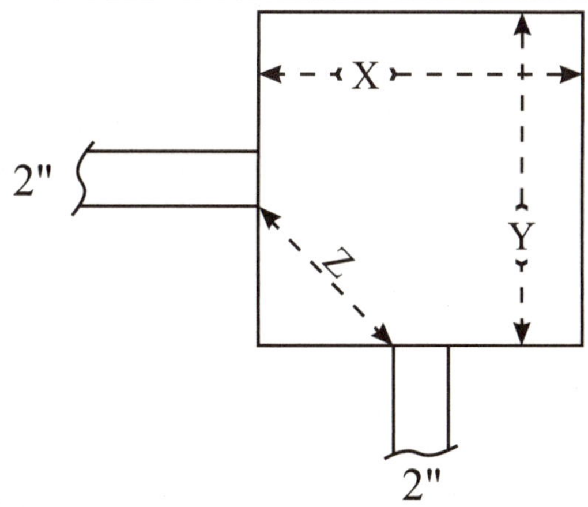

FIGURE 8-11

The minimum dimension for X is 6 × 2 = 12 + 0 = 12″
The minimum dimension for Y is 6 × 2 = 12 + 0 = 12″
The minimum dimension for Z is 6 × 2 = 12″

Up to this point, pull and junction boxes have been either straight pulls or angle pulls. If the box contains both straight and angle pulls, calculate by both methods. After the calculations are finished, compare and choose the largest dimension of either the left side or the right side. Also compare and choose the largest dimension of either the top or the bottom.

Remember, straight pull dimensions are calculated by multiplying the trade size of the largest raceway by eight. Don't make the mistake of adding anything else to that figure.

EXAMPLE

What is the minimum dimension of X and Y in the drawing below?

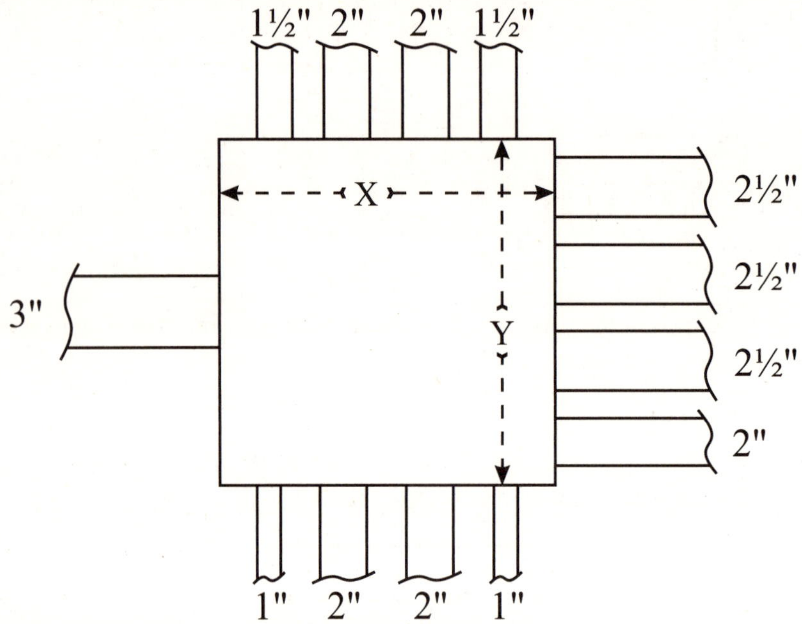

FIGURE 8-12

Straight pulls for X = 8 × 3 = 24″
Angle pulls for X =
Left side: 6 × 3 = 18 + 0 = 18″
Right side: 6 × 2½ = 15 + 2½ + 2½ + 2 = 22″

Straight pulls for Y = 8 × 2 = 16″
Angle pulls for Y =
Top: 6 × 2 = 12 + 2 + 1½ + 1½ = 17″
Bottom: 6 × 2 = 12 + 2 + 1 + 1 = 16″
The minimum dimension for X is 24″ and the minimum dimension for Y is 17″.

END-OF-CHAPTER PROBLEMS

1. What is the minimum dimension for X in the drawing below?
   a) 12″
   b) 14″
   c) 16″
   d) 20″

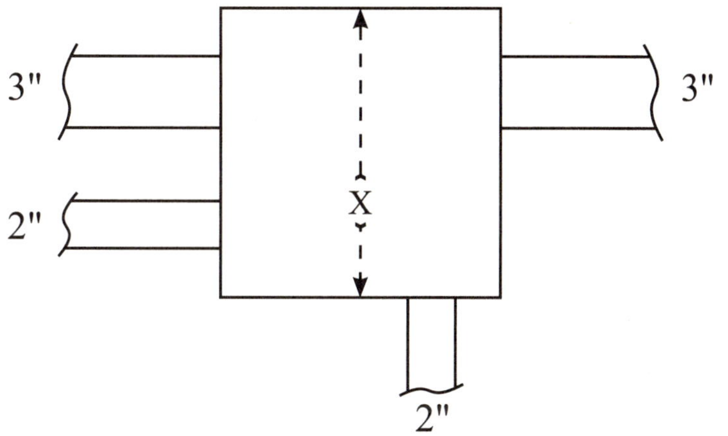

FIGURE 8-13

2. What minimum size octagon box is required for two 14-2 with ground nonmetallic-sheathed cables, one 12-2 with ground nonmetallic-sheathed cable passing through the box, three cable clamps, one hickey, and two 16-AWG fixture wires?
   a) 12.5 in.^{3}
   b) 15.5 in.^{3}
   c) 18 in.^{3}
   d) 21.5 in.^{3}
3. What is the maximum number of 12-2 with ground nonmetallic-sheathed cables permitted in an 18-cubic-inch device box if a duplex receptacle is also in the box?
   a) 1
   b) 2
   c) 3
   d) 4

4. An existing 24-cubic-inch junction box already contains two 12-2 with ground nonmetallic-sheathed cables. How many 14-2 with ground nonmetallic-sheathed cable can be installed in this box?
   a) 2
   b) 3
   c) 4
   d) 5
5. A pull box has two 2-inch conduits entering on the same wall of the box. Conductors will enter the box from one conduit and will exit through the other. The conductors will be larger than 6-AWG conductors. What is the minimum distance between these conduits?
   a) 12 inches
   b) 14 inches
   c) 18 inches
   d) 20 inches
6. What is the maximum number of 14-2 with ground nonmetallic-sheathed cables permitted in an 18-cubic-inch device box if a single-pole switch is also in the box?
   a) 1
   b) 2
   c) 3
   d) 4
7. What is the maximum number of 14-AWG conductors permitted in a 60-cubic-inch junction box?
   a) 15
   b) 24
   c) 26
   d) 30
8. What is the minimum dimension for X in the drawing below?
   a) 12″
   b) 14″
   c) 20″
   d) 24″

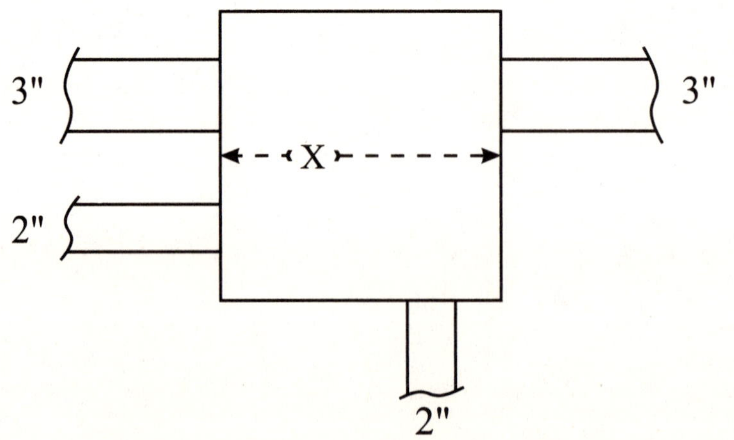

FIGURE 8-14

9. What is the maximum number of 12-AWG conductors permitted in a 4 × 2⅛-inch square metal box?
   a) 9
   b) 12
   c) 13
   d) 15
10. How many conductors in a junction box shall be reduced for one switch, two internal cable clamps, and three grounding conductors?
   a) 3
   b) 4
   c) 5
   d) 6
11. A pull box has one 2-inch conduit entering on one wall and one 2½-inch conduit entering the box on an adjacent wall. Conductors will enter the box from one conduit and will exit through the other. The conductors will be larger than 6-AWG conductors. What is the minimum distance between these raceway entries?
   a) 12 inches
   b) 15 inches
   c) 20 inches
   d) 22 inches
12. For calculating the maximum number of conductors permitted in an outlet box, the volume allowance required for each 12-AWG conductor is _____ cubic inches.
   a) 3
   b) 2.5
   c) 2.25
   d) 2
13. Six 12-AWG conductors will be spliced with wire connectors in a metal box. The box will have no grounding conductor, no clamps, and no device. A flat blank cover will be installed on the box. What is the minimum size octagon box required?
   a) 4 × 1¼-inch
   b) 4 × 1½-inch
   c) 4 × 2-inch
   d) 4 × 2⅛-inch
14. An existing 27-cubic-inch nonmetallic junction box already contains six 12-AWG conductors. How many 10-AWG conductors may be added to this junction box?
   a) 5
   b) 6
   c) 7
   d) 8

Motor Controls

CHAPTER 9

Basic Motor Controls

Motor controls can be as simple as a single-pole switch installed to start and stop a motor. Motor controls can also be very complex with a multitude of pushbuttons, switches, pilot lights, relays, and overloads. While some motor control drawings on exams show a disconnecting means for the motor, most do not. Switch types may include limit, flow, pressure or vacuum, float, foot, and temperature actuated. Switches can be normally open; normally closed; normally open, held closed; or normally closed, held open. Although motors can be single-phase or three-phase voltage, control circuits are usually single-phase. A drawing may show lines 1, 2, and 3 supplying a three-phase motor, but only two of the lines supplying the control circuit. Most motor control drawings on exams have a stop pushbutton, a start (or forward and reverse) pushbutton, a coil to energize the motor contacts, an auxiliary normally open contact, and overloads.

A coil is one of the items typically found in a motor control circuit. A coil is part of a motor contactor. When the coil on the contactor is energized, normally open contacts will close. Coils are drawn as a circle with the letter M inside. On most exams, the coil voltage is not provided, nor is it needed. On a basic motor control drawing, when the coil is energized, the contacts supplying power to the motor close and the motor starts. Overloads are typically shown on the right side of the control circuit. It is not uncommon to see one, two, or three overloads. A single-pole switch could be installed to energize the coil.

Figure 9-1 shows a basic motor control circuit. A single-pole switch is used to energize the coil. As drawn, the switch is open, the M coil is not energized, the M contacts are open, and the motor is not energized.

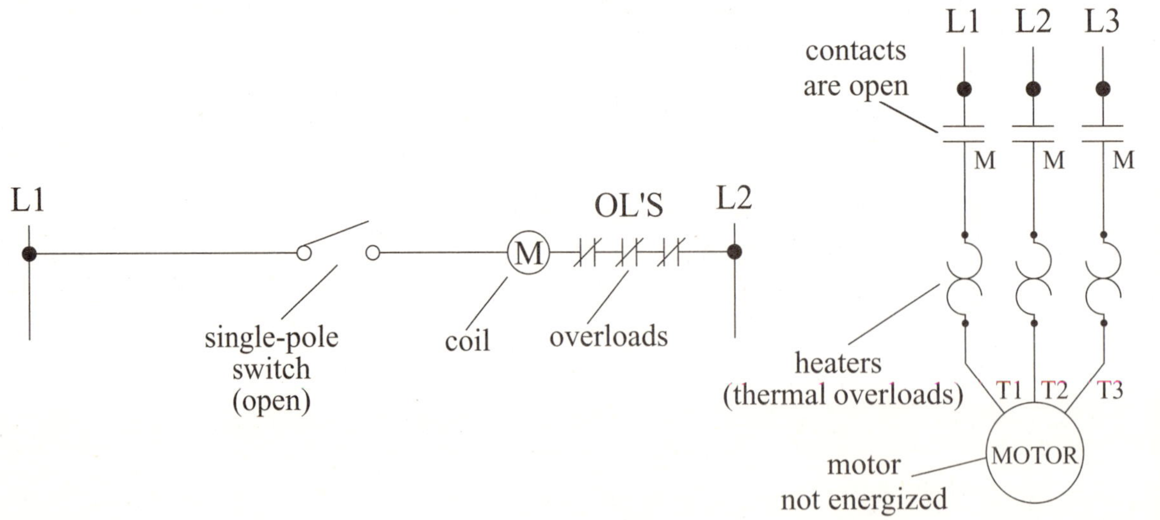

FIGURE 9-1

Some motor-control items are designated as normally open or normally closed. For example, normally open contacts are contacts that are open when the circuit is not energized. When the circuit is energized, the contacts close. When the circuit is deenergized, the contacts return to the open position. Normally closed contacts are closed until the circuit is energized, and at that time the contacts open. Overloads look like normally closed contacts, but they are not. Overloads protect the motor from an overload condition. **Figure 9-2** shows the same basic motor control circuit. As drawn, the switch is closed, the M coil is energized, the M contacts are closed, and the motor is energized.

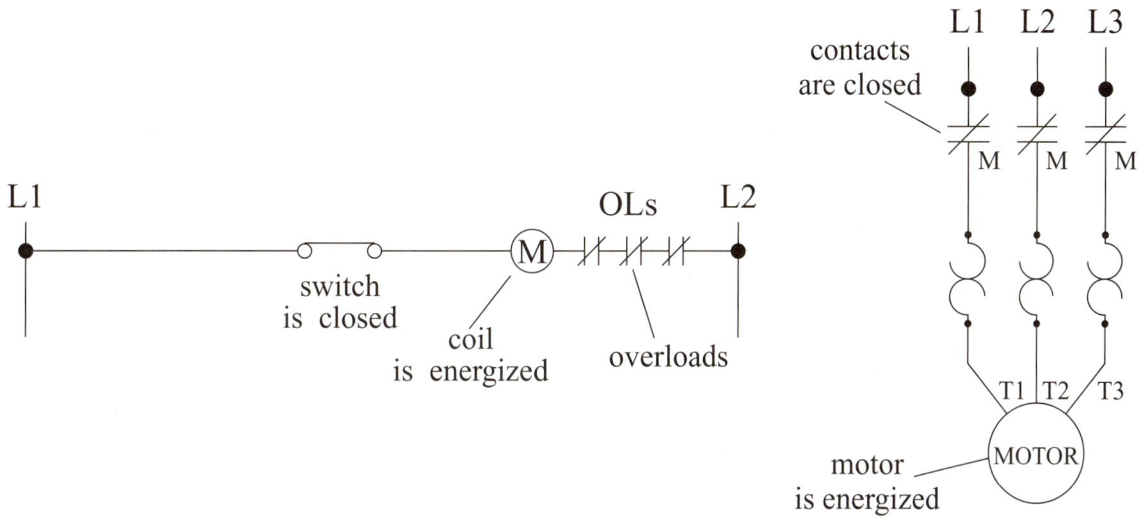

FIGURE 9-2

Start and Stop Pushbuttons

Exam questions do not always show a diagram of the circuit supplying the motor. Motor control drawings can have a two-wire control device or start and stop pushbuttons. Usually, stop pushbuttons are normally closed and start pushbuttons are normally open. Start and stop pushbuttons are typically drawn as momentary contacts. For example, after a normally open start button is pressed and released, it returns to the open position. A normally open auxiliary contact, installed in parallel with the start button, keeps the coil in the control circuit energized and the motor running. If the auxiliary contact is not installed, the coil loses power as soon as the start button is released. The motor will run only when the start button is pushed and is held in the closed position.

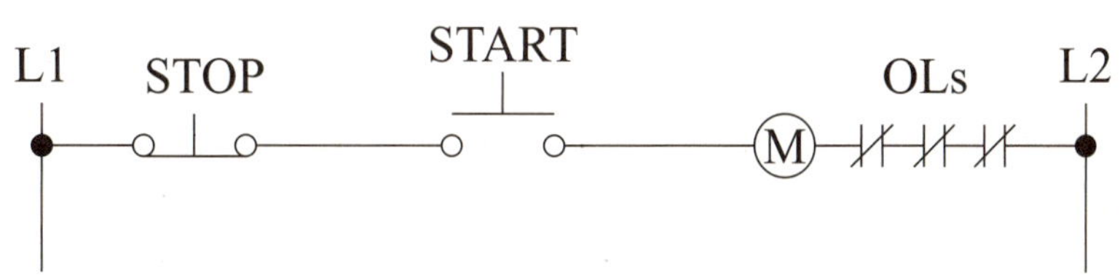

FIGURE 9-3

Without an auxiliary contact in the control circuit, the motor will not operate when the normally open start pushbutton is released. (As drawn in Figure 9-13, this start button is a jog button.)

An auxiliary contact installed in parallel with the start pushbutton can keep the motor running after the start button is released. Although the auxiliary contact under the start button is usually labeled M (below), it is sometimes labeled HC, for holding contact. In motor control drawings, it is not unusual to see one, two, or three overloads.

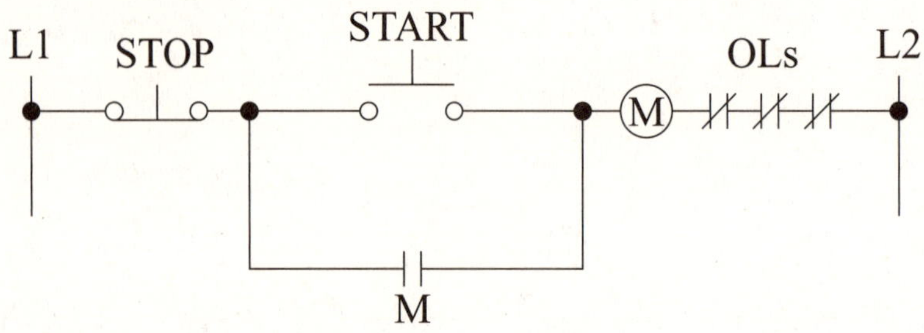

FIGURE 9-4

Sometimes additional or remote start pushbuttons are needed in a motor control circuit. Additional start buttons are usually connected in parallel with the existing start button.

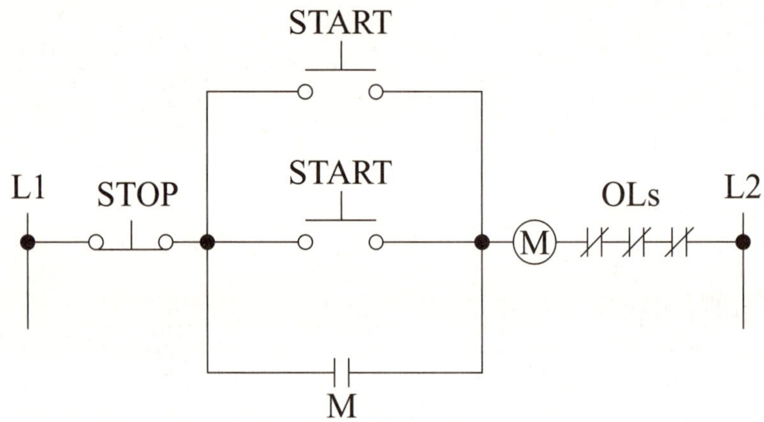

FIGURE 9-5

Sometimes additional or remote stop pushbuttons are needed in a motor control circuit. Additional stop buttons are usually connected in series to the existing stop button.

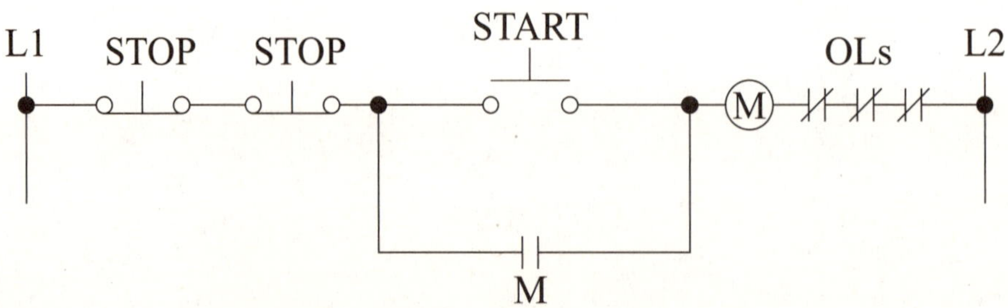

FIGURE 9-6

Two-Wire Control Circuits

Control circuits can be either two-wire or three-wire. Below is a basic diagram of a two-wire control circuit.

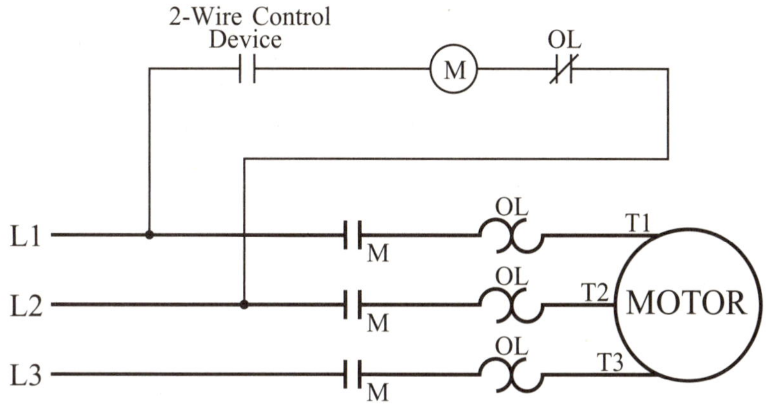

FIGURE 9-7

The following is another drawing of a two-wire control circuit. This drawing shows a wiring diagram of a starter operated from a two-wire control device. Examples of two-wire control devices include liquid level switches, limit switches, pressure and vacuum switches, temperature actuated switches, and flow switches.

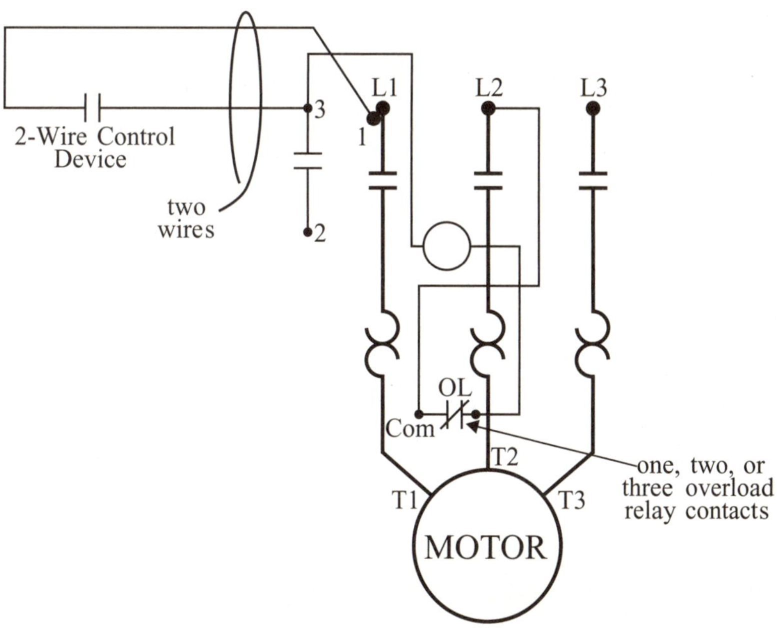

FIGURE 9-8

NOTE

The purpose of a sump pump is to remove water that has risen to a predetermined level. If the motor in Figure 9-9 is installed as a sump pump and wire A is not installed per instructions, the control circuit will not function properly: the motor will start pumping water only when the container is empty.

The pump in the wiring diagram below is controlled by a float switch, which is a two-wire control device. The float switch contains both normally open and normally closed contacts. As drawn in Figure 9-9, the control circuit is connected through the normally open contact of the float switch. The dashed line is not an existing conductor. It shows where a conductor could be installed to change the operation of the pump. When water in the tank drops, the float attached to the float switch will also drop. When the water reaches a predetermined low or empty level, the normally open contact in the float switch will close. At that time, the M coil energizes, the contacts supplying power to the motor close, and the motor starts pumping water back into the tank. When water in the tank rises, the float attached to the float switch will also rise. When the water reaches a predetermined full level, the contact that was closed (because of the dropped float) will open. When the contact in the float switch opens, the M coil will no longer be energized, the contacts supplying power to the motor will open, and the motor will stop pumping water back into the tank. With this type of installation, the auxiliary contact (between 2 and 3) is not needed because the M relay coil will function directly from the float switch.

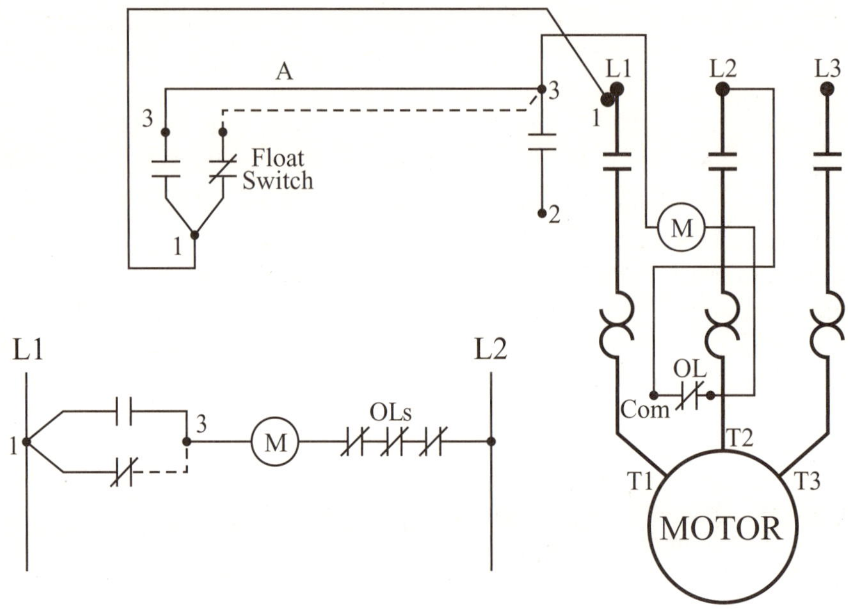

FIGURE 9-9 The float switch diagram, as shown, is in a full tank. The motor can operate as a sump pump by removing wire A and connecting it to the dotted line.

The next diagram shows the motor control circuit in Figure 9-9 wired to operate as a sump pump. Wire A has been removed from its original position and installed where the dashed line was located in the previous drawing. Now, the normally closed float switch is in series with the M relay coil and the motor overloads that are in the control circuit. When the tank is empty, the float is in the low position. When water in the tank rises, the float attached to the float switch will also rise. When the water reaches a predetermined full level, the contact that was open (because the float was down) will close. At that time, the M coil energizes, the contacts supplying power to the motor close, and the motor starts pumping water out of the tank. When water in the tank drops, the float attached to the float switch will also drop. When the water reaches a predetermined low or empty level, the normally closed contact in the float switch will open. When the contact in the float switch opens, the M coil will no longer be energized, the contacts supplying power to the motor will open, and the motor will stop pumping water out of the tank.

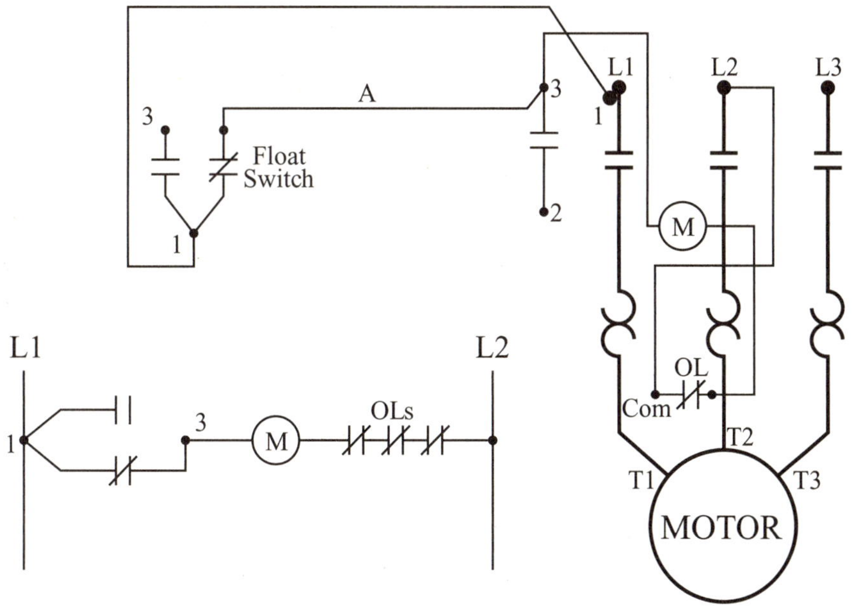

FIGURE 9-10 The float switch diagram, as shown, is in a full tank.

Three-Wire Control Circuits

Below is a basic diagram of a three-wire control circuit. Instead of a two-wire control device, start and stop pushbuttons control the operation of the motor. Pushbuttons are normally open, normally closed, or a combination of the two. Here, the stop button is normally closed and the start button is normally open. An auxiliary M contactor is in parallel with the start pushbutton. This contactor keeps the motor energized after the start button has been released.

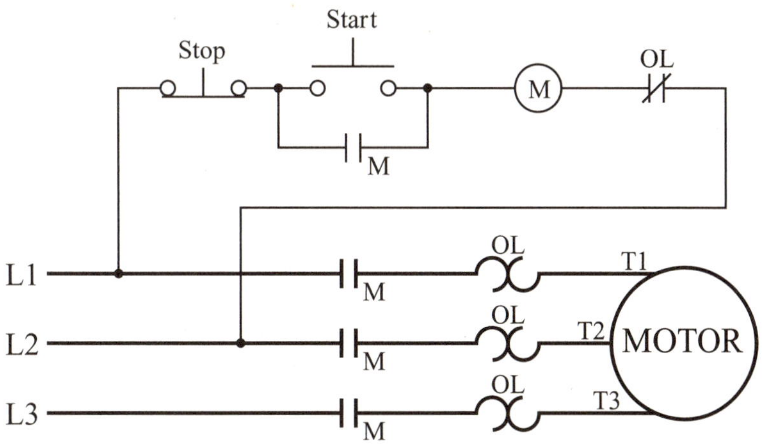

FIGURE 9-11

The following diagram is another drawing of a three-wire control circuit. This drawing shows a wiring diagram of a starter operated from a start–stop pushbutton. Three wires are needed between the starter and the start–stop pushbutton. The auxiliary contact is on the starter between 2 and 3.

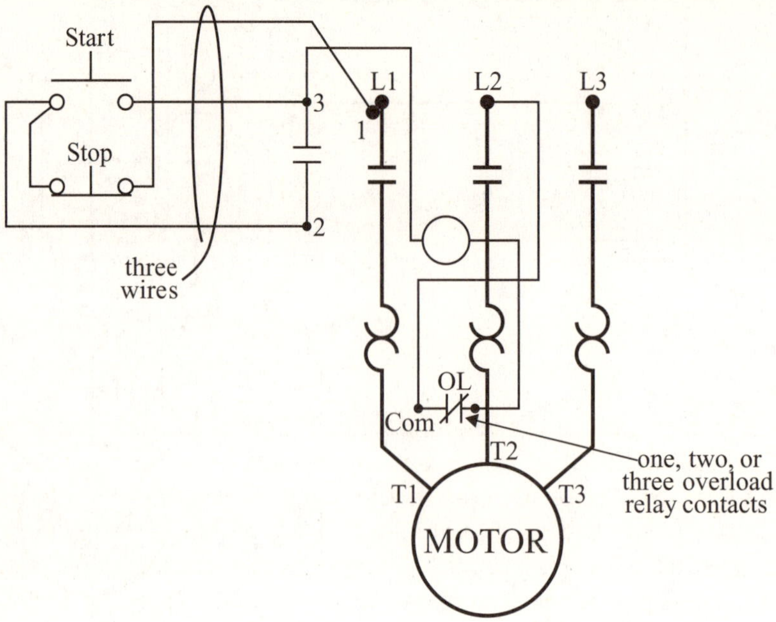

FIGURE 9-12

Jog–Run Selector Switch

Some motor control circuits have a jog–run selector switch. Jogging is momentary operation of the motor from its resting position. With jogging it is possible to operate the motor in small increments. When the selector switch is in the run position, the motor operates normally because the HC contact is in parallel with the start pushbutton. Note that the auxiliary contact in Figure 9-13 is labeled HC for holding contact. When the selector switch is in the jog position, the HC contact is disconnected from the circuit. The motor will operate only when the start pushbutton is pressed.

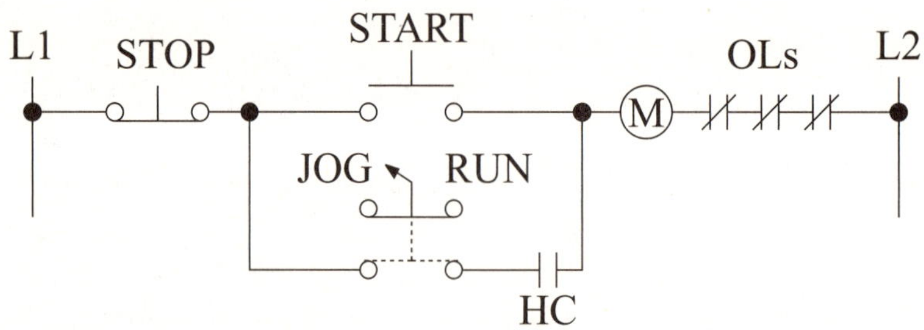

FIGURE 9-13

Reversing Starters

Reversing starters typically have three pushbuttons: forward, reverse, and stop. The dashed line under the normally open forward pushbutton and connecting to the normally closed pushbutton indicates that the two buttons are mechanically connected. The normally open and normally closed contacts of the reverse pushbutton are also mechanically connected. The normally closed contact that is part of the reverse pushbutton acts as a stop when the motor is running forward. Pressing the reverse pushbutton (when the motor is running forward) breaks the circuit supplying the forward coil and then makes the circuit supplying the reverse coil. As drawn in Figure 9-14, there is no limit switch and therefore it is not necessary to press the stop pushbutton before changing the direction of the motor.

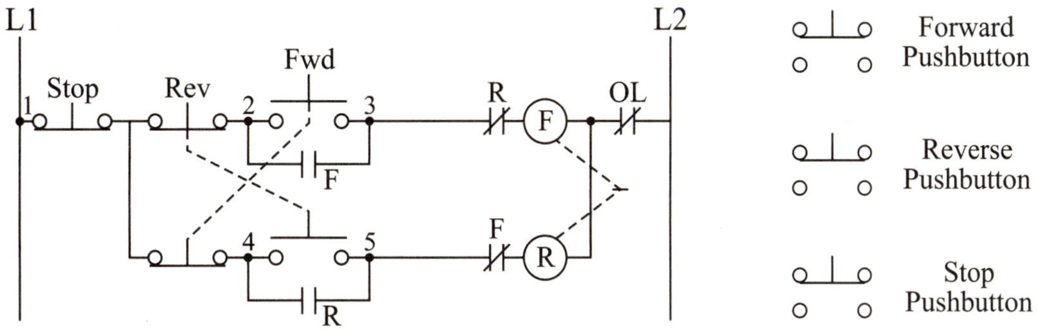

FIGURE 9-14

Pilot lights are sometimes drawn into the control circuit. Forward and reverse pilot lights on a reversing starter indicate whether the motor is in the forward or the reverse running position. Figure 9-15 shows a motor control circuit with pilot lights. When the forward pushbutton is pressed, the F (forward) coil and the forward pilot light are energized. When the reverse pushbutton is pressed, the R (reverse) coil and the reverse pilot light are energized. Normally closed contacts keep both coils from being energized at the same time. The right side of this drawing shows the terminations for the pushbuttons.

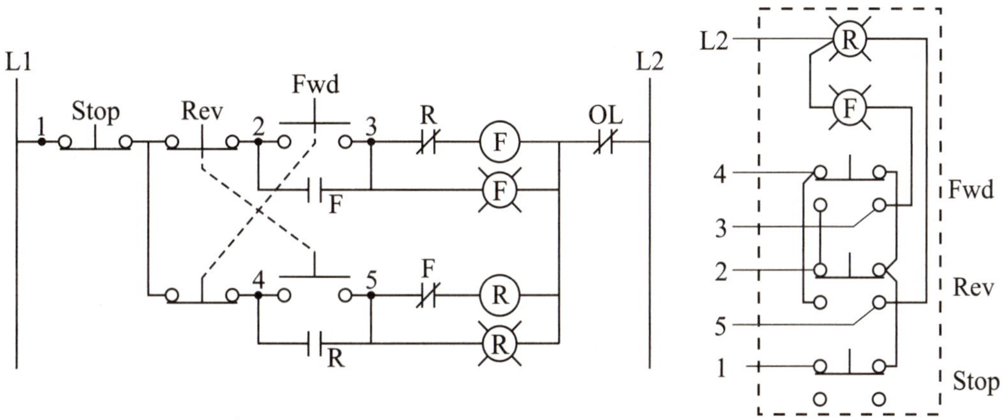

FIGURE 9-15

Similar to the forward–reverse motor control circuit is the fast–slow motor control circuit. Figure 9-16 shows a motor control with fast and slow pushbuttons. This diagram has only one pilot light, which is energized with either coil.

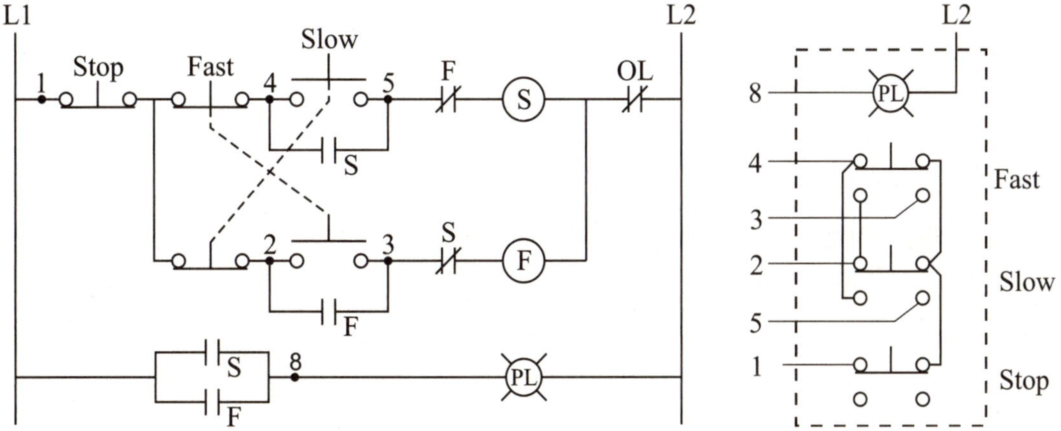

FIGURE 9-16

Sometimes it is necessary to change the direction of rotation on a three-phase motor. The direction can be reversed by swapping or interchanging any two of the three lines supplying power to the motor. Figure 9-17 shows the power circuit and control circuit of a reversing starter. Follow the lines supplying power to the motor and determine which lines are interchanged. When the normally open forward contacts are closed, L1 connects to T1, L2 connects to T2, and L3 connects to T3. When the normally open reverse contacts are closed, L1 now connects to T3, L2 again connects to T2, and L3 now connects to T1. Therefore in this diagram, L1 and L3 are interchanged to reverse the motor.

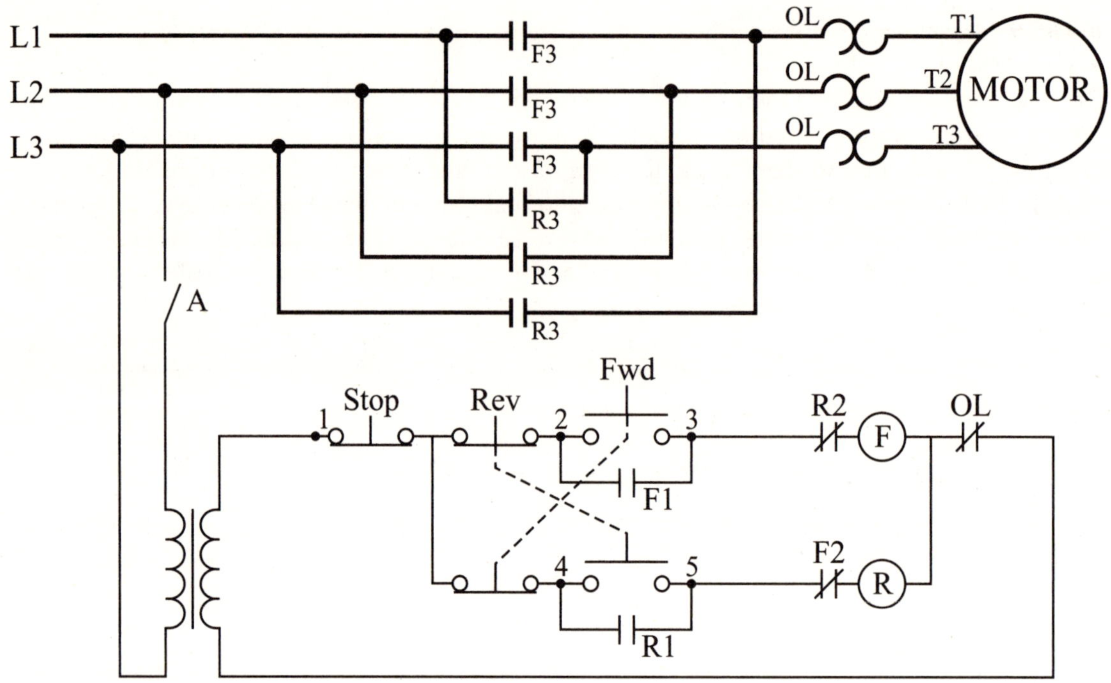

FIGURE 9-17

While reversing starters typically interchange Lines 1 and 3, different lines could be interchanged on an electrical exam. Since A is a disconnect, A must be closed before the motor can run in either direction. Pressing the forward pushbutton energizes the F coil. The F coil will close the normally open F1 and F3 contacts and will also open the normally closed F2 contact. The F2 contact is in the control circuit as a safety measure. When the F coil is energized, the R coil cannot energize because the circuit is open at the F2 contact. Whereas F1 must close before the forward pushbutton is released, it is not mandatory for F2 to open. As drawn, it is not necessary to press the stop button before reversing the motor. To reverse the motor from the forward direction, simply press the reverse pushbutton.

Resistor Starters

An electrical exam question may provide a drawing and ask for the type of starter in the motor control circuit. The two most common types on electrical exams are full-voltage and resistor. To determine the starter type, look at the circuit supplying power to the motor. All the motor control drawings up to this point are full-voltage starters. A resistor starter is a type of reduced voltage starter and is easy to identify because resistors are in the circuit supplying power to the motor. Figure 9-18 is an example of a resistor starter.

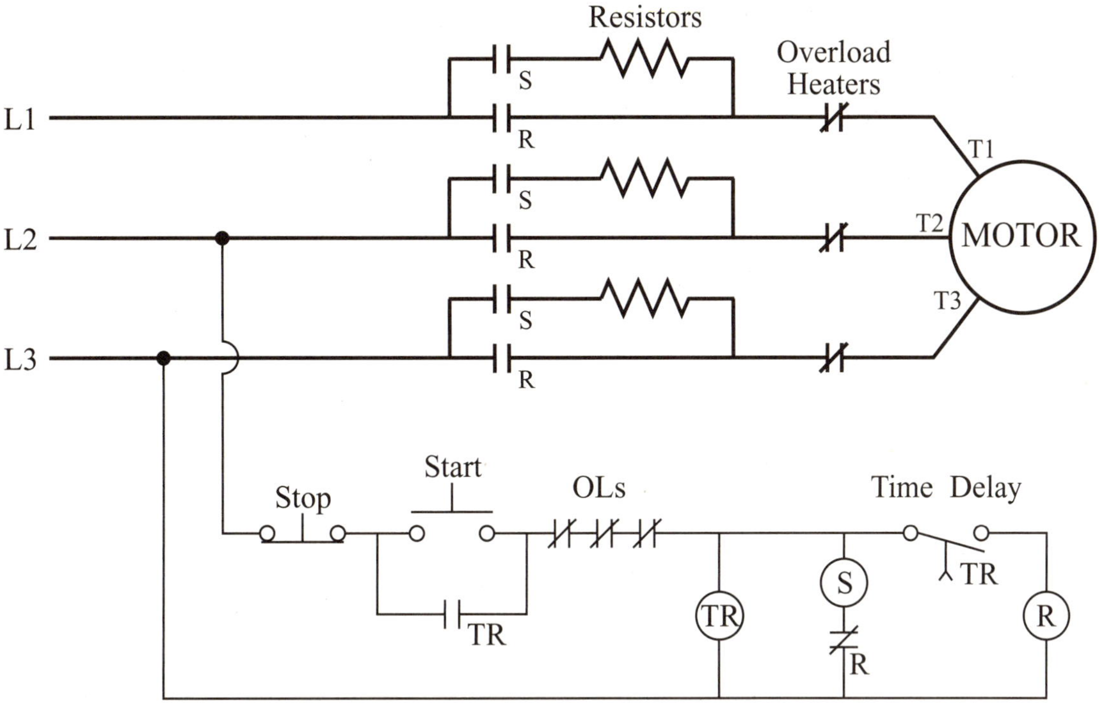

FIGURE 9-18

The resistors are in the power circuit, not the control circuit. The resistors are in series with the normally open S contacts and are in parallel with the normally open R contacts. The resistors are also in series to the overload heaters. When Start is pressed, the TR and S coils become energized. When the TR coil is energized, the time delay begins its cycle. This time delay device is a normally open contact that closes after it times out. Because the S coil is energized, the normally open S contacts close. When the S contacts close, the motor is energized, but because of the resistors in the circuit, voltage to the motor is reduced. When the time delay contact closes, the R coil becomes energized. The R coil will close the normally open R contacts and will also open the normally closed R contact. At this point, the motor will be supplied with full voltage. Also, the S coil will no longer be energized because the normally closed R contact has opened its circuit.

END-OF-CHAPTER PROBLEMS

Questions 1 through 4 refer to the following diagram:

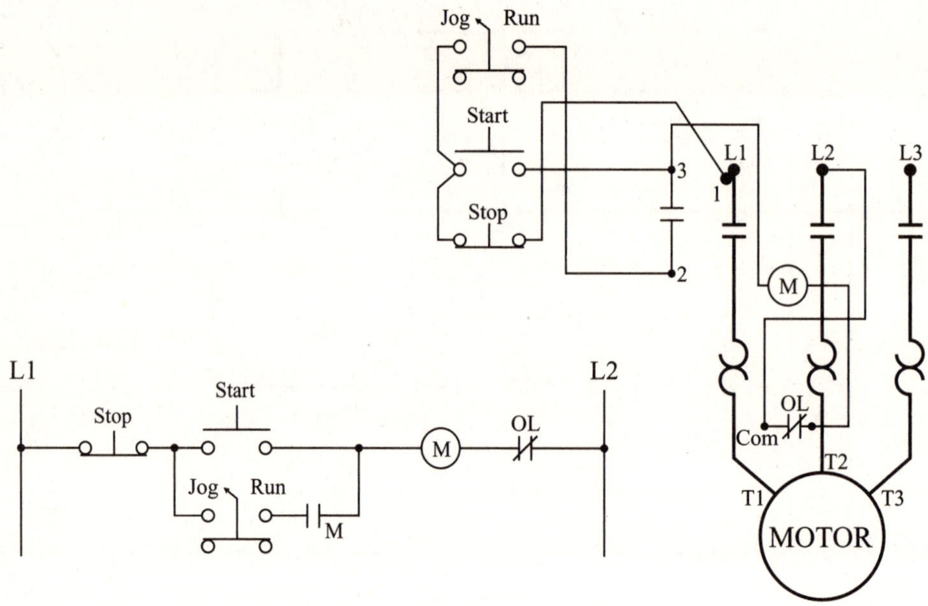

FIGURE 9-19

1. This is a(n) _____ starter.
   a) autotransformer
   b) full voltage
   c) resistor
   d) reactor
2. When the control circuit is in the run mode, which is the proper sequence of operation?
   a) Start is pushed, the normally open M contacts close, and the M relay coil is deenergized.
   b) Start is pushed; the M relay coil is energized after the normally open M contacts close.
   c) Start is pushed, energizing the M relay coil; the normally open M contacts open.
   d) Start is pushed, energizing the M relay coil; the normally open M contacts close.
3. Which of the following statements is true?
   a) The overload is in series with the M relay coil.
   b) The auxiliary M contact is in parallel with the M relay coil.
   c) The stop button is in parallel with the control circuit.
   d) The heaters are in series with the M relay coil.
4. When the control circuit is in the jog mode, which is the proper sequence of operation?
   a) Start is pushed, energizing the M relay coil; the normally open M contacts open.
   b) Start is pushed; the M relay coil is energized after the normally open M contacts close.
   c) Start is pushed, energizing the M relay coil; the jog switch times out and opens the circuit to the M relay coil.
   d) Start is pushed, energizing the M relay coil; the auxiliary M contact is bypassed in the jog mode; the M relay coil remains energized only when the start button is in the closed position.

Questions 5 through 9 refer to the following diagram:

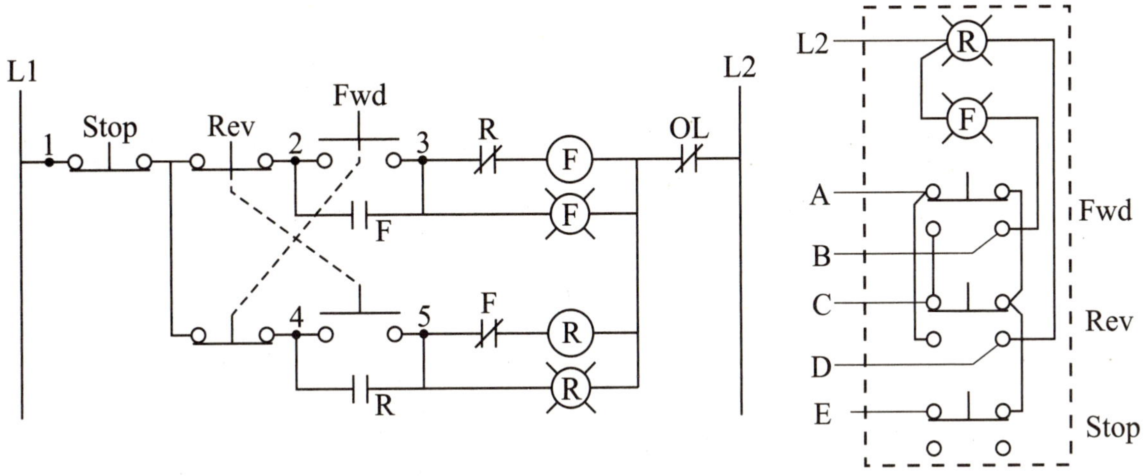

FIGURE 9-20

5. The letter B on the pushbutton station represents the number _____ on the control circuit diagram.

a) 2

b) 4

c) 1

d) 3

6. The letter E on the pushbutton station represents the number _____ on the control circuit diagram.

a) 3

b) 2

c) 4

d) 1

7. The letter C on the pushbutton station represents the number _____ on the control circuit diagram.

a) 2

b) 5

c) 1

d) 4

8. The letter A on the pushbutton station represents the number _____ on the control circuit diagram.

a) 5

b) 3

c) 4

d) 2

9. The letter D on the pushbutton station represents the number _____ on the control circuit diagram.

a) 2

b) 5

c) 3

d) 4

Questions 10 through 14 refer to the following diagram:

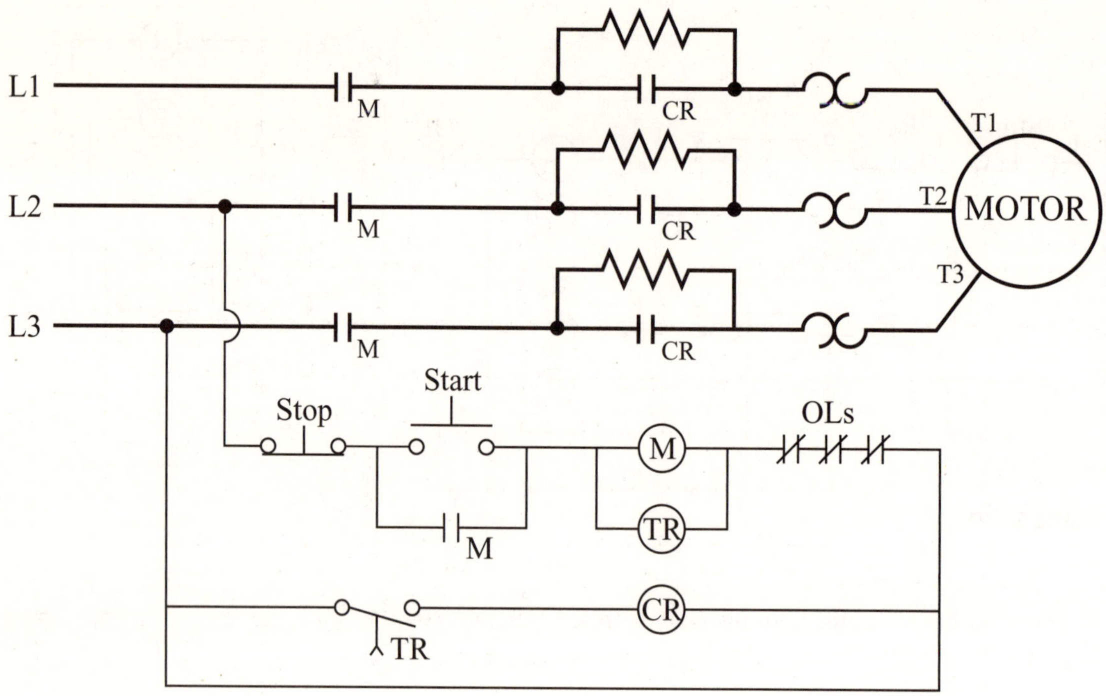

FIGURE 9-21

10. This is a(n) _____ starter.
   a) reactor
   b) resistor
   c) full voltage
   d) autotransformer
11. When the motor is at full speed, _____.
   a) all normally open M contacts are open
   b) all normally open CR contacts are open
   c) all normally open CR contacts are closed
   d) all normally open OLs are open.
12. The resistors are _____.
   a) in parallel with the CR contacts
   b) in parallel with the M contacts
   c) in parallel with the OLs
   d) in series with the CR contacts
13. Which of the following statements is true?
   a) When normally open M contacts close, the CR coil is energized.
   b) CR contacts close before M contacts close.
   c) Normally open M contacts open when the TR coil is energized.
   d) When the CR coil is energized, the motor will run at full speed.

14. A remote start station needs to be connected in _____.

a) series

b) parallel

c) series-parallel

d) Line 3

Questions 15 through 20 refer to the following diagram:

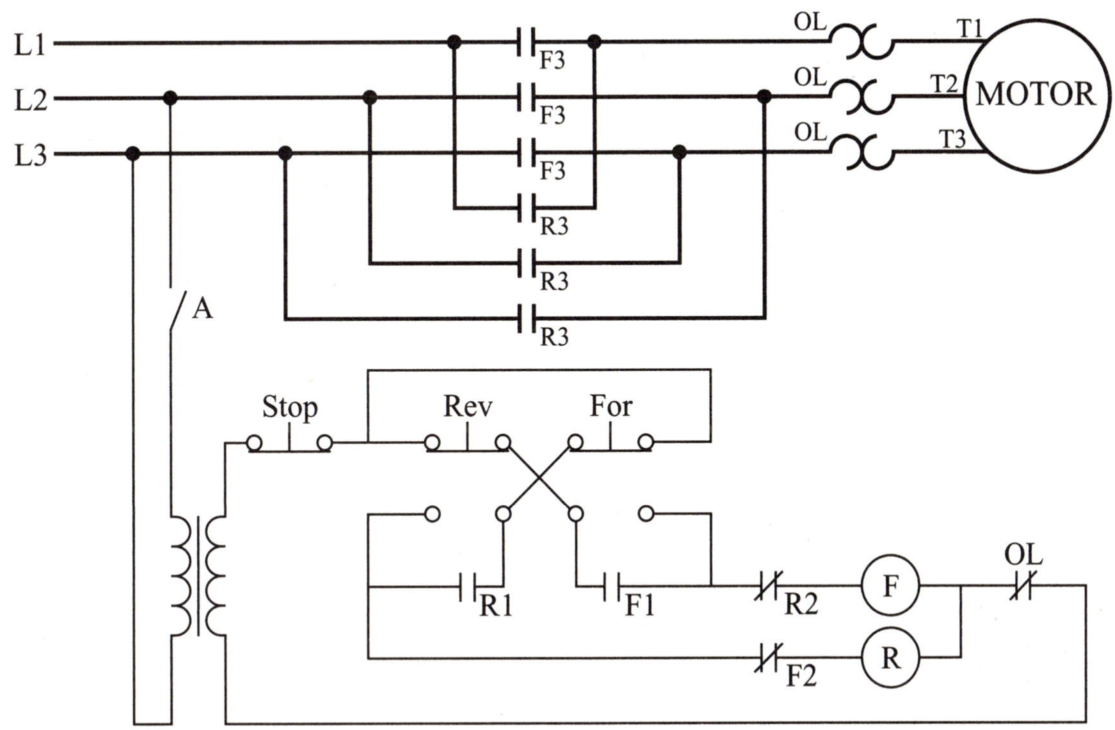

FIGURE 9-22

15. This is a(n) _____ starter.

a) reactor

b) resistor

c) full voltage

d) autotransformer

16. To reverse the motor, _____.

a) L1 and L3 are interchanged

b) L1 and L2 are interchanged

c) L2 and L3 are interchanged

d) the field is reversed

17. For the motor to run in the forward direction, which of the following statements is true?

a) F1 must close before For is released.

b) F1 must close before F2 opens.

c) F2 must open before For is released.

d) F2 must open and F1 must close before For is released.

18. If the motor is running forward, which of these statements is true?
   a) A is closed, R1 is closed, and F2 is open.
   b) A is closed, F1 is closed, and F2 is open.
   c) A is closed, For is released, and F2 is closed.
   d) A is closed, and F1, F2, and F3 are closed.
19. To reverse the motor from the forward position, ____.
   a) R2 must be open
   b) F1 must be open
   c) the stop button must be pressed before pressing the reverse button
   d) the reverse button may be pressed with the motor running forward
20. A remote stop switch can be wired in _____.
   a) L1
   b) L2
   c) L3
   d) the control circuit

Motor Load Calculations

CHAPTER 10

Article 430 of the *NEC* covers single- and multimotor calculations. There are six steps in calculating motor circuits: full-load currents, motor overload protection, branch-circuit conductor size, branch-circuit protection, feeder conductor size, and feeder protection.

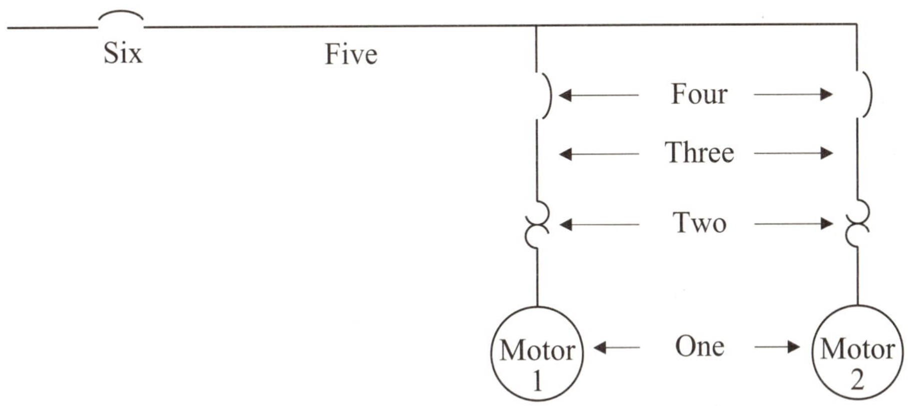

FIGURE 10-1

Table 10-1 Motor Load Calculation Step and *Code* References

One: Motor full-load current (FLC)	Tables 430.247, 430.248, and 430.250	
Two: Motor overload protection (heaters and overloads)	**Use the nameplate FLC if it is given.**	
	Standard or minimum size 430.32(A)(1)	
	Motors with a service factor of 1.15 or greater	FLC × 125%
	Motors with a temperature rise of 40° C or less	FLC × 125%
	All other motors	FLC × 115%
	Maximum size 430.32(C)	
	Motors with a service factor of 1.15 or greater	FLC × 140%
	Motors with a temperature rise of 40° C or less	FLC × 140%
	All other motors	FLC × 130%
	Ultimate trip current (or thermal protection) 430.32(A)(2)	
Three: Branch-circuit conductor size	430.22(A)	FLC × 125%
	Select the conductor size from Table 310.16	
Four: Branch-circuit protection	Table 430.52. The maximum rating or setting of branch-circuit short-circuit and ground-fault protective devices. Where the values as determined by Table 430.52 do not correspond to the standard ampere ratings in 240.6(A), the next higher standard size, rating, or setting is permitted. [430.52(C)(1) Ex. 1]	
Five: Feeder conductor size	430.24. Use this section when there is more than one motor (generally for feeders). Multiply the largest motor FLC by 125% and add the FLC of all other motors in the group. Note that the motor group is the maximum FLC of motors connected to each conductor.	
Six: Feeder protection	430.62(A). The feeder protective device must not be larger than the largest rating or setting of the branch-circuit short-circuit protective device for any motor of the group, plus the sum of the FLCs of the other motors of the group (on the same phase).	

1. Motor Full-Load Current

The first step is to find the full-load current (FLC) of the motor. Tables 430.247, 430.248, and 430.250 are used to find the FLC of dc, single-phase, and three-phase motors. Also use these tables when calculating branch-circuit conductors, branch-circuit protection, feeder conductors, and feeder protection.

EXAMPLE

What is the full-load current of a 10-hp, 208-volt, three-phase motor?

Step 1. Look at Table 430.250 in the 208-volt column.

Step 2. The FLC of a 10-hp, 208-volt, three-phase motor is 30.8 amperes.

PROBLEMS

Find the full-load currents for the following motors:

1. A 1½-hp, 240-volt, three-phase motor
2. A 5-hp, 120-volt, dc motor
3. A 3-hp, 220-volt, single-phase motor
4. A 40-hp, 230-volt, three-phase synchronous motor
5. A 5-hp, 120-volt, single-phase motor
6. A 75-hp, 208-volt, three-phase induction motor
7. A 75-hp, 240-volt, dc motor

2. Motor Overload Protection

Motor overload protection protects the operation of the motor in excess of normal full-load current rating. Be careful not to confuse motor overload protection with branch-circuit protection. Field terms for motor overload protection include the terms heaters and overloads. When sizing motor overload protection, use the nameplate rating of the motor if it is given. Read and highlight 430.6(A)(2). On electrical exams, motor overload protection may be referred to as running over-current protection, running overload protection, or heaters and overloads.

EXAMPLES

What minimum size motor overload protective device is required for a 5-hp, 230-volt, three-phase motor?

Step 1. The full-load current of the motor is 15.2 amperes. [Table 430.250]

Step 2. The motor in the question above does not have a service factor or a temperature rise.

Step 3. Multiply the FLC by 115% (all other motors).

15.2 × 115% = 17.48-ampere overload protective device.

What is the minimum size motor overload protection for the same motor with a 40° C temperature rise?

Step 1. The full-load current of the motor is 15.2 amperes. [Table 430.250]

Step 2. Multiply the FLC of the motor by 125% because of the temperature rise.

15.2 × 125% = 19-ampere overload protective device.

What is the maximum size motor overload protection for a 3-hp, 208-volt, single-phase motor with a service factor of 1.2?

Step 1. The full-load current of the motor is 18.7 amperes. [Table 430.248]

Step 2. Because of the service factor, multiply the FLC of the motor by 140%. This question is asking for the maximum size overloads. [430.32(C)]

18.7 × 140% = 26.18-ampere overload protective device.

What are the minimum and maximum size overloads (heaters) for a 5-hp, 208-volt, three-phase motor with a nameplate rating of 14 amperes?

Step 1. The full-load current of the motor is 14 amperes (nameplate rating).

Step 2. Because the nameplate is given, multiply 14 amperes by 115%.

14 × 115% = 16.1 ampere overloads (minimum).

Step 3. Multiply 14 amperes by 130% to find the maximum size overloads.

14 × 130% = 18.2-ampere overloads (maximum).

3. Branch-Circuit Conductor Size

Branch-circuit conductors are the conductors supplying power to the motor. Conductors supplying a single motor that is used in a continuous duty application must have an ampacity of at least 125 percent of the motor's full-load current rating as determined by 430.6(A)(1). [430.22(A)] The motor's full-load current rating used in this calculation must be from Tables 430.248 through 430.250. Do not use the nameplate rating of the motor even if it is given. After multiplying the motor's full-load current rating by 125 percent, select a conductor from Table 310.16.

NOTE

The voltages listed in Tables 430.248 and 430.250 have the following ranges: 110-115-120, 220-230-240, 440-460-480, and 550-575-600.

EXAMPLES

What size TW conductors are required to feed a 5-hp, 240-volt, single-phase motor?

Step 1. The full-load current is 28 amperes. [Table 430.248]

Step 2. Multiply the FLC of the motor by 125%. [430.22]

28 × 125% = 35 amperes

Step 3. Select a copper conductor from Table 310.16, in the 60° C (140° F) column, that has an ampacity rating of at least 35 amperes.

Step 4. The minimum size copper conductors required are 8-AWG TW conductors.

What size THW conductors are required to feed a 7 ½-hp, 208-volt, three-phase motor?

Step 1. The full-load current is 24.2 amperes. [Table 430.250]

Step 2. Multiply the FLC of the motor by 125%. [430.22]

24.2 × 125% = 30.25 amperes

Step 3. Select a copper conductor from Table 310.16, in the 75° C (167° F) column, that has an ampacity rating of at least 31 amperes.

Step 4. The minimum size copper conductors required are 10-AWG THW conductors.

NOTE

When sizing conductors for motors, the asterisk (*) in Table 310.16 will not apply. [240.4(G)]

NOTE

Sometimes there will be extra, unnecessary information in questions. The question above is asking only for the minimum ampacity rating in order to size the conductor.

EXAMPLE

What is the minimum load current rating, in amperes, for the branch-circuit conductors for a 25-hp, 230-volt, three-phase, synchronous motor with autotransformer starting? The motor has a code letter A, will be protected with time-delay fuses, and has a service factor of 1.1.

Step 1. The FLC of this motor is 53 amperes. [Table 430.250, Synchronous Type]

Step 2. Multiply 53 amperes by 125%. [430.22]

53 × 125% = 66.25 amperes

4. Branch-Circuit Protection

Branch-circuit conductors must be protected against overcurrent due to short circuits or grounds. The motor branch-circuit short-circuit and ground-fault protective device must be sized in accordance with 430.52. Use Table 430.52 to size nontime-delay fuses, dual element (time-delay) fuses, and inverse time breakers. Although the table provides percentages for instantaneous trip breakers, it is rare to have an exam question ask for this type of branch-circuit protection. When exam questions ask for motor branch-circuit overcurrent protection, find the maximum rating or setting allowed. Where the values as determined by Table 430.52 do not correspond to the standard ampere ratings in 240.6(A), the next higher standard size, rating, or setting is permitted. [430.52(C)(1) Ex. 1]

NOTE

When sizing branch-circuit protection, if the setting does not correspond to a standard size fuse or breaker found in 240.6(A), the next higher size fuse or breaker shall be permitted.

NOTE

The asterisk (*) in Table 310.16, referenced in 240.4(D), states that the overcurrent protection shall not exceed 20 amperes for 12-AWG conductors. This asterisk does not apply to motors. Article 430 is used to calculate overcurrent protection for motors as found in 240.4(G).

EXAMPLES

What is the maximum size inverse time circuit breaker allowed for branch-circuit protection on a 10-hp, 230-volt, three-phase motor?

Step 1. The full-load current of the motor is 28 amperes. [Table 430.250]

Step 2. The maximum percentage rates for motor branch-circuit short-circuit protective devices are found in Table 430.52. First, find the motor in the Type of Motor column. Next, follow straight across the page to the correct fuse or breaker column. Finally, choose the corresponding percentage rate. The percentage rate for this motor is 250%.

Step 3. Multiply 28 amperes by 250%.

28 × 250% = 70 amperes

The maximum size inverse time circuit breaker, for branch-circuit protection, allowed on a 10-hp, 230-volt, three-phase motor is a 70-ampere breaker.

What are the maximum size dual element fuses allowed for branch-circuit protection on a 3-hp, 208-volt, single-phase motor?

Step 1. The full-load current is 18.7 amperes. [Table 430.248]

Step 2. The percentage rate is 175%. [Table 430.52]

Step 3. Multiply the full-load current by the percentage rate.

18.7 × 175% = 32.725 amperes

The maximum dual element fuses for branch-circuit protection allowed for a 3-hp, 208-volt, single-phase motor are 35-ampere fuses.

The minimum size 75° C branch-circuit conductors for this motor are 12-AWG copper conductors (18.7 × 125% = 23.375 amperes). Table 310.16 states that a 12-AWG THW has a current rating of 25 amperes.

5. Feeder Conductor Size

Conductors supplying two or more motors must be sized in accordance with 430.24. Although this requirement pertains to branch-circuit, feeder, and service conductors, only feeder conductors will be covered in this chapter. Multiply the full-load current (FLC) rating of the highest rated motor by 125% and add the FLC ratings of all other motors in the group. Note that the motor group is the maximum FLC of motors connected to each conductor. Always select the smallest size conductors allowed per the *Code*.

NOTE

Always use the largest FLC, which is not necessarily the largest horsepower.

EXAMPLES

What size THW feeder conductors are required to feed a 5-hp, 230-volt, single-phase motor and a 3-hp, 115-volt, single-phase motor?

Step 1. The FLC of a 5-hp, 230-volt motor is 28 amperes. [Table 430.248]

Step 2. The FLC of a 3-hp, 115-volt motor is 34 amperes. [Table 430.248]

Step 3. The motor drawing the highest full-load current is the 3-hp motor (34 amperes).

Step 4. Multiply the full-load current of that motor by 125%.

$34 \times 125\% = 42.5$

Step 5. Add to that amount the full-load current of the other motor.

$42.5 + 28 = 70.5$ amperes

Step 6. Look in Table 310.16 to find a THW conductor with a rating of at least 71 amperes.

Step 7. The feeder conductors required for these motors are 4-AWG THW copper conductors.

What size THW feeder conductors are required to feed a 10-hp, 208-volt, 3-phase motor and three 5-hp, 208-volt, 3-phase motors?

Step 1. The FLC of a 10-hp, 208-volt, 3-phase motor is 30.8 amperes. [430.250]

Step 2. The FLC of a 5-hp, 208-volt, 3-phase motor is 16.7 amperes (each). [430.250]

Step 3. The motor drawing the highest full-load current is the 10-hp motor (30.8 amperes).

Step 4. Multiply the full-load current of that motor by 125%.

$30.8 \times 125\% = 38.5$

Step 5. Add to that amount the current ratings of the other motors.

$38.5 + 16.7 + 16.7 + 16.7 = 88.6$ amperes

Step 6. Look in Table 310.16 to find a THW conductor with a rating of at least 89 amperes.

Step 7. The feeder conductors required for these motors are 3-AWG THW copper conductors.

Sometimes there will be motors on the feeder conductors with different characteristics, such as 3-phase and single-phase motors on a 3-phase feeder. It may not be necessary to include the full-load current ratings of all the motors. In accordance with 430.24, conductors supplying two or more motors must have an ampacity not less than 125 percent of the FLC rating of the highest rated motor plus the sum of the FLC ratings of all the other motors in the group. It is permissible to balance the motors as evenly as possible per phase before determining the number of motors. The best way to start this type of motor calculation is to draw a simple diagram. Balance the motors as evenly as possible between phases and base the calculation on the phase with the highest amount of current.

EXAMPLE

What size THW feeder conductors are required to feed a 5-hp, 208-volt, three-phase motor and three 2-hp, 120-volt, single-phase motors? If the motors are balanced, each phase will have one 5-hp motor and one 2-hp motor.

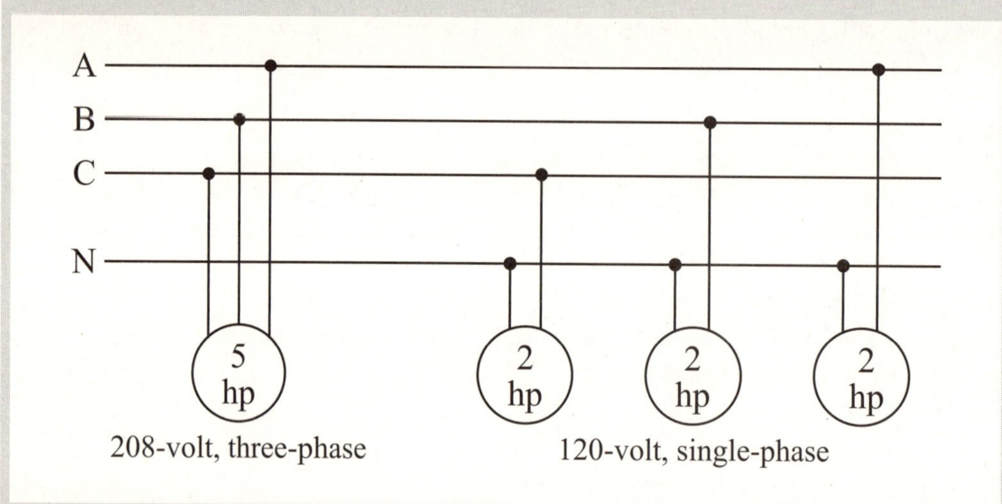

FIGURE 10-2

Step 1. The FLC of the 5-hp motor is 16.7 amperes on each phase.

Step 2. The FLC of each 2-hp motor is 24 amperes on one phase.

Step 3. The motor drawing the highest full-load current is the 2-hp motor (24 amperes).

Step 4. Multiply the full-load current of that motor by 125%.

$24 \times 125\% = 30$

Step 5. Add to that amount the current rating of the other motor on the same phase.

$30 + 16.7 = 46.7$ amperes

Step 6. The feeder conductors required for these motors are 8-AWG THW copper conductors.

NOTE

There are only two motors per conductor, not four.

6. Feeder Protection

Feeder conductors must also be protected against overcurrent due to short circuits or grounds. A feeder supplying specific fixed motor loads and consisting of conductor sizes based on 430.24 must be calculated in accordance with 430.62(A). The rating or setting of the feeder overcurrent protective device must not be larger than the largest rating or setting of the branch-circuit short-circuit and ground-fault protective device for any motor, plus the sum of the full-load currents of the other motors of the group. [430.62(A)]

EXAMPLE

What size inverse time breaker is required for the feeder protective device for three 3-hp, 230-volt, three-phase motors? Each of the motor branch circuits is protected with a 25-ampere breaker.

Step 1. The FLC for each motor is 9.6 amperes.

Step 2. The branch-circuit protection is given in this problem (25 amperes).

Step 3. Add the rating of the largest branch-circuit protective device to the current ratings of the other motors.

$25 + 9.6 + 9.6 = 44.20$ = 40-ampere breaker

NOTE

The feeder must not be larger than 44.2; therefore the maximum size inverse time breaker is 40 amperes.

Sometimes the branch-circuit protection must be calculated before the feeder protection can be calculated. Remember, single-phase motors on three-phase feeders can be balanced, as evenly as possible, between phases.

EXAMPLE

What is the maximum feeder overcurrent protection for the following motors? Use inverse time breakers.

- One 15-hp, 3-phase, 208-volt motor
- Three 5-hp, 1-phase, 120-volt motors

If the motors are balanced, each phase will have one 15-hp motor and one 5-hp motor.

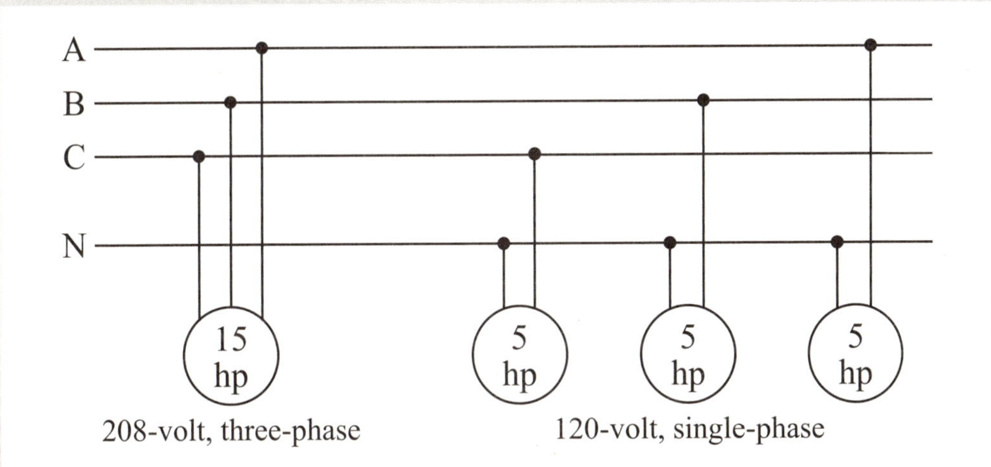

FIGURE 10-3

Step 1. The FLC of the 15-hp motor is 46.2 amperes.
Step 2. The FLC of the 5-hp motors is 56 amperes each.
Step 3. Find the largest branch-circuit protective device. Compare all motors.
46.2 × 250% = 115.5 = 125-ampere breaker
56 × 250% = 140 = 150-ampere breaker
Step 4. Add the largest branch-circuit protection to the other motors in the group (on the same phase).
150 + 46.2 = 196.2 (There are only two motors on each phase.)
Step 5. The maximum feeder overcurrent protection for these motors is a 175-ampere breaker.

NOTE

The feeder cannot be larger than 196.2 amperes. The next size standard breaker less than 196.2 amperes is a 175-ampere breaker. [240.6(A)]

END-OF-CHAPTER PROBLEMS

1. What is the maximum feeder overcurrent protection for the following motors? Use time-delay fuses.
   - One 20-hp, 3-phase, 208-volt motor
   - Three 5-hp, 1-phase, 120-volt motors

   a) 200 amperes
   b) 175 amperes
   c) 150 amperes
   d) 125 amperes

2. What is the minimum and maximum motor overcurrent protection for a 5-hp, three-phase, 208-volt motor?
   a) 19.2 amperes, 21.7 amperes
   b) 19.2 amperes, 23.4 amperes
   c) 20.1 amperes, 21.7 amperes
   d) 20.9 amperes, 23.4 amperes
3. What size branch-circuit protection is required for a 30-hp, 480-volt, 3 phase, squirrel-cage, autotransformer starting motor protected by dual element fuses?
   a) 60 amperes
   b) 70 amperes
   c) 80 amperes
   d) 100 amperes
4. What is the maximum feeder overcurrent protection for the following motors? Use an inverse time breaker.
   - One 20-hp, three-phase, 230-volt motor
   - Three 3-hp, single-phase, 230-volt motors

   a) 125 amperes
   b) 150 amperes
   c) 175 amperes
   d) 200 amperes
5. What is the branch-circuit protection for a 10-hp, 3-phase, 230-volt motor? Use an inverse time breaker.
   a) 30 amperes
   b) 35 amperes
   c) 70 amperes
   d) 80 amperes
6. The volt-amps of a 10-hp, 230-volt, three-phase motor are closest to______.
   a) 11,150
   b) 10,000
   c) 7460
   d) 6440
7. What size conductors are required to feed a 1½-hp, 230-volt, single-phase motor? The motor is 150 feet from the source. Do not exceed *NEC* recommendations.
   a) 8-AWG THW
   b) 10-AWG THW
   c) 12-AWG THW
   d) 14-AWG THW
8. What is the minimum load current rating for the branch-circuit conductors for a 3-hp, 1-phase, 208-volt motor?
   a) 13.3 amperes
   b) 18.7 amperes
   c) 23.4 amperes
   d) 25 amperes

9. What is the load per phase of a 12-kW, 3-phase, 240-volt motor?
   a) 4000 watts
   b) 6928 watts
   c) 12,000 watts
   d) 20,784 watts
10. What is the wattage per phase of a 1½-hp, 230-volt, 3-phase motor?
   a) 2390 watts
   b) 1380 watts
   c) 797 watts
   d) 460 watts
11. The smallest size feeder to one 5-hp, three-phase, 208-volt motor and three 2-hp, single-phase, 120-volt motors would require a minimum ampacity of _____.
   a) 100 amperes
   b) 85 amperes
   c) 65 amperes
   d) 50 amperes
12. What is the full-load current drawn by a 7½-hp, three-phase, 208-volt motor?
   a) 16.7 amperes
   b) 22 amperes
   c) 24.2 amperes
   d) 44 amperes
13. The branch-circuit protection for one 2-hp, single-phase, 120-volt motor using an inverse time circuit breaker would be _____.
   a) 70 amperes
   b) 60 amperes
   c) 50 amperes
   d) 40 amperes
14. A 5-hp, three-phase, 208-volt motor will have the branch circuit sized by _____.
   a) 15.2 amperes
   b) 16.72 amperes
   c) 19 amperes
   d) 20.9 amperes
15. A 5-hp, three-phase, 208-volt motor is on the same feeder as three 2-hp, 1-phase, 120-volt motors. The branch-circuit protection on each motor is an inverse time circuit breaker. The 5-hp branch-circuit protection is a 40-ampere breaker. The feeder protection should be a(n) _____-ampere inverse time breaker.
   a) 50
   b) 60
   c) 70
   d) 80

16. The feeder conductors are required to be _____ THW conductors for the four motors in the previous question.
   a) 10-AWG
   b) 8-AWG
   c) 6-AWG
   d) 4-AWG
17. What is the load per phase of a 9-kW, 3-phase, 208-volt motor?
   a) 3000 watts
   b) 5196 watts
   c) 9000 watts
   d) 15,588 watts
18. What minimum size conductors are required to feed a 1-phase, 1½-hp motor? The conductors are in a conduit, in free air, 150 feet from the source. The voltage on the motor is 115 volts.
   a) 12-AWG THW
   b) 10-AWG THW
   c) 8-AWG THW
   d) 6-AWG THW
19. What is the full-load current drawn by a 7½-hp, 3-phase, 230-volt motor?
   a) 15.2 amperes
   b) 16.72 amperes
   c) 22 amperes
   d) 24.2 amperes
20. What is the output wattage of a 3-phase, 208-volt, 5-hp motor?
   a) 3730 watts
   b) 6024 watts
   c) 6660 watts
   d) 7266 watts
21. What is the maximum feeder overcurrent protection for the following motors? Use an inverse time circuit breaker.
   - One 20-hp, 3-phase, 208-volt motor
   - Three 3-hp, 1-phase, 120-volt motors

   a) 150 amperes
   b) 175 amperes
   c) 200 amperes
   d) 275 amperes
22. A single-phase motor draws 23 amperes at 240 volts. The feeder is 175′ long, in conduit, in free air. What minimum size uncoated stranded THW copper conductors are recommended by the *National Electrical Code?*
   a) 12 AWG
   b) 10 AWG
   c) 8 AWG
   d) 4 AWG

23. What minimum size equipment grounding conductor is required for a 175-ampere motor?

a) 10 AWG
b) 8 AWG
c) 6 AWG
d) 4 AWG

24. What branch-circuit protection is required for a three-phase, 3-hp, 208-volt motor? Use an inverse time circuit breaker.

a) 15 ampere
b) 20 ampere
c) 25 ampere
d) 30 ampere

25. What is the maximum feeder overcurrent protection for the following motors? Use an inverse time circuit breaker.

- One 20-hp, 3-phase, 208-volt motor
- Three 2-hp, 1-phase, 120-volt motors

a) 125 amperes
b) 150 amperes
c) 175 amperes
d) 200 amperes

26. The full-load current of a three-phase, 25-hp, synchronous, 460-volt motor at unity power factor is _____.

a) 26 amperes
b) 34 amperes
c) 53 amperes
d) 68 amperes

27. A 15-hp, 3-phase, 240-volt, motor running overload protection shall be set at approximately _____ amperes.

a) 42
b) 48
c) 52.5
d) 58.8

28. What is the minimum and maximum motor overcurrent protection for a 7.5-hp, three-phase, 208-volt motor?

a) 30.25 amperes, 33.88 amperes
b) 30.25 amperes, 31.46 amperes
c) 27.83 amperes, 30.25 amperes
d) 27.83 amperes, 31.46 amperes

29. What is the feeder ampacity for one three-phase, 3-hp, 208-volt motor and two three-phase, 2-hp, 208-volt motors?

a) 21 amperes
b) 28 amperes
c) 32 amperes
d) 35 amperes

30. What size heaters are recommended for a 3-phase, 5-hp, 208-volt motor with a nameplate rating of 14 amperes?

a) 18.2 amperes
b) 17.5 amperes
c) 16.1 amperes
d) 14.0 amperes

31. The ampacity of a feeder for one three-phase, 3-hp, 208-volt motor and one three-phase, 2-hp, 208-volt motor should be at least _____ amperes.

a) 18
b) 19
c) 21
d) 22.63

32. A 50-hp, 480-volt, three-phase motor has a nameplate rating of 61 amperes. What is the maximum branch-circuit rating if this motor is protected by time-delay fuses?

a) 100 amperes
b) 110 amperes
c) 125 amperes
d) 150 amperes

33. What ampere feeder is required for one 15-hp, two 7½-hp, and three 3-hp motors? All motors are 208-volt, three-phase.

a) 130 amperes
b) 140 amperes
c) 150 amperes
d) 175 amperes

34. What is the voltage drop on a 6-AWG conductor feeding a 15-hp, 480-volt, three-phase motor? The length is 150 feet and the conductors are in a rigid metal conduit.

a) 1.55 volts
b) 2.68 volts
c) 3.09 volts
d) 3.34 volts

35. The branch-circuit protection for one 7½-hp, three-phase, 230-volt motor using an inverse time circuit breaker would be _____.

a) 30 amperes
b) 35 amperes
c) 55 amperes
d) 60 amperes

36. What ampere feeder is required for one 25-hp, one 30-hp, and one 40-hp motor? All motors are 440-volt, three-phase, synchronous-type motors.

a) 109 amperes
b) 126 amperes
c) 139 amperes
d) 220 amperes

37. If a 100-ampere inverse time breaker will serve as the branch-circuit protection for a single motor, what minimum size grounding conductor is required?

a) 10 AWG

b) 8 AWG

c) 6 AWG

d) 3 AWG

38. What size THW copper feeder conductors are required for a 10-hp, 480-volt, three-phase motor with 14 amperes full-load current and a 30-hp, 480-volt, three-phase motor with 40 amperes full-load current?

a) 1/0 AWG

b) 4 AWG

c) 6 AWG

d) 8 AWG

39. If two 20-hp, 3-phase, 208-volt motors are fed by the same feeder, what would be the minimum current in amperes required for the feeder?

a) 100

b) 130

c) 150

d) 175

40. What size branch-circuit protection is required for a 30-hp, 230-volt, 3-phase, squirrel-cage, autotransformer starting motor protected by nontime-delay fuses?

a) 150 amperes

b) 175 amperes

c) 200 amperes

d) 250 amperes

CHAPTER 11

One-Family Dwelling Load Calculations

Branch-circuit, feeder, and service load calculations are given in Article 220 of the *National Electrical Code.* This is the first of three chapters explaining load calculation provisions. While load calculation requirements may be difficult to understand as they appear in Article 220, they are easy to understand as they appear in this manual. This manual divides load calculations into three types: one-family, multifamily, and commercial. Each type is divided into a series of easy-to-understand steps. Carefully study the steps for calculating one-family dwellings because the steps for calculating multifamily dwellings are almost identical.

Calculation Steps

1. General Lighting and Receptacle Loads

The minimum lighting and receptacle loads for listed occupancies are found in Table 220.12. The residential unit load per square foot is located across from dwelling units in Table 220.12. The minimum lighting load for a dwelling unit is 3 volt-amperes (VA) per square foot.

According to *NEC* 220.12, "the floor area for each floor shall be calculated from the outside dimensions, dwelling unit, or other area involved. For dwelling units, the calculated floor area shall not include open porches, garages, or unused or unfinished spaces not adaptable for future use." [220.12]

NOTE

In dwelling units, all general-use receptacle outlets of 20-ampere rating or less "are included in the general lighting load calculations of 220.12. No additional load calculations shall be required for such outlets." [220.14(J)]

EXAMPLE

What is the general lighting load, before any derating, for a house with the outside dimensions of 32′ × 54′ on the first floor and 21′ × 42′ on the second floor?

a) 882 VA b) 1728 VA c) 2610 VA d) 7830 VA

Step 1. Find the total area of the house. Multiply the dimensions of each floor, then add the two.

32 × 54 = 1728 sq. ft. (first floor)

21 × 42 = 882 sq. ft. (second floor)

1728 + 882 = 2610 total sq. ft.

Step 2. Multiply the total square footage by 3 volt-amperes.

2610 × 3 = 7830 volt-amperes

The minimum general lighting load for this house is 7830 VA. Answer d) is correct.

EXAMPLE

A homeowner wants to add ten general-purpose receptacles. What is the additional load to the service?

a) 5400 VA b) 1800 VA c) 180 VA d) 0 VA

As specified in 220.14(J), no additional load calculation is required for these outlets. Answer d) is correct.

2. Small-Appliance Branch Circuits

Each small-appliance branch circuit required must be calculated at 1500 volt-amperes. These loads can be included with the general lighting load and subjected to the demand factors provided in Table 220.42. [220.52(A)] At least two small-appliance branch circuits are required. [210.11(C)(1)]

EXAMPLE

What is the minimum calculated load, before derating, for a dwelling unit kitchen with five appliance branch circuits?

a) 1500 VA b) 3000 VA c) 4500 VA d) 7500 VA

Multiply the number of small-appliance branch circuits by 1500. If the question does not provide the number, insert two small-appliance branch circuits.

5 × 1500 = 7500 volt-amperes

The minimum load for five appliance branch circuits is 7500 VA. Answer d) is correct.

3. Laundry Branch Circuits

Each laundry branch circuit must be calculated at no less than 1500 volt-amperes. These loads can be included with the general lighting load and are subjected to the demand factors provided in Table 220.42. [220.52(B)] At least one laundry branch circuit is required. [210.11(C)(2)]

EXAMPLE

What is the minimum calculated load, before derating, for a dwelling unit laundry room with three laundry branch circuits?

a) 1500 VA b) 3000 VA c) 4500 VA d) 7500 VA

Multiply the number of laundry branch circuits by 1500. If the question does not provide the number, insert one laundry branch circuit.

3 × 1500 = 4500 volt-amperes

The minimum load for three laundry branch circuits is 4500 volt-amperes. Answer c) is correct.

4. Lighting Load Demand Factors

Lighting load demand factors are in Table 220.42. The demand factors shall apply to that portion of the total branch-circuit load calculated for general illumination.

EXAMPLE

What is the lighting demand load for a house with the outside dimensions of 52′ × 40′ on the first floor and 26′ × 20′ on the second floor?

a) 4435 VA b) 4680 VA c) 6255 VA d) 12,300 VA

Step 1. To find the general lighting load, start by calculating the total floor area and multiplying by 3 volt-amperes per square foot.

52 × 40 = 2080 ft^2 (first floor)

26 × 20 = 520 ft^2 (second floor)

2080 + 520 = 2600 ft^2 (total)

2600 × 3 = 7800 volt-amperes

Step 2. Add the small appliance branch circuits.

1500 × 2 = 3000 volt-amperes

Step 3. Add the laundry circuit.

1500 × 1 = 1500 volt-amperes

continues

NOTE

When calculating the total lighting load, include a minimum of two small appliance branch circuits and one laundry circuit.

continued

Step 4. Add together Steps 1, 2, and 3 for the total volt-amperes before derating.

7800 + 3000 + 1500 = 12,300 volt-amperes

Step 5. Apply the demand factors of Table 220.42.

$$\begin{array}{r} 12{,}300 \\ \underline{-\,3000} \\ 9300 \end{array}$$

3000 × 100% = 3000

9300 × 35% = 3255

3000 + 3255 = 6255 volt-amperes

The general lighting demand load for the above house is 6255 volt-amperes. Answer c) is correct.

5. Fastened-in-Place Appliances

For the appliance load, it is "permissible to apply a demand factor of 75 percent to the nameplate rating load of four or more appliances fastened in place." [220.53] Do not include electric ranges, clothes dryers, space-heating equipment, or air conditioning equipment.

EXAMPLE

What is the appliance load contribution, in volt-amperes, for a house that has a 1.2-kVA dishwasher, a 0.76-kVA garbage disposer, a 1.1-kVA trash compactor, and a 4.5-kVA water heater?

a) 5.67 **b)** 7.56 **c)** 5670 **d)** 7560

Step 1. Notice that the question is asking for the answer to be given in volt-amperes. Convert all of the loads to volt-amperes by multiplying kVA by 1000.

Table 11-1 Converting Loads to Volt-Amperes

Dishwasher	1.2 × 1000 = 1200
Garbage disposer	0.76 × 1000 = 760
Trash compactor	1.1 × 1000 = 1100
Water heater	4.5 × 1000 = 4500

Step 2. Find the total load for the fastened-in-place appliances.

1200 + 760 + 1100 + 4500 = 7560 volt-amperes

Step 3. Since there are four fastened-in-place appliances, multiply the load by 75 percent.

7560 × 75% = 5670 volt-amperes

The calculated load for these appliances is 5670 volt-amperes. Answer c) is correct.

6. Clothes Dryer Load

According to the *NEC*, "the load for household electric clothes dryers . . . shall be either 5000 watts (volt-amperes) or the nameplate rating, whichever is larger." [220.54] A dryer is not required for a load calculation. If a dryer is not given in a problem, then assume no dryer and continue to the seventh step. The neutral demand on a dryer is 70 percent for feeders and services. [220.61(B)]

Explanations and examples of household electric clothes dryer calculations are covered in Chapter 4.

EXAMPLE

What is the one-family load contribution for a 4.5-kW clothes dryer?

a) 4.5 watts b) 5.0 watts c) 4500 watts d) 5000 watts

The minimum load (using the standard method) for a clothes dryer is 5000 watts. Answer d) is correct.

7. Household Cooking Appliances

Requirements for household electric ranges and other cooking appliances are in 220.55 and Table 220.55. Ranges and other cooking appliances are not required for a load calculation. If a range is not given in a problem, then assume no range and continue to the next step. The neutral demand on a range is 70 percent for feeders and services. [220.61(B)] Explanations and examples of household electric cooking equipment calculations are covered in Chapter 4.

8. Heating and Air Conditioning Loads

"Where it is unlikely that two or more noncoincident loads will be in use simultaneously, it shall be permissible to use only the largest load(s)." [220.60] The heating system and the air conditioning system of a dwelling unit are noncoincident loads. Since the air handler works with either the heating or the air conditioning, it should be calculated separately and added into the calculated load.

A heat pump with supplementary heat is not considered a noncoincident load. Add the compressor full load current (FLC) with the maximum amount of heat that can be energized with the compressor running.

9. Twenty-Five Percent of Largest Motor

As specified in 220.50 and 430.24, the largest motor must be multiplied by 125 percent. However, all of the motors should already be included before this step. Therefore, only the additional 25 percent of the largest motor is needed to complete this one-family dwelling calculation. In most cases, the largest motor is the air conditioner compressor. If the air conditioner load was omitted, then use the next largest motor.

NOTE

Table 310.15 (B)(6) does not apply to any single-phase dwelling service over 400 amperes.

10. Total Calculated Load

Add Steps 4 through 9 together to find the total calculated load in volt-amperes. Find the amperage by dividing the volt-amperes by the voltage. Find the conductor size in Table 310.15(B)(6).

11. Neutral Loads

Size the neutral conductor by calculating the total neutral loads. For ranges, wall-mounted ovens, counter-mounted cooking equipment, and clothes dryers, the neutral load is 70%. [220.61(B)] The neutral service or main power feeder conductor can be smaller than the ungrounded conductors, provided the requirements of 215.2, 220.61, and 230.42 are met. [310.15(B)(6)] The neutral cannot be smaller than the required grounding electrode conductor specified in Table 250.66, but it is not required to be larger than the largest ungrounded service-entrance phase conductor. [250.24(C)(1)]

Calculating Circuits in Dwellings

Exam questions will sometimes ask for the minimum number of circuits in dwellings. The questions will ask for the minimum number of 15- or 20-ampere branch circuits. A question might provide the square-foot dimension or the volt-amperes allocated for general lighting.

NOTE

The general lighting and receptacle circuits in dwelling occupancies are not considered continuous loads. It is not necessary to multiply the load by 125 percent or to multiply the overcurrent protection by 80 percent.

EXAMPLE

A dwelling has 2500 square feet of living area. How many 15-ampere circuits are required for this dwelling?

a) 2 b) 3 c) 5 d) 8

Step 1. Multiply 2500 sq. ft. by 3 volt-amperes per square foot as specified in Table 220.12.

2500 × 3 = 7500 total volt-amperes allocated for general lighting

Step 2. Divide the volt-amperes by 120 volts. This will give the current required for general purpose lighting and receptacles.

7500 ÷ 120 = 62.5 amperes

Step 3. Divide the total current by the rating (15 or 20 amperes) of the circuit.

62.5 ÷ 15 = 4.17

Step 4. Since at least 4.17 circuits are required, round up to 5.

The minimum number of circuits is five. Answer c) is correct.

A question may ask for the minimum number of 20-ampere circuits in a dwelling. One-family dwellings require at least two 20-ampere small appliance branch circuits, one 20-ampere laundry branch circuit, and one 20-ampere bathroom branch circuit. These four 20-ampere circuits must be added to the calculated number of 20-ampere general lighting circuits.

EXAMPLE

A dwelling has 3600 volt-amperes allocated for general lighting and receptacles. How many 20-ampere branch circuits are required for this dwelling?

a) 6 b) 5 c) 3 d) 2

Step 1. Divide the volt-amperes by 120 volts. This will give the current required for general purpose lighting and receptacles.

3600 ÷ 120 = 30 amperes

Step 2. Divide the total current by 20 amperes.

30 ÷ 20 = 1.5 = 2

Two branch circuits are required for general lighting and receptacles.

Step 3. Add two small appliance branch circuits, one laundry branch circuit, and one bathroom branch circuit to the calculated number of general lighting and receptacle circuits.

2 + 2 + 1 + 1 = 6

The minimum number of circuits is six. Answer a) is correct.

Standard Method: One-Family Dwellings

1. General Lighting and Receptacle Loads [Table 220.12]

3 volt-amperes × square footage (outside dimensions)

Do not include porches, garages, or unused or unfinished spaces not adaptable for future use.

2. Small Appliance Branch Circuits [220.52(A)]

2 × 1500 = 3000 volt-amperes (minimum)

3. Laundry Branch Circuits [220.52(B)]

1 × 1500 = 1500 volt-amperes (minimum)

4. General Lighting Load [Table 220.42]

Add Steps 1, 2, and 3 to find the lighting demand load.

5. Fastened-in-Place Appliances [220.53]

With four or more fastened-in-place appliances, apply a demand factor of 75% to the nameplate rating load. Do not include electric ranges, clothes dryers, space-heating equipment, or air conditioning equipment.

6. Clothes Dryer [220.54]

Include a load of 5000 volt-amperes or the nameplate rating, whichever is larger. The neutral demand is 70% for feeders. [220.61(B)]

7. Ranges [220.55]

Use Table 220.55 and notes.

The neutral demand is 70% for feeders. [220.61(B)]

8. Heat or Air Conditioner [220.60]

Compare the heat and A/C and omit the smaller.

Include the air handler on either one.

9. Largest Motor [220.50 and 430.24]

Multiply the largest motor by 25% and add to the load.

10. Size the Feeder and/or Service Conductors [Table 310.15(B)(6)]

Add Steps 4 through 9 together to find the total calculated load in volt-amperes.

11. Size the Neutral Conductor [220.61(B)]

The neutral conductor can be smaller than the ungrounded conductors, provided the requirements of 215.2, 220.61, and 230.42 are met. [310.15(B)(6)] The neutral cannot be smaller than the required grounding electrode conductor specified in Table 250.66, but it is not required to be larger than the largest ungrounded service-entrance phase conductor. [250.24(C)(1)]

EXAMPLE

Single-phase, 115/230 volt
1800 square foot house (outside dimensions)
12-kW range, 115/230 volt
5-kW water heater, 230 volt
5-hp air conditioner motor, 230 volt
½-hp air handler, 115 volt
15-kW heat, 230 volt
4-kW clothes dryer, 115/230 volt
1.5-kW dishwasher, 115 volt
¼-hp garbage disposer, 115 volt
½-hp trash compactor, 115 volt

Table 11-2 Calculating Volt-Amperes

General lighting and receptacles	Volt-amperes
3 VA per sq. ft.	1800 × 3 = 5400
Small appliance	1500 × 2 = 3000
Laundry	1500 × 1 = 1500
Total:	9900

3000 × 100% = 3000
9900 - 3000 = 6900
6900 × 35% = 2415
3000 + 2415 = 5415 VA

continues

continued

Table 11-3 One-Family Load Calculation (Standard Method)

	Service	Neutral
General lighting and receptacles	5415	5415
Water heater 5000 × 75%	3750	0
Dishwasher 1500 × 75%	1125	1125
Disposer 5.8 × 115 × 75%	500	500
Compactor 9.8 × 115 × 75%	845	845
Dryer 4 kW = 5 kW minimum	5000	3500
Range 12 kW = 8 kW [Table 220.55]	8000	5600
Heat (omit A/C)	15,000	0
Air handler 9.8 × 115	1127	1127
Largest motor 9.8 × 115 × 25%	282	282
Total	**41,044**	**18,394**

Service conductors
41,044 ÷ 230 = 178.5 amperes

200-ampere service [Table 310.15(B)(6)] = 2/0-AWG THW Cu

Neutral conductors
18,394 ÷ 230 = 80 amperes

80 amperes [Table 310.16] = 4-AWG THW Cu

Grounding electrode conductor [Table 250.66]
4-AWG copper

Optional Method: One-Family Dwellings

There are two methods for calculating feeder and service loads. The first method (discussed before the optional method) is in Part III of Article 220. The first method is often referred to as the standard method. The second method is the optional method and is in Part IV of Article 220. Unless an exam question specifically requires finding the load by using the optional method, use the standard method. As stated in 220.82(A), the minimum size service is 100 amperes when calculating by the optional method.

1. General Lighting and Receptacle Loads [220.82(B)(1)]

3 volt-amperes × square footage (outside dimensions)

Do not include porches, garages, or unused or unfinished spaces not adaptable for future use.

2. Small Appliance Branch Circuits [220.82(B)(2)]

2 × 1500 = 3000 volt-amperes (minimum)

3. Laundry Branch Circuit [220.82(B)(2)]

1 × 1500 = 1500 volt-amperes (minimum)

4. Fastened-in-Place Appliances [220.82(B)(3)]

Use the nameplate rating of all fastened-in-place appliances, including ranges, ovens, cook tops, clothes dryers, and water heaters.

5. Motors (not air conditioning or heat) [220.82(B)(4)]
Use the nameplate rating and convert to volt-amperes.

6. Total the Volt-Amperes from Steps 1, 2, 3, 4, and 5 [220.82(B)]
Multiply the first 10,000 volt-amperes by 100%.

Then, multiply the remainder of the volt-amperes by 40%.

7. Use the Largest of the Following [220.82(C)]

- 100% of air conditioning and cooling
- 100% of a heat pump (without supplemental heating)
- 100% of electric thermal storage systems
- 100% of a heat pump compressor and 65% of supplemental electric heating
- 65% of three or fewer space-heating units
- 40% of four or more separately controlled space-heating units

8. Size the Feeder Conductors
Add Steps 6 and 7 to find the total calculated load in volt-amperes.

NOTE

There is no optional method for calculating the neutral conductor.

EXAMPLE

Use the optional method to find the minimum size service and service conductors for this one-family dwelling. The source voltage is 120/240 single-phase.

1900 square feet of living area
14-kW range, 120/240 volt
4.5-kW water heater, 240 volt
4-hp air conditioner, 240 volt
1200-watt dishwasher, 120 volt
⅓-hp garbage disposer, 120 volt
4000-watt electric clothes dryer, 120/240 volt
18-kW heating, (separate controls) in six rooms, 240 volt

General lighting and receptacle
1900 (ft^2) × 3 (volt-amperes) = 5700 volt-amperes

Table 11-4 One-Family Load Calculation (Optional Method)

	Service
General lighting and receptacle	5700
Small appliance	3000
Laundry	1500
Range	14,000
Water heater	4500
Dishwasher	1200
Garbage disposer 7.2 × 120	864
Clothes dryer	4000
Total	**34,764**

continues

continued

The first 10,000 volt-amperes × 100% = 10,000
The remainder 24,764 VA × 40% = 9906
Heat (omit air conditioning) 18,000 × 40% = 7200
Total = 27,106 volt-amperes

Service conductors 27,106 ÷ 240 = 113 amperes
125-ampere service = 2-AWG THW Cu
The minimum size service required, using the optional method, is 125 amperes.

END-OF-CHAPTER PROBLEMS

1. A one-family dwelling has 6000 watts of heat and 4000 watts of air conditioning. What is the demand load contribution for the heat and air? Use the optional method.
   a) 10,000 watts
   b) 6000 watts
   c) 4000 watts
   d) 3900 watts
2. A homeowner wants to add four general-purpose receptacles. What is the additional load to the service?
   a) 720 volt-amperes
   b) 1500 volt-amperes
   c) 3000 volt-amperes
   d) None of these
3. What size THW conductors are required to feed a 13.55-kW range in a one-family dwelling? The voltage is 120/240 volts.
   a) 10 AWG
   b) 8 AWG
   c) 6 AWG
   d) 4 AWG
4. A dwelling has a 175-ampere service that is fed with THW copper conductors. What is the minimum size copper grounding electrode conductor?
   a) 8 AWG
   b) 6 AWG
   c) 4 AWG
   d) 1/0 AWG
5. The neutral of a 5-kW, residential, 240-volt clothes dryer would contribute _____ amperes to the service neutral size.
   a) 14.58
   b) 20.83
   c) 29.17
   d) 41.67

6. A house has four 6-ampere, 230-volt room air conditioners and 1-kW heaters in each of five rooms. What is the kW load before demand factors?
   a) 2 kW
   b) 3.25 kW
   c) 5 kW
   d) 5.52 kW
7. An owner wishes to add 9 additional receptacle outlets to the drawings of his residence. What load will this add to the service?
   a) 0 watts
   b) 27 watts
   c) 900 watts
   d) 1620 watts
8. A one-family dwelling has 6000 watts of heat and 3000 watts of air conditioning. What is the demand load contribution for the heat or air? Use the optional method.
   a) 3000 watts
   b) 3900 watts
   c) 6000 watts
   d) 6900 watts
9. A one-family dwelling has a total calculated load of 67.5 kVA at 120/240 volts. Using the optional method, what minimum size service disconnect is required?
   a) 100 amperes
   b) 125 amperes
   c) 150 amperes
   d) 200 amperes
10. A one-family dwelling has a general lighting load of 6000 volt-amps. What is the minimum number of 15-amp branch circuits required?
    a) 3
    b) 4
    c) 5
    d) 6
11. What is the neutral load in amperes for a 12-kW residential range that is rated 115/230 volts?
    a) 24.35
    b) 34.78
    c) 36.52
    d) 52.17
12. What is the feeder neutral load for two 4-kW wall ovens and one 5.1-kW counter-mounted cooking unit in the kitchen of a one-family dwelling?
    a) 8.04 kW
    b) 7.21 kW
    c) 5.88 kW
    d) 5.04 kW

Multifamily-Dwelling Load Calculations

CHAPTER 12

Standard Method: Multifamily Dwellings

Multifamily-dwelling load calculations are almost identical to the one-family dwelling load calculations described in the previous chapter. See Chapter 11 for examples of each individual step.

1. **General Lighting and Receptacle Loads [Table 220.12]**

   3 volt-amperes × square footage × total number of units

   Do not include porches, garages, or unused or unfinished spaces not adaptable for future use.

2. **Small Appliance Branch Circuits [220.52(A)]**

   2 × 1500 × total number of units

   3000 volt-amperes (minimum per unit)

3. **Laundry Branch Circuit(s) [220.52(B)]**

   1 × 1500 × total number of units

   1500 volt-amperes (minimum per unit)

   If laundry facilities are on the premises, omit this step. [210.52(F) Ex. 1]

4. **General Lighting Load [Table 220.42]**

   Add Steps 1, 2, and 3 together and find the lighting demand load.

5. **Fastened-in-Place Appliances [220.53]**

   With four or more fastened-in-place appliances, apply a demand factor of 75% to the nameplate rating load, then multiply by the total number of units. Do not include electric ranges, dryers, space-heating equipment, or air conditioning equipment.

6. **Clothes Dryer [220.54]**

   Include a load of 5000 volt-amperes or the nameplate rating, whichever is larger.

   The neutral demand is 70% for feeders. [220.61(B)]

7. **Ranges [220.55]**

   Use Table 220.55 and notes.

   The neutral demand is 70% for feeders. [220.61(B)]

8. **Heat or Air Conditioner [220.60]**

   Compare the heat and air conditioning and omit the smaller.

   Include the air handler on either one.

9. **Largest Motor (one motor only) [220.50 and 430.24]**

   Multiply the largest motor by 25% and add to the load.

10. **Size the Feeder and/or Service Conductors**

    Add Steps 4 through 9 to find the total calculated load in volt-amperes.
    Single-phase I = VA ÷ E
    Three-phase I = VA ÷ (E × 1.732)

11. **Size the Neutral Conductor [220.61(B)]**

    Single-phase I = VA ÷ E
    Three-phase I = VA ÷ (E × 1.732)
    Multiply unbalanced load over 200 amperes by 70%. The load cannot be smaller than the grounding conductor. The neutral cannot be smaller than 12½% of the phase wires larger than 1100-kcmil copper or 1750 aluminum. [250.24(C)(1)]

EXAMPLE

An 18-unit condominium has a service voltage of 120/240-volt single-phase. The service to each unit shall be 120/240-volt single-phase. Find the calculated load on the service and the neutral. Each unit has the following:

1000 ft^2 floor area
4.5-kW water heater, 240 volt
3-hp air conditioner, 240 volt
4.5-kW heat, 240 volt
8-kW range, 120/240 volt
4.5-kW dryer, 120/240 volt
½-hp air handler, 120 volt

1000 × 3 × 18 = 54,000
Small appliance × 18 = 54,000
Laundry × 18 = 27,000
135,000 VA
-3,000 × 100% = 3,000
132,000
-117,000 × 35% = 40,950
15,000 × 25% = 3,750
General lighting and receptacles 47,700 VA

Table 12-1 Multifamily Load Calculation (Standard Method)

	Service	Neutral
General lighting and receptacles	47,700	47,700
Water heater 4500 × 75% × 18	60,750	0
Dryer 5 kW × 18 × 40% [Table 220.54]	36,000	25,200
Ranges [220.55]	33,000	23,100
Heat (omit air conditioning) 4500 × 18	81,000	0
Air handler 9.8 × 120 × 18	21,168	21,168
Largest motor 9.8 × 120 × 25%	294	294
Total	**279,912**	**117,462**

Service conductors

279,912 ÷ 240 = 1166 amperes

The minimum size service required for this condominium is 1200 amperes.

Neutral conductors

117,462 ÷ 240 = 489 amperes

489 amps
-200 amps × 100% = 200 amps
289 amps × 70% = +202 amps
402 amperes total neutral load

NOTE

If the load is over 800 amperes, the ampacity of the conductors should be rated at, or above, the overcurrent protective device. [240.4(B)]

Optional Method: Multifamily Dwellings

Multifamily-dwelling calculations by the optional method are similar to the optional method for one-family dwelling load calculations described in the previous chapter. In accordance with 220.84(A), three conditions must be met:

1. No dwelling unit can be supplied by more than one feeder.
2. Each dwelling unit must be equipped with electric cooking equipment (unless meeting the exception).
3. Each dwelling unit must be equipped with either electric space-heating or air conditioning, or both.

It is not permissible to calculate feeder and service neutral conductors by the optional method. It is also not permissible to calculate house loads by the optional method. The house load must be calculated by Part III of Article 220.

1. **General Lighting and Receptacle Loads [220.84(C)(1)]**

   3 volt-amperes × square footage × total number of units

   Do not include porches, garages, or unused or unfinished spaces not adaptable for future use.
2. **Small Appliance Branch Circuits [220.84(C)(2)]**

   2 × 1500 × total number of units
3. **Laundry Branch Circuit [220.84(C)(2)]**

   1 × 1500 × total number of units

   If laundry facilities are on premises, omit this step. [210.52(F) Ex. 1]
4. **Fastened-in-Place Appliances [220.84(C)(3)]**

   Nameplate × number of units

   Use the nameplate rating of all the appliances, including the clothes dryer.
5. **Motors (not air conditioning or heat) [220.84(C)(4)]**

   Multiply the nameplate rating in volt-amperes by the number of units.
6. **Heat or Air Conditioner [220.84(C)(5)]**

   Largest of the heat or the air conditioner × number of units
7. **Total the Volt-Amperes**

   Add Steps 1 through 6.
8. **Demand Factor [Table 220.84]**

   Find the demand factor percent from the number of dwelling units and multiply by Step 7.
9. **Add the House Load [220.84(B)]**

   Calculate the house load (if any) by Part III and add to Step 8.
10. **Size the Feeder Conductors**

    Single-phase: I = VA ÷ E

    Three-phase: I = VA ÷ (E × 1.732)

    There is no optional method for calculating the neutral conductor.

EXAMPLE

Use the optional method to find the service demand load for a 16-unit apartment complex with a service voltage of 120/208, three-phase. The service to each unit is 120/208 volt, single-phase. A common laundry facility is available to all tenants. Each unit has the following:

1050 ft^2 floor area

11-kW range, 120/208 volt

5-kW water heater, 208 volt

1/3-hp garbage disposer, 120 volt

3-hp air conditioner, 208 volt

10-kW heat, 208 volt

1.5-kW dishwasher, 120 volt

General lighting and receptacles 1050 × 3 = 3150 × 16 = 50,400 total volt-amperes

Table 12-2 Multifamily Load Calculation (Optional Method)

	Service
General lighting and receptacles	50,400
Small appliance 1500 × 2 × 16	48,000
Laundry [210.52(F) Ex. 1]	0
Range 11,000 × 16	176,000
Water heater 5000 × 16	80,000
Disposer 7.2 × 120 × 16	13,824
Dishwasher 1500 × 16	24,000
Heat (omit air conditioning) 10,000 × 16	160,000
Total	**552,224 VA**

Table 220.84 demand factor for 16 units is 39%.

552,224 × 39% = 215,367 volt-amperes

215,367 ÷ (208 × 1.732) = 598 amperes

The minimum size service required, using the optional method, is 600 amperes.

END-OF-CHAPTER PROBLEMS

1. If 3000 watts are allocated for lighting on 15-ampere circuits and no laundry facilities are available in an apartment complex, the minimum number of 20-ampere circuits required per apartment would be _____. The apartment service is 120/240 volts.
   a) 6
   b) 5
   c) 4
   d) 3
2. A building has 40 apartments. Each apartment has a 12-kW electric range on 230/115 volts. What is the minimum service demand load for all 40 ranges?
   a) 40 kW
   b) 55 kW
   c) 105.6 kW
   d) 480 kW
3. A quadruplex has a 125-ampere panel in each unit. What ampere size feeder is required for all 4 units? Use the optional method.
   a) 125
   b) 225
   c) 350
   d) 500
4. What is the service demand load for thirty 4.5-kW clothes dryers? Use the optional method.
   a) 45 kW
   b) 44.55 kW
   c) 40.5 kW
   d) 135,000 watts
5. A four-unit apartment has the following ranges: a 16-kW, a 14-kW, a 10-kW, and an 8-kW. What is the kW demand load that these ranges add to the service?
   a) 17 kW
   b) 17.85 kW
   c) 18.7 kW
   d) 45 kW
6. What is the demand load for a 15-unit apartment building with a 3.5-kW cooktop in each unit? The service is 115/230 volts.
   a) 30 kW
   b) 21 kW
   c) 16.8 kW
   d) None of these

7. An apartment has 3000 watts allocated for general lighting and receptacles. How many 20-ampere circuits are required for this apartment? There is no laundry on the premises.
   a) 6
   b) 5
   c) 4
   d) 2
8. What is the feeder demand for a twenty-eight unit apartment building that has a 14-kW range in each unit? The service is 115/230 volt.
   a) 53.8 kW
   b) 47.3 kW
   c) 43 kW
   d) 10.3 kW
9. A total of ten 6-kW dishwashers are in a ten-unit apartment building. What is the kW demand load using the standard calculation?
   a) 3 kW
   b) 35 kW
   c) 40 kW
   d) 45 kW
10. What is the demand on the house load for an apartment building with a laundry that consists of ten 1.5-kW washers and ten 5-kW clothes dryers?
    a) 15 kW
    b) 25 kW
    c) 40 kW
    d) 65 kW
11. A 50-unit apartment building has a 12-kW electric range in each unit. What is the load on the neutral? The service is 120/240 volt.
    a) 43.75 amperes
    b) 156 amperes
    c) 182 amperes
    d) 260 amperes
12. Each apartment has 4800 watts allocated for general lighting on 20-ampere circuits. No laundry facilities are available in the apartment complex. What is the minimum number of 20-ampere circuits required for each apartment? The service is 240/120 volt, single-phase.
    a) 2
    b) 4
    c) 5
    d) 6

13. There is a 6-kW water heater in each unit of a 60-unit apartment building. What is the calculated load added to the service for these water heaters? Do not use the optional method.
   a) 70 kW
   b) 234 kW
   c) 270 kW
   d) 360 kW
14. A 12-unit apartment building has a 12-kW electric range in each unit. What is the load on the neutral? The service is 115/230 volt.
   a) 18.9 amperes
   b) 82 amperes
   c) 117 amperes
   d) 438 amperes
15. A 55-unit apartment building has a 14-kW range in each unit. What is the calculated load for these ranges? The service is 115/230 volt.
   a) 51 kW
   b) 66.3 kW
   c) 72.9 kW
   d) 79.5 kW
16. An all-electric apartment complex has a 4.5-kVA clothes dryer in each of four units. Using the optional method, these four dryers have a demand load of _____ kVA.
   a) 8.1
   b) 9
   c) 18
   d) 20
17. What is the feeder demand load for a 12-unit apartment building with 800 ft^2 in each unit? Each apartment also has a 12-kW range and a 1.2-kW dishwasher. There is no laundry on the premises. The voltage is 120/240 volts.
   a) 48–56 kVA
   b) 56–64 kVA
   c) 64–72 kVA
   d) 72–80 kVA
18. What is the service demand load for an apartment complex with the following ranges? The service is 120/240 volt single phase.
   - Five 8-kW ranges
   - Four 12-kW ranges
   - Seven 16-kW ranges

   a) 13.75 kW
   b) 31 kW
   c) 32.55 kW
   d) 34.1 kW

Commercial Load Calculations

CHAPTER 13

Step-by-Step Commercial Load Calculations

1. General Lighting Loads by Occupancies

Find the type of occupancy from *NEC* Table 220.12. When there are assembly halls and auditoriums, multiply their square-foot area by 1 volt-ampere per square foot. Multiply the square-foot area of halls, corridors, closets, and stairways by 0.5 volt-amperes per square foot. Multiply the square-foot area of storage spaces by 0.25 volt-amperes per square foot.

Table 13-1 Examples of Unit Loads for Different Occupancy Types

Occupancy Type	Unit Load (in volt-amperes per square foot)
Restaurants	2
Banks	3½
Hospitals	2
Lodge rooms	1½
Storage warehouses	¼
Storage spaces	¼
Schools	3
Beauty shops	3

2. Other Loads

Find the volt-amperes for other lighting and outlet loads in accordance with 220.14. Include show window lighting 220.14(G) and 220.43(A) (where required) and a sign and outline lighting circuit 220.14(F) (where required).

EXAMPLE

What are the calculated loads for the following items?

Five heavy-duty lampholders

5 × 600 = 3000 VA

Fifty receptacles

50 × 180 = 9000 VA

One sign outlet

1200 VA

30′ of show window lighting

30 × 200 = 6000 VA

40′ of multioutlet assembly

40 ÷ 5 = 8 × 180 = 1440 VA

3. Continuous and Noncontinuous Lighting Loads

Separate the continuous loads from the noncontinuous loads. Multiply the noncontinuous loads by 100%. Multiply the continuous loads by 125%.

EXAMPLE

What is the calculated load for store lighting 90 kVA, show window 35 feet, parking lot lighting 6 kVA, and general purpose receptacles 3 kVA?

Store lights

90,000 × 125% = 112,500

Show window lighting

35 × 200 × 125% = 8750

Parking lot lighting

6000 × 125% = 7500

Receptacles

3000 × 100% = 3000

Total volt-amperes

112,500 + 8750 + 7500 + 3000 = 131,750

4. Lighting Load Demand Factors

Apply the lighting load demand factors to hospitals, hotels and motels (including apartment houses without provision for cooking by tenants), and storage warehouses. [Table 220.42]

NOTE

Only noncontinuous loads may be derated in Table 220.42. Any load that is a continuous load must be multiplied by 125% and not derated.

EXAMPLE

What is the lighting demand load for patients' rooms, in a hospital, where the room dimensions total 45,000 square feet?

a) 18,000 VA b) 28,000 VA c) 45,000 VA d) 90,000 VA

Step 1. Find the total volt-amperes. Multiply the square-foot area by the volt-amperes per square foot for the listed occupancy.

45,000 × 2 (volt-amperes per square foot) = 90,000

Step 2. Derate according to Table 220.42.

First 50,000 × 40% = 20,000

Remaining 40,000 × 20% = 8000

20,000 + 8000 = 28,000

The total lighting demand load for these hospital patients' rooms is 28,000 volt-amperes. Answer b) is correct.

5. Receptacle Load Demand Factors

The receptacle load for other than dwelling units can be reduced by demand factors in Table 220.44:

- The first 10 kVA or less at 100 percent
- The remaining kVA multiplied by 50 percent

EXAMPLE

What is the demand load for 80 receptacles in an office building?

a) 180 VA b) 7200 VA c) 12,200 VA d) 14,400 VA

Step 1. Since receptacle outlets are calculated at 180 volt-amperes each, multiply the number of receptacles by 180.

80 (receptacles) × 180 (volt-amperes) = 14,400 volt-amperes

Step 2. The first 10,000 or less is calculated at 100%.

10,000 × 100% = 10,000

Step 3. Find the remaining volt-ampere load over 10,000 and multiply by 50%.

14,400 - 10,000 = 4400

4400 × 50% = 2200

Step 4. Add the results from Steps 2 and 3.

10,000 + 2200 = 12,200

The demand load for 80 receptacles in an office building is 12,200 volt-amperes. Answer c) is correct.

6. Heating or Air Conditioning

Compare the heat and air conditioning and omit the smaller. [220.60] Include the air handler on either one.

7. Kitchen Equipment

Using Table 220.56, calculate the load for commercial cooking equipment which has either thermostatic control or intermittent use as kitchen equipment. This table does not apply to space heaters or ventilating or air conditioning equipment.

NOTE

The feeder demand load for kitchen equipment must not be smaller than the sum of the two largest kitchen equipment loads.

EXAMPLE

A school has twenty 8-kW ranges in a home economics classroom. What is the minimum load to the service?

a) 35 kW b) 44.8 kW c) 56 kW d) 160 kW

Step 1. Find the demand load from Columns B and C of Table 220.55.

B = 20 × 8 × 28% = 44.8

C = 35 kW

Step 2. Compare the two and select the smaller number, 35 kW. Answer a) is correct.

Since the demand in Column C can be found most quickly, look at Column C first. Then look at the answers to see if that number happens to be the smallest number in the answers. If the demand in Column C is the smallest of the answers, there is no need to calculate Column B.

Notice that in the previous example, the Column C answer of 35 kW is found first. Looking at the answers, 35 kW is the smallest number. Therefore, the Column B number must be larger than 35 kW. Select answer a) and move on to the next question. If the demand in Column C is not the smallest of the answers, Column B must be calculated. Compare Column B and Column C and select the lower kW.

NOTE

For household cooking equipment rated over 1¾ kW and used in instructional programs, use Table 220.55.

8. All Other Loads

Add all other noncontinuous loads into the calculation at 100%. Multiply all other continuous loads (operating for 3 hours or more) by 125%. [215.3]

9. Largest Motor

Multiply the largest motor (one motor only) by 25% and add to the load. [220.50 and 430.24]

10. Calculate the Service Size

Size the service by adding Steps 3 through 8 together to find the total calculated load:

Single-phase I = VA ÷ E

Three-phase I = VA ÷ (E × 1.732)

11. Calculate the Neutral Load

Calculate the neutral load. A further demand factor of 70% shall be permitted for any neutral load over 200 amperes. [220.61(B)]

EXAMPLE

What is the neutral demand load, in amperes, for a 30,000 square foot store? The service is 120/208 volt three-phase.

a) 83 amperes b) 219 amperes c) 250 amperes d) 279 amperes

Step 1. Multiply the store's square-foot area by 3 because the general lighting load for a store is 3 volt-amperes per square foot.

30,000 × 3 = 90,000 VA

Step 2. Since store lighting will be on for more than 3 hours, multiply the general lighting load by 125 percent.

90,000 × 125% = 112,500 VA

Step 3. Find the current by dividing the volt-ampere load by the voltage of the service.

112,500 ÷ (208 × 1.732) = 312.3 amperes

Step 4. Since this neutral load is more than 200 amperes, apply the 70% demand factor to the load in excess of 200 amperes.

$$\begin{array}{rcr} 312.3 & & \\ \underline{-200.0} \times 100\% & = & 200 \\ 112.3 \times 70\% & = & \underline{79} \\ & & 279 \end{array}$$

The neutral demand load for this 30,000 square foot store is 279 amperes. Answer d) is correct.

Commercial Load Calculations

1. **General Lighting Loads by Occupancies [Table 220.12]**

   Multiply the square foot area by the volt-amperes per type of occupancy.

2. **Other Loads [220.14]**

   Include a sign outlet where required. [220.14(G) and 220.43(A)]

   Include show window lighting where required. [220.14(F)]

3. **Continuous and Noncontinuous Lighting Loads**

   Keep the noncontinuous loads at 100%.

   Multiply the continuous loads by 125%.

4. **Lighting Load Demand Factors [Table 220.42]**

   Apply the demand factors to hospitals, hotels, motels, apartment houses (without provision for cooking), and storage warehouses.

5. **Receptacle Load Demand Factors [Table 220.44]**

   If the receptacle load is more than 10 kVA, apply the demand factors from Table 220.44.

6. **Heating or Air Conditioning [220.60]**

   Compare the heat and air conditioning and omit the smaller.

7. **Kitchen Equipment [Table 220.56]**

   The feeder demand load for kitchen equipment must not be smaller than the sum of the two largest kitchen equipment loads.

   For household cooking equipment rated over 1¾ kW and used in instructional programs, use Table 220.55.

8. **All Other Loads [215.2(A) and 215.3]**

   Add all other noncontinuous loads into the calculation at 100%.

   Multiply all other continuous loads (operating for 3 hours or more) by 125%.

9. **Largest Motor [220.50 and 430.24]**

   Multiply the largest motor by 25% and add to the load.

10. **Calculate the Service Size**

    Add Steps 3 through 9 for the total volt-ampere load.

    Single-phase I = VA ÷ E

    Three-phase I = VA ÷ (E × 1.732)

11. **Calculate the Neutral Load**

    A further demand factor of 70% shall be permitted for any neutral load over 200 amperes. [220.61(B)]

Optional Method Load Calculation for a School

An optional method for calculating school feeders or services is in 220.86. This method is in lieu of calculating by Part III of Article 220. The total connected load includes the entire interior and exterior lighting, power, water heating, cooking, other loads, and the larger of the air conditioning load or space heating load within the building or structure. The school must be equipped with electric space heating, air conditioning, or both.

The neutral load must be calculated by Part III of Article 220. Portable classroom buildings cannot be calculated by the optional school method.

> **NOTE**
>
> Calculating the total connected load may not be necessary during an exam. The total connected load is usually given if the exam question is asking for the optional load calculation of a school. Two numbers are typically provided: the total connected load and the square-foot area of the school.

1. Total Connected Load

Find the total connected load. When calculating the total connected load of a school, use only 100% of the load. Do not use 125% for continuous loads. Do not increase the largest motor by 25%. Do not derate the receptacle load. Do not derate cooking equipment.

2. Average Volt-Amperes Per Square Foot

Find the average volt-amperes per square foot. Divide the total connected load by the total square-foot area of the school.

3. Demand Factors

Apply the demand factors (percents) from Table 220.86 to the volt-amperes found in Step 2, and add them together to find the new volt-amperes per square foot.

4. Total Load After Demand Factors

Multiply the volt-amps found in Step 3 by the total square foot area of the school.

EXAMPLE

Using the optional method, what is the kVA demand load for a school with a total connected load of 500 kVA? The area of the school is 20,000 square feet.

Step 1. The total connected load is given in this problem (500,000 volt-amperes).

Step 2. Calculate the average volt-amperes per square foot.

500,000 ÷ 20,000 = 25 volt-amperes per square foot

Step 3. Apply the demand factors (percents) from Table 220.86.

		25				
The first 3 at 100%	=	-3	× 100%	=	3	
		22				
The next 17 at 75%	=	-17	× 75%	=	12.75	
The remainder at 25%	=	5	× 25%	=	1.25	
					17 volt-amperes per square foot	

Step 4. Multiply the volt-amps by the total square foot area.

17 × 20,000 = 340,000 (volt-amperes) ÷ 1000 = 340 kVA

The demand load for this school, using the optional method, is 340 kVA.

END-OF-CHAPTER PROBLEMS

1. What is the calculated load for a store front that has 85 linear feet of show window? There are 32 medium-base lamps supplying the show window space.
   a) 8000 VA
   b) 17,000 VA
   c) 25,000 VA
   d) 33,500 VA
2. A school has twenty 16′ × 32′ classrooms, 2000 ft^2 halls, and an office that is 60′ × 18′. What is the continuous demand load?
   a) 16–26 kVA
   b) 26–36 kVA
   c) 36–46 kVA
   d) 46–56 kVA
3. What is the additional demand load for a motel with 80 receptacles in public areas (not counting room receptacles)?
   a) 180 VA
   b) 12.2 kVA
   c) 14.4 kVA
   d) 18 kVA
4. What is the demand load for a school with an area of 24,000 sq. ft. and 408 kW? Use the optional method.
   a) 170 kVA
   b) 324 kVA
   c) 408 kVA
   d) None of these
5. What is the demand load in amperes for a commercial store with these 120-volt loads?
   - One 1-hp, 120-volt motor
   - Two 1920-VA lights, 120 volts
   - One 16-ampere receptacle circuit, 120 volts

   a) 76 amperes
   b) 64 amperes
   c) 48 amperes
   d) 80 amperes
6. An office has a lighting load of 72 kW that is continuous. The voltage is 120/240. What is the neutral demand load?
   a) 375 amperes
   b) 350 amperes
   c) 325 amperes
   d) 322.5 amperes

7. An office building has an area of 40,000 square feet, and the number of receptacles is unknown. How many 2-wire, 20-ampere, 120-volt circuits are required?
   a) 120
   b) 90
   c) 60
   d) 40
8. What is the service demand load for a 50-unit motel with 15′ × 30′ rooms and 4000 sq. ft. of halls?
   a) 20,800 VA
   b) 22,000 VA
   c) 22,500 VA
   d) 47,000 VA
9. What is the demand load for a school that has a total connected load of 160,000 watts and 20,000 square feet? Use the optional method.
   a) 106 kVA
   b) 112 kVA
   c) 120 kVA
   d) 135 kVA
10. A motel has 2000 sq. ft. of halls and stairs. How much lighting load must be added?
    a) 4 kVA
    b) 2 kVA
    c) 1.25 kVA
    d) 1 kVA
11. There are forty-six 250-watt fluorescent lighting fixtures on a 120-volt, 3-phase, 4-wire balanced service. How many 20-ampere circuits are required? The load is continuous.
    a) 5
    b) 6
    c) 7
    d) 8
12. The neutral of a 6-kW commercial 240-volt, 1-phase, electric clothes dryer would be sized on the basis of ______.
    a) 50 amperes
    b) 25 amperes
    c) 21 amperes
    d) 15 amperes
13. What is the demand load for a school that has a total connected load of 680 kW and 20,000 square feet? Use the optional method.
    a) 680 kVA
    b) 385 kVA
    c) 170 kVA
    d) 19.25 kVA

14. What is the minimum size grounding conductor for a 100-ampere inverse time circuit breaker?
    a) 6 AWG
    b) 8 AWG
    c) 10 AWG
    d) 12 AWG
15. A store front has a total of 24 linear feet of show window. Twelve linear feet are on each side of the front door. How many receptacle outlets are required?
    a) None
    b) One
    c) Two on each side
    d) One on each side
16. An office has 40 receptacles. What is the volt-ampere load for theses 40 receptacles?
    a) 180 VA
    b) 3600 VA
    c) 7200 VA
    d) 40,000 VA
17. What is the service demand load for a mobile home park that has 200 mobile homes? The calculated demand load on each trailer is 60 amperes at 240/120 volts.
    a) 633.6 kVA
    b) 704 kVA
    c) 1180.8 kVA
    d) 2880 kVA
18. What size equipment grounding conductor is required for a circuit being fed with a 150-ampere overcurrent protective device?
    a) 8 AWG
    b) 6 AWG
    c) 4 AWG
    d) Cannot be determined
19. What is the demand load for a school with an area of 20,000 square feet and a total connected load of 206.6 kW? Use the optional method.
    a) 680 kVA
    b) 385 kVA
    c) 170 kVA
    d) 19.25 kVA
20. What size equipment grounding conductor is required for a circuit being fed with an 80-ampere overcurrent protective device?
    a) 4 AWG
    b) 6 AWG
    c) 8 AWG
    d) 10 AWG

21. If a motel has an 80 × 100-foot assembly hall and a 30 × 40-foot restaurant, the lighting load would be _____.
    a) 0–10,000 volt-amperes
    b) 10,000–20,000 volt-amperes
    c) 20,000–30,000 volt-amperes
    d) 30,000–40,000 volt-amperes
22. A developer is building a marina. Twenty boats require 120-volt, 20-ampere receptacles; seventeen of the spaces require 120-volt, 30-ampere receptacles; and seven of the spaces require 120-volt, 50-ampere receptacles. What is the minimum ampacity for a 120-volt feeder?
    a) 630 amperes
    b) 730 amperes
    c) 1260 amperes
    d) 1460 amperes
23. What load shall be added to the service for an office building with 200 receptacles?
    a) 15 kVA
    b) 18 kVA
    c) 23 kVA
    d) 36 kVA
24. What is the demand load for a school with an area of 20,000 square feet and a total connected load of 206.6 kW? Use the optional method.
    a) 8.5 kVA
    b) 10.33 kVA
    c) 96.8 kVA
    d) 170 kVA
25. A bank has an area of 2400 square feet and no receptacles are shown on the drawings. What is the receptacle demand load?
    a) 2400 VA
    b) 3000 VA
    c) 8400 VA
    d) 10,800 VA
26. What is the load current rating for a feeder supplying a continuous lighting load of 39 amperes and a single motor that draws 24 amperes?
    a) 63 amperes
    b) 69 amperes
    c) 72.75 amperes
    d) 78.75 amperes
27. A 480/277-volt feeder in Building 1 supplies power to a 400-ampere panel in Building 2. The branch circuits in the second building are mostly 480-volt motor loads. What is the minimum size copper grounded conductor that must be run to the second building?
    a) 3 AWG
    b) 1/0 AWG
    c) 3/0 AWG
    d) 500 kcmil

28. A commercial building has a 200-ampere service fed with THW conductors. What size grounding electrode conductor is required?
   a) 8 AWG
   b) 6 AWG
   c) 4 AWG
   d) 2 AWG

29. A store front has 15 linear feet of show window. How many receptacle outlets are required?
   a) 3
   b) 2
   c) 1
   d) 0

30. Tap conductors are needed for a 1200-ampere THW copper service. The tap conductors have a length of six feet and will not extend beyond the panelboard. If the calculated load is 150 amperes, what minimum size conductors are required?
   a) 1/0 THW
   b) 2/0 THW
   c) 3/0 THW
   d) 4/0 THW

31. The service conductors on a three-phase service are 500-kcmil copper conductors. Only three-phase loads will be installed in the building. What is the minimum size copper grounded conductor that must be installed?
   a) 1/0 AWG
   b) 2/0 AWG
   c) 3/0 AWG
   d) No neutral is required.

32. Paralleled 500-kcmil THWN conductors are protected by an 800-ampere overcurrent protection device. The paralleled conductors are in a trough and will be tapped. The tap conductors will terminate a single circuit breaker and their length is 20 feet. What is the smallest size tap conductors permitted?
   a) 3/0 THW
   b) 4/0 THW
   c) 300-kcmil THW
   d) 500-kcmil THW

33. What is the minimum bending radius for a 500-kcmil THW conductor?
   a) 6.3 inches
   b) 7.6 inches
   c) 7.75 inches
   d) 8 inches

Efficiency and Power Factor

CHAPTER 14

Efficiency

Efficiency and power factor calculations on electrical exams are not as difficult as they first appear. Formulas for different types of efficiency and power factor calculations are discussed in this chapter. Some formulas have many factors and some have few. One type of efficiency calculation has only three factors—two known and one unknown. Three factors associated with transformer questions are input, output, and efficiency. Although there are formulas for these types of calculations, using the formula is not even necessary. The following examples will show how this is possible. These questions will not be on any electrical exam, but they help illustrate the simplicity of solving some types of efficiency questions.

EXAMPLES

The input of the pump below is 100 gallons per minute (gpm) and the output is 100 gpm. What is the efficiency of this pump?

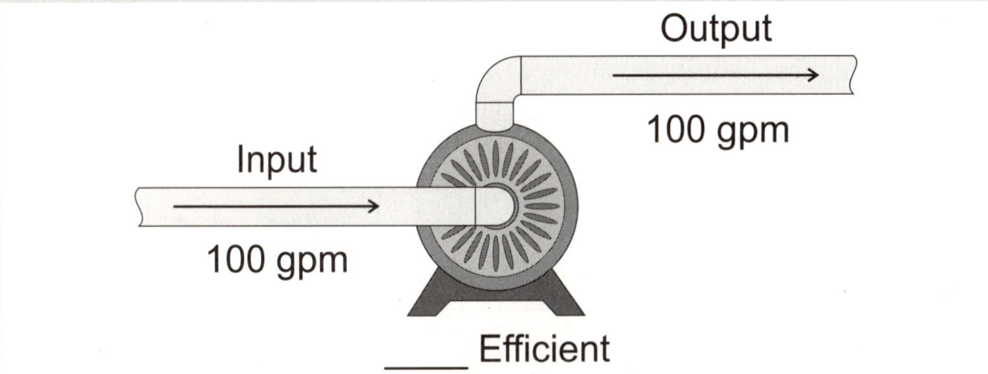

FIGURE 14-1

Answer: 100%

The pump in the previous example sprang a 5-gpm leak. The input is still 100 gpm. What is the gpm output of this pump?

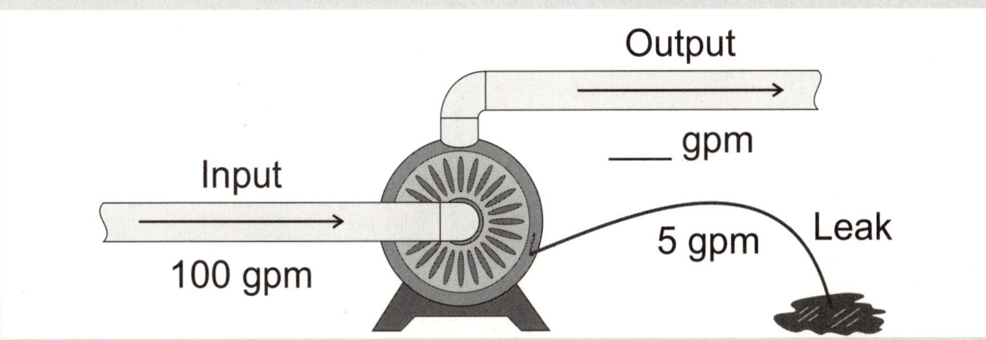

FIGURE 14-2

Answer: 95 gpm

continues

continued

Since the pump in the previous example has an input of 100 gpm and an output of only 95 gpm, what is the efficiency of this pump now?

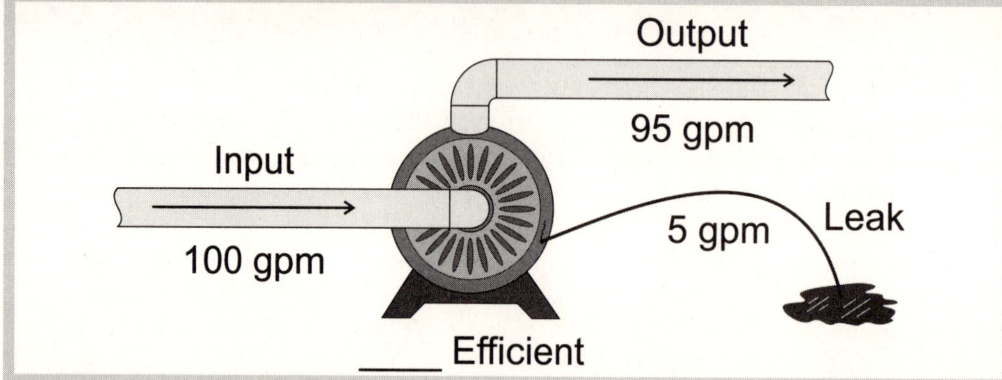

FIGURE 14-3

Answer: 95%

Transformer efficiency questions are not any more difficult than the above examples. The three factors associated with these types of questions are input, output, and efficiency. Efficiency is the ratio of the input to the output and is expressed as a percentage or as a decimal.

Formulas for calculating the efficiency of transformers are given in **Table 14-1.**

Table 14-1 Transformer Efficiency Formulas
$\text{Efficiency} = \dfrac{\text{Output}}{\text{Input}}$
$\text{Input} = \dfrac{\text{Output}}{\text{Efficiency}}$
Output = Input × Efficiency

A few basic rules apply to transformer efficiency questions:

- Input is always greater than output, therefore input is the larger number and output is the smaller number.
- Efficiency is always less than 100%.

EXAMPLES

What is the efficiency of a single-phase transformer with an input of 10,000 watts and an output of 9600 watts? The power factor is 100%.

a) 0.96% b) 9.6% c) 96% d) 104%

Step 1. Draw the problem.

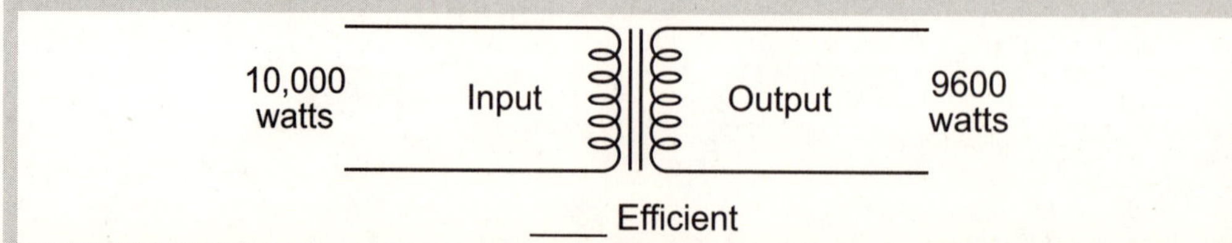

FIGURE 14-4

Step 2. Divide the output watts by the input watts.

$$\text{Efficiency} = \frac{\text{Output}}{\text{Input}} = \frac{9600}{10{,}000} = 0.96 = 96\% \text{ efficient}$$

Answer c) is correct.

Forgetting this formula during an exam is not critical. This question can be solved just by knowing efficiency is always less than 100%. Start by dividing one number by the other number. If the numbers are reversed, the answer will be more than 100%.

10,000 ÷ 9600 = 1.0417 = 104%

Since efficiency cannot be more than 100%, this is incorrect. Note that 104% is one of the incorrect answers given with the question.

Swap the two numbers and divide again.

9600 ÷ 10,000 = 0.96 = 96%

What is the volt-ampere (VA) output of a 45-kVA transformer that is 98% efficient? The power factor is 100%.

a) 43,200 VA **b)** 44,100 VA **c)** 45,918 VA **d)** 46,875 VA

Step 1. Draw the problem.

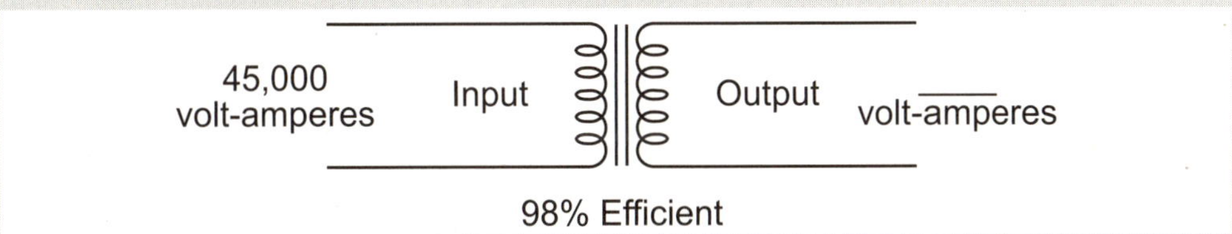

FIGURE 14-5

Step 2. Multiply input volt-amperes by efficiency percent.

Output = Input × Efficiency = 45,000 × 98% = 44,100 VA

Answer b) is correct.

This question can be solved just by knowing input is always greater than output. Since the answer will be smaller than 45,000, multiply 45,000 by 98%.

45,000 × 98% = 44,100 VA

What is the input (in watts) of a three-phase transformer that has an output of 75,000 watts and an efficiency of 96%? The power factor is 100%.

a) 72,000 watts **b)** 73,500 watts **c)** 76,531 watts **d)** 78,125 watts

Step 1. Draw the problem. Since watts are given instead of volts and amperes, the problem can be drawn as a single-phase transformer.

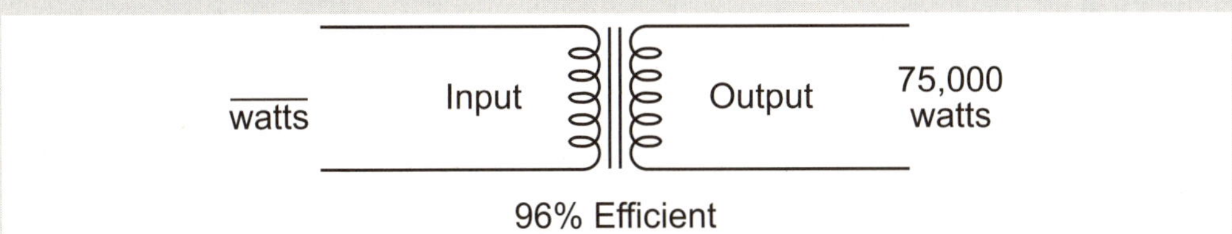

FIGURE 14-6

Step 2. Divide output watts by efficiency percent.

$$\text{Input} = \frac{\text{Output}}{\text{Efficiency}} = \frac{75{,}000}{96\%} = 78{,}125 \text{ watts}$$

Answer d) is correct.

This question can be solved just by knowing input is always greater than output. Since the answer will be greater than 75,000, divide 75,000 by 96%.

75,000 ÷ 96% = 78,125 watts

Power Factor

Power factor is the ratio between true power (measured in watts) and apparent power (measured in volt-amperes). A purely resistive alternating-current (ac) circuit has a unity (100%) power factor. Direct-current (dc) circuits also have a unity (100%) power factor. In a dc circuit, or in a purely resistive ac circuit, the voltage and current are in phase with each other. This is true power measured in watts.

The Ohm's law formula for true power is Watts (power) = E (volts) × I (amperes)

Apparent power exists when voltage and current are out of phase with each other. Apparent power is also referred to as volt-amperes, VA, kVA, floating power, or reactive power. Inductance and capacitance will cause a circuit to be out of phase.

Power factor is expressed as a percentage or as a decimal. Unity power factor is another way of expressing 100% power factor. Power factor cannot be greater than 100%.

The formula for power factor is

$$\text{Power Factor} = \frac{\text{True Power (watts)}}{\text{Apparent Power (volt amperes)}} = \frac{\text{Watts}}{\text{VA}}$$

EXAMPLE

What is the power factor of a three-phase circuit that has an apparent power of 15,000 VA but a true power of only 13,800 watts?

a) 0.92% **b)** 87% **c)** 92% **d)** 109%

Divide watts (13,800) by volt-amperes (15,000).

$$\text{PF} = \frac{\text{Watts}}{\text{VA}} = \frac{13{,}800}{15{,}000} = 0.92 = 92\%$$

Answer c) is correct.

This question can be solved by knowing that power factor cannot be greater than 100%. Start by dividing one number by the other number. If the numbers are reversed, the answer will be more than 100%.

15,000 ÷ 13,800 = 1.087 = 109%

Since power factor cannot be more than 100%, this is incorrect. Note that 109% is one of the incorrect answers given with the question.

Swap the two numbers and divide again.

13,800 ÷ 15,000 = 0.92 = 92%

NOTE

Apparent power (VA) is always more than true power (watts).

The formula for apparent power is

$$\text{Apparent Power} = \frac{\text{True Power}}{\text{Power Factor}} = \frac{\text{Watts}}{\text{PF}}$$

The formulas for true power in single-phase circuits and three-phase circuits are in **Table 14-2.**

Table 14-2 True Power Formulas

Circuit	Formula
Single-phase	True power (watts) = E × I × PF
Three-phase	True power (watts) = E × 1.732 × I × PF

EXAMPLE

What is the kW of a 208-volt, three-phase circuit that has a current of 119.5 amperes and a power factor of 93%?

a) 23 b) 25 c) 40 d) 43

Multiply all the factors given in the problem. Do not forget to multiply the voltage by 1.732 because it is three-phase.

Watts = 208 × 1.732 × 119.5 × 93% = 40,037 ÷ 1000 = 40 kW

Answer c) is correct.

Efficiency and Power Factor

Some formulas contain both efficiency and power factor figures. Motor efficiency questions may contain power factor percentages with alternating-current motors. If power factor is not given in the problem, then assume 100% power factor.

The formulas for calculating the efficiency of motors are in **Table 14-3.**

Table 14-3 Motor Efficiency Formulas

Motor	Efficiency Formula
Direct current	$\text{Efficiency} = \dfrac{\text{hp} \times 746}{E \times I}$
Single-phase ac	$\text{Efficiency} = \dfrac{\text{hp} \times 746}{E \times I \times PF}$
Three-phase ac	$\text{Efficiency} = \dfrac{\text{hp} \times 746}{E \times 1.732 \times I \times PF}$

EXAMPLE

What is the efficiency of a 15-hp, 230-volt, three-phase motor that draws 41.3 amperes and has a power factor of 97%?

a) 68% b) 70% c) 72% d) 121%

Select the three-phase ac formula and insert the known factors.

$$\text{Eff.} = \frac{15 \times 746}{230 \times 1.732 \times 41.3 \times 0.97} = 0.7012 = 70\% \text{ efficient}$$

Answer b) is correct.

Some questions ask for efficiency after giving watts, voltage, current, and power factor. Formulas for solving these questions are similar to the ones for solving motor efficiency questions. The formulas are identical except for the top line. In the formulas in Table 14-4, watts replaces hp × 746. This is understandable because wattage of a motor is found by multiplying horsepower by 746.

Use the formulas in **Table 14-4** to find efficiency when watts, voltage, current, and power factor are known.

Table 14-4 Formulas to Find Efficiency

Single-phase	$\text{Efficiency} = \dfrac{\text{Watts}}{E \times I \times PF}$
Three-phase	$\text{Efficiency} = \dfrac{\text{Watts}}{E \times 1.732 \times I \times PF}$

The last set of formulas in this chapter solve for current. These are the same formulas that are in Table 14-4 except current and efficiency are transposed.

Use the formulas in **Table 14-5** to find current when watts, voltage, efficiency, and power factor are known.

Table 14-5 Formulas to Find Current

Single-phase	$I = \dfrac{\text{Watts}}{E \times \text{Efficiency} \times PF}$
Three-phase	$I = \dfrac{\text{Watts}}{E \times 1.732 \times \text{Efficiency} \times PF}$

EXAMPLE

What is the current of a 22-kW rated load that has an efficiency of 97% and a power factor of 88%? The source voltage is 208-volt, three-phase.

a) 61 amperes b) 63 amperes c) 69 amperes d) 72 amperes

Select the three-phase ac formula and insert the known factors.

$$I = \frac{22{,}000}{208 \times 1.732 \times 0.97 \times 0.88} = 71.54 = 72 \text{ amperes}$$

Answer d) is correct.

END-OF-CHAPTER PROBLEMS

1. A three-phase wattmeter has a reading of 12,442 watts. An ammeter has a reading of 36 amperes per phase. The voltage is 208/120 volts, three-phase. What is the power factor?
   a) 0.55
   b) 0.96
   c) 1.04
   d) Unity
2. What is the efficiency of a 20-hp, 240-volt, three-phase motor that is drawing 48 amperes? The circuit has a power factor of 97%.
   a) 133.7%
   b) 103%
   c) 86%
   d) 77.09%
3. A three-phase wattmeter has a reading of 14,965 watts. An ammeter has a reading of 36 amperes per phase. The voltage is three-phase, 240/120 volts. What is the power factor?
   a) 0.52
   b) 0.55
   c) 0.96
   d) Unity
4. A single-phase wattmeter has a reading of 10,000 watts. An ammeter has a reading of 50 amperes per phase. The voltage is 230/115 volts, single-phase. What is the power factor?
   a) 115%
   b) 94%
   c) 87%
   d) 78%
5. What is the input current of a 1½-hp, 230-volt, single-phase motor that has an efficiency of 85%?
   a) 5.7 amperes
   b) 4.9 amperes
   c) 4.1 amperes
   d) 3.2 amperes
6. A three-phase wattmeter has a reading of 12,960 watts. An ammeter has a reading of 36 amperes per phase. The voltage is three-phase, 208/120 volts. What is the power factor?
   a) 0.99%
   b) 57.7%
   c) 96%
   d) Unity

7. If a wattmeter reads 15,000 watts connected to a 3-phase, 208/120-volt load and the power factor measures 0.98, the current will be _____ amperes.

a) 40.8
b) 41.67
c) 42.5
d) 73.59

8. If a wattmeter reads 15,000 watts connected to a three-phase, 208/120-volt load, and the power factor measures 0.96, the current will be _____ amperes.

a) 72.12
b) 44.72
c) 43.37
d) 41.67

9. A 1½-hp, 120-volt motor has an efficiency of 90% and a power factor of 85%. What is the full load current of this motor?

a) 7.1 amperes
b) 12.2 amperes
c) 15.3 amperes
d) 20 amperes

10. What size conductors are required for a 3-hp, 230-volt, single-phase motor that is 90% efficient and has a power factor of 88%?

a) 14 AWG
b) 12 AWG
c) 10 AWG
d) 8 AWG

11. A 5-hp, 200-volt, three-phase motor has an efficiency of 92% and a power factor of 90%. What is the full load current of this motor?

a) 8.92 amperes
b) 10.77 amperes
c) 13.02 amperes
d) 22.52 amperes

12. What is the power factor of a 208-volt, three-phase circuit that draws 33 amperes when the load is 10,860 watts?

a) 63%
b) 91%
c) 109%
d) 158%

13. What is the VA input to a fully loaded 5-hp, three-phase, 208-volt motor?

a) 3500–4000
b) 4500–5000
c) 5100–5900
d) 6000–6100

14. What is the efficiency of a motor that draws 4 kVA and delivers 4 hp?
   a) 75%
   b) 83%
   c) 100%
   d) 134%
15. What is the efficiency of a 5-hp, 230-volt, single-phase motor that consumes 6440 volt-amperes?
   a) 16%
   b) 58%
   c) 70%
   d) 173%
16. A 15-hp, 208-volt, three-phase motor has a power factor of 0.97. What will be the reading on the wattmeter?
   a) 11,536
   b) 16,145
   c) 16,644
   d) 17,159
17. What is the efficiency of a 20-hp, 208-volt, three-phase motor? Assume unity power factor.
   a) 60%
   b) 70%
   c) 80%
   d) 90%

Transformers

CHAPTER
15

Single-Phase Transformers

A transformer is used to change voltages in ac systems. An isolation-type transformer has two sets of windings that are electrically insulated from each other. The input side is called the primary. The output side is called the secondary.

A transformer ratio is the number of turns in the primary compared to the number of turns in the secondary. A step-down transformer with a 2-to-1 ratio has twice the number of turns in the primary as it does in the secondary.

EXAMPLE

If a transformer has 100 turns and 240 volts in the primary, what will be the voltage in the secondary if the secondary winding has 50 turns?

Step 1. Draw the problem.

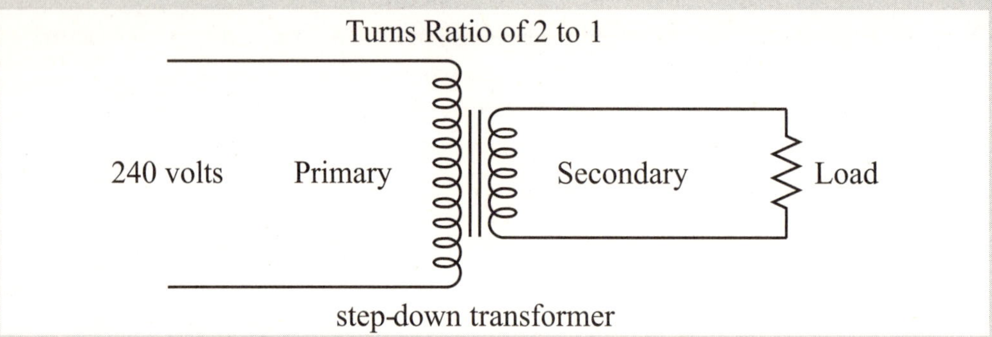

FIGURE 15-1

Because the primary has twice the number of turns as the secondary, the turns ratio of this transformer is 2 to 1. The voltage is directly proportional to the turns ratio. Since the secondary has half the number of turns as the primary, the secondary voltage will be half the primary voltage.

Step 2. Divide the primary voltage by two.

$240 \div 2 = 120$ volts

The secondary voltage of this transformer is 120 volts.

Figure 15-1 is an example of a step-down transformer with a turns ratio of 2 to 1. The first number in a transformer ratio refers to the primary winding. The second number in a step-down transformer will be one. A step-up transformer has more turns in the secondary than in the primary. The first number in a step-up transformer will be one. See **Figure 15-2** for an example of a step-up transformer. This transformer has 240 volts in the primary and 480 volts in the secondary. The turns ratio of this transformer is 1 to 2.

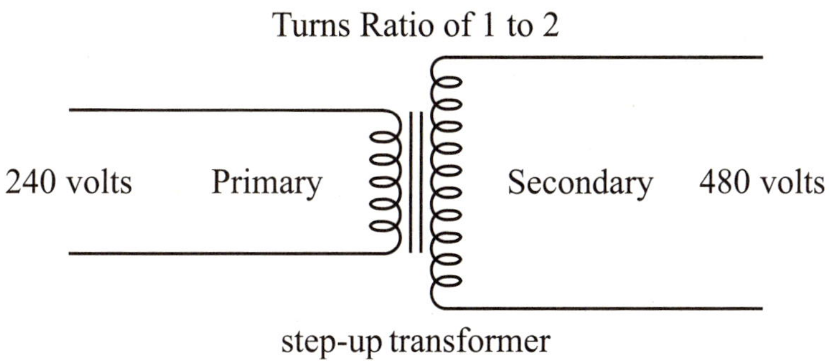

FIGURE 15-2

An autotransformer has a primary and secondary that uses the same set of windings.

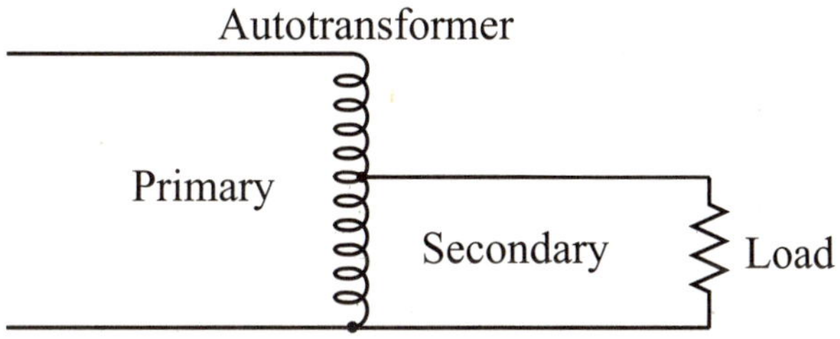

FIGURE 15-3

The following formula can be used to solve certain types of transformer exam questions. The formula can be transposed to fit the need of the problem.

$$\frac{T_P}{T_S} = \frac{E_P}{E_S} = \frac{I_S}{I_P}$$

Formula definitions:

T_P = number of turns in the primary

T_S = number of turns in the secondary

E_P = voltage in the primary

E_S = voltage in the secondary

I_P = current in the primary

I_S = current in the secondary

Variations of the formula are shown in **Table 15-1**.

Table 15-1 Variations to the Transformer Formula

Condition	Formula
Find turns in the primary when T_S, E_P, and E_S are known	$T_P = \frac{E_P \times T_S}{E_S}$
Find turns in the secondary when T_P, E_P, and E_S are known	$T_S = \frac{E_S \times T_P}{E_P}$
Find turns in the primary when T_S, I_P, and I_S are known	$T_P = \frac{I_S \times T_S}{I_P}$
Find turns in the secondary when T_P, I_P, and I_S are known	$T_S = \frac{I_P \times T_P}{I_S}$
Find voltage in the primary when E_S, T_P, and T_S are known	$E_P = \frac{T_P \times E_S}{T_S}$
Find voltage in the secondary when E_P, T_P, and T_S are known	$E_S = \frac{T_S \times E_P}{T_P}$
Find voltage in the primary when E_S, I_P, and I_S are known	$E_P = \frac{I_S \times E_S}{I_P}$
Find voltage in the secondary when E_P, I_P, and I_S are known	$E_S = \frac{I_P \times E_P}{I_S}$
Find the current in the primary when I_S, T_P, and T_S are known	$I_P = \frac{T_S \times I_S}{T_P}$
Find the current in the secondary when I_P, T_P, and T_S are known	$I_S = \frac{T_P \times I_P}{T_S}$
Find the current in the primary when I_S, E_P, and E_S are known	$I_P = \frac{E_S \times I_S}{E_P}$
Find the current in the secondary when I_P, E_P, and E_S are known	$I_S = \frac{E_P \times I_P}{E_S}$

EXAMPLE

A transformer secondary delivers 15 amperes at 120 volts. The current in the primary is 5 amperes. What is the voltage in the primary?

a) 120 volts **b)** 240 volts **c)** 360 volts **d)** 480 volts

Step 1. Find the formula that starts with E_P. The formula needs to contain current in the primary I_P, current in the secondary I_S, and voltage in the secondary E_S.

$$E_P = \frac{I_S \times E_S}{I_P}$$

Step 2. Replace the letters with the given values and solve the problem.

$$E_P = \frac{15 \times 120}{5} = \frac{1800}{5} = 360 \text{ volts}$$

The voltage in the primary is 360 volts. Answer c) is correct.

If a transformer is 100% efficient, the primary volt-amperes will be equal to the secondary volt-amperes. To calculate the volt-amperes in the primary when the efficiency is less than 100%, use the following formula:

$$\text{Primary VA} = \frac{\text{Secondary VA}}{\text{Efficiency}}$$

NOTE

In a transformer, when efficiency is less than 100%, the primary volt-amperes will always be larger than the secondary volt-amperes.

EXAMPLE

What are the primary volt-amperes in a three-phase transformer where the secondary has 4000 volt-amperes? The transformer is 80% efficient.

a) 3200 VA b) 4000 VA c) 5000 VA d) 8660 VA

Step 1. Replace the letters with the values given and solve the problem.

$$\text{Primary VA} = \frac{4000}{0.80} = 5000 \text{ volt-amperes}$$

There are 5000 volt-amperes in the primary. Answer c) is correct.

To calculate the volt-amperes in the secondary when the primary volt-amperes and the efficiency are known, use this formula:

$$\text{Secondary VA} = \text{Primary VA} \times \text{Efficiency}$$

NOTE

In a transformer, when efficiency is less than 100%, the secondary volt-amperes will always be smaller than the primary volt-amperes.

EXAMPLE

What are the secondary volt-amperes in a transformer where the primary has 4000 volt-amperes? The transformer is 75% efficient.

a) 3000 VA b) 4000 VA c) 5333 VA d) 6928 VA

Step 1. Replace the letters with the values given and solve the problem.

Secondary VA = 4000 × 75% = 3000 volt-amperes

There are 3000 volt-amperes in the secondary. Answer a) is correct.

To calculate the efficiency of a transformer when the primary and secondary volt-amperes are known, use this formula:

$$\text{Efficiency} = \frac{\text{Secondary VA}}{\text{Primary VA}}$$

NOTE

In a transformer problem, efficiency will never be more than 100%.

EXAMPLE

What is the efficiency of a three-phase transformer that has 6250 volt-amperes in the primary and 5000 volt-amperes in the secondary?

a) 125% b) 0.8% c) 90% d) 80%

Step 1. Replace the letters with the values given and solve the problem.

$$\text{Efficiency} = \frac{5000}{6250} = 0.8 = 80\%$$

The efficiency of this transformer is 80%. Answer d) is correct.

Drawing the problem on a sheet of scratch paper will help determine the answer. When asked to find current, if watts (volt-amperes) and voltage are given, use Ohm's law. Calculate the current in a transformer by dividing the watts (volt-amperes) by the voltage.

EXAMPLE

What is the current of the primary and the secondary of a transformer that has a load of 4.5 kVA? The secondary voltage is 240 volts, the ratio of the transformer is 2 to 1, and the efficiency is 90%.

Step 1. Draw the problem.

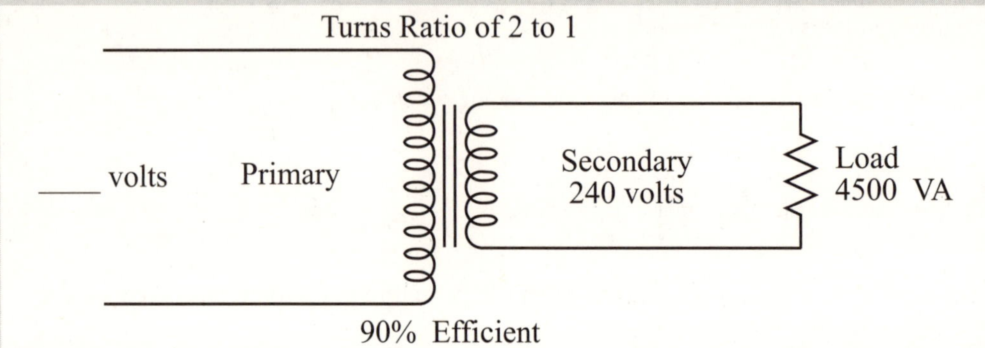

FIGURE 15-4

Step 2. The primary voltage can be calculated from the ratio given. The primary voltage is two times larger than the secondary voltage ($240 \times 2 = 480$). The primary voltage is 480 volts.

Step 3. Calculate the primary volt-amperes.

$$\text{Primary VA} = \frac{\text{Secondary VA}}{\text{Efficiency}} = \frac{4500}{90\%} = 5000 \text{ volt-amperes}$$

Step 4. Calculate the current in the secondary.

$$\text{Secondary Current} = \frac{\text{Secondary VA}}{\text{Secondary Volts}} = \frac{4500}{240} = 18.75 \text{ amperes}$$

Step 5. Calculate the current in the primary.

$$\text{Primary Current} = \frac{\text{Primary VA}}{\text{Primary Volts}} = \frac{5000}{480} = 10.42 \text{ amperes}$$

Three-Phase Transformers

A three-phase transformer is simply three single-phase transformers in one housing. In fact, three single-phase transformers can even be connected to form a three-phase transformer bank.

The two most common methods of connecting single-phase transformers to form a three-phase transformer bank are delta connected and wye connected. These two types of connections create four combinations of transformers.

- Delta to delta
- Delta to wye
- Wye to delta
- Wye to wye

There are certain properties that remain constant on both delta and wye transformers. The formulas on the following pages are used to solve certain transformer problems. Take transformer problems one step at a time. A question is asking for information on either a delta transformer or a wye transformer.

Wye Transformer

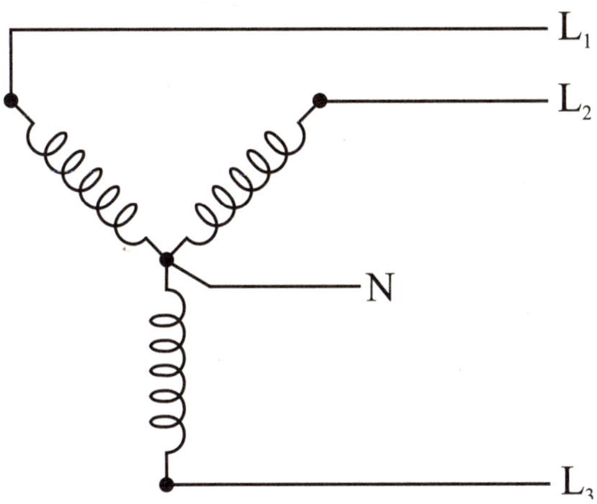

FIGURE 15-5

- Phase current is equal to line current.
- Phase voltage will be smaller than line voltage.
- Phase voltage is equal to line voltage divided by 1.732.
- Line voltage will be larger than phase voltage.
- Line voltage is equal to phase voltage multiplied by 1.732.

Table 15-2 Formulas for a Wye Transformer
$I_L = I_P$
$E_P = E_L \div 1.732$
$E_L = E_P \times 1.732$
$VA = E_P \times I_P \times 3$
$VA = E_L \times I_L \times 1.732$
$I_P = \frac{VA}{E_P \times 3}$
$I_L = \frac{VA}{E_L \times 1.732}$

Delta Transformer

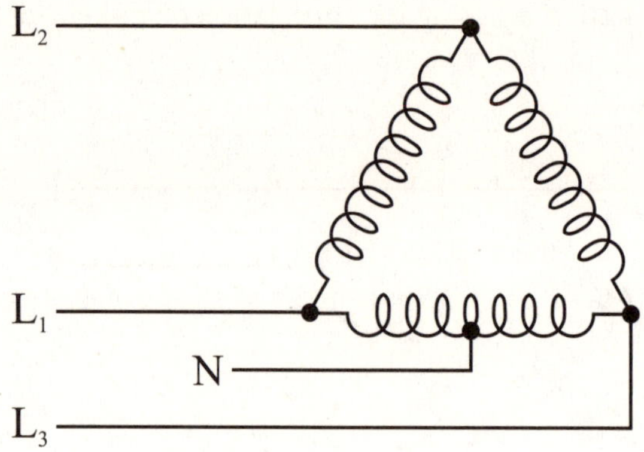

FIGURE 15-6

- Phase voltage is equal to line voltage.
- Phase current will be smaller than line current.
- Phase current is equal to line current divided by 1.732.
- Line current will be larger than phase current.
- Line current is equal to phase current multiplied by 1.732.

Table 15-3 Formulas for a Delta Transformer
$E_L = E_P$
$I_P = I_L \div 1.732$
$I_L = I_P \times 1.732$
$VA = E_P \times I_P \times 3$
$VA = E_L \times I_L \times 1.732$
$I_P = \frac{VA}{E_P \times 3}$
$I_L = \frac{VA}{E_L \times 1.732}$

The type of transformer is not always given because it may not be necessary to solve the problem. The last four formulas for delta transformers are the same as the last four formulas for wye transformers.

EXAMPLE

The line current of a 3-phase load is 10 amperes on a 208/120-volt transformer. What is the total wattage?

a) 4157 watts **b)** 3600 watts **c)** 2400 watts **d)** 1200 watts

Step 1. Find the formula. In this question, two voltages are given. Unless phase and line voltage are the same, line voltage will be larger than phase voltage. Therefore, in this question, line voltage is 208 volts.

$VA = E_L \times I_L \times 1.732$

Step 2. Replace the letters with the values given and solve the problem.

$VA = 208 \times 10 \times 1.732 = 3602.56$

The total wattage is approximately 3600 watts. Answer b) is correct.

When a question asks for neutral current, use the formula for phase current. Only loads that use a neutral are counted. A load that does not use a neutral will not contribute to the neutral current.

EXAMPLE

The secondary load on a 3-phase, 480-208/120-volt transformer is nine 3000-watts of 120-volt lighting and nine 3000-watts of single-phase, 208-volt loads. The neutral must have a load current rating of at least _____ amperes.

a) 50 b) 75 c) 86.5 d) 150

Step 1. Find the total neutral load. The 208-volt loads do not contribute to the neutral current.

9 × 3000 = 27,000 watts on the neutral

Step 2. Find the formula and replace the letters with the values given. Solve the problem.

$$I_P = \frac{VA}{E_P \times 3} = \frac{27{,}000}{120 \times 3} = 75 \text{ amperes}$$

There will be 75 amperes on the neutral conductor. Answer b) is correct.

A delta-delta transformer has the same voltage potential for line voltage as it does for phase voltage. A question may give the voltage of a delta-delta transformer as 230/115 volts. Phase and line voltage are both 230 volts. There is a grounded (neutral) tap in the center of one of the windings. The potential between the neutral and either line on that winding will be 115 volts because it is one half of the full winding. The third conductor coming from the transformer is known as the high leg. The voltage potential from this conductor to the neutral conductor will be higher than the other two conductors.

The high leg can be calculated by multiplying the single-phase center-tapped voltage by 1.732. The high leg to ground on a 230/115-volt transformer is approximately 200 volts because 115 × 1.732 = 199.18.

EXAMPLE

A three-phase delta-delta transformer has a 240/120-volt secondary. What is the voltage between the high leg and ground?

a) 120 volts b) 190 volts c) 208 volts d) 240 volts

Step 1. Multiply the single-phase center-tapped voltage by 1.732.

120 × 1.732 = 207.84 = 208 volts

Answer c) is correct.

When comparing transformer ratios, always compare phase voltage to phase voltage. In a delta transformer, line and phase are the same value. In a wye transformer, line and phase are different values. If a delta-wye transformer question is asking for the turns ratio, make sure both primary and secondary are phase voltage. The primary will always be the first number in the ratio.

EXAMPLE

What is the turns ratio of a three-phase delta-wye transformer with a voltage of 480-208/120 volts?

a) 1 to 1 b) 2.3 to 1 c) 3 to 1 d) 4 to 1

Phase voltage in the delta is 480 volts. Phase voltage in the wye is 120 volts, not 208.

480 ÷ 120 = 4

The ratio is 4 to 1. Answer d) is correct.

Exercise caution when calculating the turns ratio in a delta-delta transformer. A delta-delta transformer with a voltage of 460-230/115 does not have a 4-to-1 ratio. Line and phase voltage in a delta transformer are the same. The number 115 does not represent phase voltage. It is a result of a ground tap in the center of one phase. Since phase voltage is 230 volts, 115 is one half of one phase. Remember: always compare phase voltage to phase voltage.

EXAMPLE

What is the turns ratio of a three-phase delta-delta transformer with a voltage of 480-240/120 volts?

a) 1 to 1 b) 2 to 1 c) 3 to 1 d) 4 to 1

Step 1. Compare primary phase voltage to secondary phase voltage.

480 compared to 240 = 2 to 1

Answer b) is correct.

Note, the 120 in this delta transformer is not phase voltage, it is one half of the phase voltage.

END-OF-CHAPTER PROBLEMS

1. What is the secondary line voltage of a delta-wye three-phase, 480-208/120-volt transformer?
   a) 120 volts
   b) 208 volts
   c) 360 volts
   d) 480 volts
2. What is the phase current of a 9000-watt load that is connected to a 208-volt, three-phase wye system?
   a) 14.4 amperes
   b) 25 amperes
   c) 43.3 amperes
   d) 75 amperes
3. What are the approximate volt-amperes per phase for the load in the following drawing? Given: A delta load with a source voltage of 208/120-volt, three-phase.
   a) 1442 VA
   b) 2497 VA
   c) 4326 VA
   d) None of these

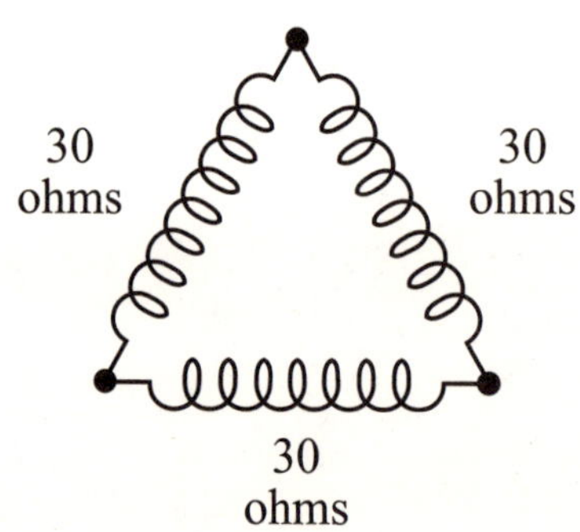

FIGURE 15-7

4. What is the line current of a 50-kVA, three-phase, 480-240/120-volt transformer?
   a) 60 amperes
   b) 104 amperes
   c) 120 amperes
   d) None of these
5. If all the load on a 3-phase, 480-208/120-volt transformer is single-phase, 120-volt, the size of the neutral on a 24-kVA transformer will have a load current rating of _____.
   a) 100 amperes
   b) 75 amperes
   c) 67 amperes
   d) 24 amperes
6. With a 3-phase, delta-wye transformer bank having a 480-volt primary and a 208/120-volt secondary, the secondary phase voltage is _____.
   a) equal to the secondary line voltage
   b) greater than the secondary line voltage
   c) less than the secondary line voltage
   d) variable
7. A three-phase, 480-208/120-volt transformer has the following loads:
   - Nine 2000-watts, 120-volts
   - Nine 2000-watts, 208-volts

   If the loads are balanced, what is the maximum current on the neutral?
   a) 150 amperes
   b) 86.5 amperes
   c) 75 amperes
   d) 50 amperes
8. The primary of a 3-phase, 480-208-volt transformer has a line current of 240 amperes. The kVA rating of this transformer would be closest to a _____ kVA transformer.
   a) 115
   b) 200
   c) 500
   d) 800
9. Three 3000-watt, 120-volt light banks are connected to a three-phase, 208/120-volt source. Each of the three light banks draws _____ amperes.
   a) 75
   b) 50
   c) 25
   d) 20
10. A three-phase, 240-volt motor has a total load of 15,000 watts. What is the wattage load on each phase of the transformer bank?
    a) 15,000 watts
    b) 5000 watts
    c) 4000 watts
    d) 3000 watts

11. A transformer has a 120-volt primary and a 480-volt secondary. There are 40 turns in the primary. How many turns are in the secondary?
    a) 10
    b) 40
    c) 120
    d) 160
12. A three-phase, 208-volt panel has a total load of 28,460 watts. Half of the load is three-phase, 208-volt; the other half is single-phase, 120-volt. What is the line current?
    a) 79 amperes
    b) 125 amperes
    c) 137 amperes
    d) 158 amperes
13. If ten single-phase, 208-volt, 5000-watt office machines were on the same phase of the same transformer, the total phase current with all the machines operating would be _____.
    a) 120 amperes
    b) 208 amperes
    c) 240 amperes
    d) 416 amperes
14. With a three-phase delta-wye transformer bank having a 480-volt primary and a 208/120-volt secondary, the line current of the secondary is 120 amperes. The primary line current will be _____.
    a) 480 amperes
    b) 120 amperes
    c) $30 \times \sqrt{3}$ amperes
    d) 30 amperes
15. A three-phase delta-delta transformer has a 230/115-volt secondary. What is the high leg voltage to ground?
    a) 115 volts
    b) 199 volts
    c) 208 volts
    d) 230 volts
16. The line current is 10 amperes on a three-phase, 208/120-volt transformer. What is the total wattage?
    a) 1200 watts
    b) 2400 watts
    c) 3600 watts
    d) 4157 watts
17. Three 1500-watt, 208-volt dryers are balanced on a 3-phase, 208/120-volt source, one connected line A-B, one B-C, and the third A-C. What is the maximum line current drawn by these three dryers?
    a) 6.25 amperes
    b) 7.2 amperes
    c) 12.5 amperes
    d) 14.4 amperes

The drawing below refers to question 18.

- 3-phase, 480-volt primary
- 3-phase, 240-volt secondary
- 1-phase, 240/120-volt secondary

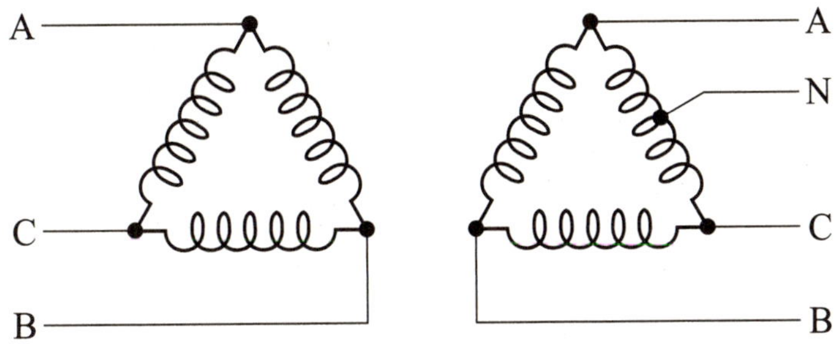

FIGURE 15-8

18. Single-phase loads total 12,000 watts on 240 volts, 6000 watts on 120 volts. Three-phase loads total 36,000 watts on 240 volts. Only 3-phase loads are connected to the high leg. The current in the high leg (B) is _____ amperes.

a) 300

b) 150

c) 86.7

d) 50

19. The secondary load of a 3-phase, 480-208/120-volt transformer is nine 3000-watts of 120-volt lighting and nine 3000-watts of 208-volt, one-phase loads. The neutral must have a load rating of at least _____ amperes.

a) 50

b) 75

c) 86.5

d) 150

20. What is the turns ratio of a three-phase, delta-delta transformer with a voltage of 480-240/120 volts?

a) 1 to 1

b) 2 to 1

c) 3 to 1

d) 4 to 1

21. If the source is a 50-kVA, three-phase, 480-208/120-volt transformer, what is the maximum primary line current expected at full load?

a) 60 amperes

b) 35 amperes

c) 110 amperes

d) 104 amperes

22. With a 3-phase delta-wye transformer bank having a 480-volt primary and a 208/120-volt secondary, the line current of the secondary is 120 amperes. The primary phase current will be _____.
    a) 480 amperes
    b) 120 amperes
    c) $30 \times \sqrt{3}$ amperes
    d) 30 amperes
23. What is the secondary phase current of a 45-kVA, three-phase, 208-volt transformer?
    a) 72 amperes
    b) 125 amperes
    c) 216 amperes
    d) 375 amperes
24. A three-phase, delta-wye transformer has a 480-volt primary and a 208/120-volt secondary. The primary line current is _____.
    I. the transformer kVA divided by (1.732 × 480)
    II. three times the primary phase current
    III. equal to the primary phase current
    a) I only
    b) II only
    c) III only
    d) I and II
25. A single-phase transformer has a rating of 150 kVA. The voltage on the secondary is 240/120 volts. The 240-volt loads are 50 kVA. The neutral must be sized to carry _____ amperes.
    a) 278
    b) 417
    c) 625
    d) 833
26. The maximum load on a three-phase, 4-wire, 208/120-volt lighting panel is 9.6 kW. The maximum load on the neutral is _____ amperes.
    a) 80
    b) 46.2
    c) 26.7
    d) 15.4
27. The total load on a center tap 240/120-volt transformer is 100 kW. The 120-volt loads are balanced, and the 240-volt loads consist of 20 kW. What is the maximum neutral current?
    a) 667 amperes
    b) 333 amperes
    c) 192 amperes
    d) 0 amperes

28. Three 1800-watt lighting loads and two 2-hp, single-phase, 240-volt pumps are in a 3-wire, 240/120-volt panel. What is the maximum neutral current in amperes?

a) 42 amperes
b) 27 amperes
c) 30 amperes
d) 15 amperes

29. The secondary of a 3-phase, 480-208/120-volt transformer has a line current of 200 amperes. The kVA rating of this transformer would be closest to a ______-kVA transformer.

a) 20
b) 30
c) 50
d) 75

30. What size primary overcurrent protection is required for a 480-208/120-volt transformer rated 15 kVA?

a) 20 amperes
b) 25 amperes
c) 30 amperes
d) 40 amperes

31. An auto-transformer fault sensing system can be accomplished by the use of two subtractive-connected donut-type current transformers installed to sense and signal when an unbalance occurs in the line current to the auto-transformer of ______ or more of rated current.

a) 25%
b) 33%
c) 50%
d) 66%

32. What is the maximum rating of overcurrent protection for a 277/480-volt transformer with a primary current of 8 amperes? There will be no overcurrent protection in the secondary.

a) 8 amperes
b) 10 amperes
c) 13 amperes
d) 15 amperes

33. What is the maximum rating of overcurrent protection for a 240-volt transformer with a primary current of 13 amperes? Overcurrent protection will be only in the primary.

a) 13 amperes
b) 15 amperes
c) 20 amperes
d) 30 amperes

34. What is the maximum rating of overcurrent protection percentage for a 600-volt, 9-ampere transformer where there is no secondary protection?

a) 115%
b) 125%
c) 150%
d) 167%

PART II General Exam Questions

Exam Number 1

Time Limit: 1 Hour

1. A battery operates on the basis of _____.
   a) piezoelectricity
   b) thermochemistry
   c) photovoltaics
   d) electrochemistry
2. Ohm's law states that the current is directly proportional to the voltage in a circuit and inversely proportional to the resistance.
   a) True
   b) False
3. A start switch is wired _____ in a motor control circuit.
   a) series-parallel
   b) parallel
   c) series
   d) none of these
4. An electric generator operates on the principle of _____.
   a) capacitive force
   b) magnetmotive force
   c) electromotive force
   d) magneto electricity
5. The main reason for laminating the core of a power transformer is to keep the _____.
   a) friction losses low
   b) turns ratio balanced
   c) eddy current loss at a minimum
   d) copper losses at a minimum
6. Two light bulbs are connected in parallel to a 120-volt source. Each lamp draws 0.62 amps. What is the total load of the circuit?
   a) 37.2 watts
   b) 74.4 watts
   c) 148.8 watts
   d) 46,451.6 watts
7. Electrical pressure is measured in _____.
   a) amperes
   b) watts
   c) volts
   d) ohms
8. AC voltages may be decreased or increased by a _____.
   a) transformer
   b) motor
   c) rectifier
   d) capacitor

9. The basic unit of electrical work is the _____.
   a) kVA
   b) watt-hour
   c) watt
   d) volt-ampere
10. A 1000 watt, _____ lamp uses electrical energy at the same rate as a 14.4-ohm resistor.
   a) 69.4-volt
   b) 110-volt
   c) 120-volt
   d) 240-volt
11. Solenoids are made of what type of magnets?
   a) Permanent
   b) Copper
   c) Reverse
   d) Electro
12. When a current leaves its intended path and returns to the source, bypassing the load, the circuit is _____.
   a) open
   b) shorted
   c) incomplete
   d) broken
13. The disadvantage of an autotransformer is _____.
   a) the size
   b) the poor efficiency due to a single winding
   c) the high cost
   d) the lack of isolation between the primary and secondary
14. The main reason for using oil in an oil circuit breaker is to _____.
   a) increase the capacity of current
   b) quench the arc
   c) decrease the resistance
   d) lubricate the points
15. _____ is/are the letter(s) used to indicate inductive reactance in a circuit.
   a) XL
   b) HZ
   c) Z
   d) XC
16. An outlet box should be fastened to a solid concrete wall by means of _____.
   a) wood plugs
   b) lag bolts
   c) toggle bolts
   d) expansion bolts

17. Heating the junction of two dissimilar alloys causes _____.
    a) fission
    b) electrons to flow
    c) arcing
    d) corrosion
18. If voltage rotation of a three-phase motor is A-B-C, then the current rotation will be _____.
    a) C-B-A
    b) A-C-B
    c) B-C-A
    d) B-A-C
19. Sixty-cycle frequency travels 180 degrees in what length of time?
    a) 1/30 sec.
    b) 1/60 sec.
    c) 1/120 sec.
    d) 1/180 sec.
20. A low-energy power circuit _____.
    a) is a signal circuit
    b) is a remote-control circuit
    c) has its power supplied by transformers and batteries
    d) none of these
21. _____ shall not be threaded.
    a) EMT
    b) IMC
    c) RMC
    d) Luminaire stems
22. A _____ is a device with electrical contacts that are opened or closed when the electrical current energizes the coils.
    a) thermostat
    b) transtat
    c) reactor
    d) relay
23. The total opposition to the flow of current in an ac circuit is expressed in ohms and is called _____.
    a) resistance
    b) impedance
    c) conductance
    d) reluctance

24. The power factor can be determined by the use of which combination of the following meters?

I. Ammeter
II. Watt meter
III. Watt-hour meter
IV. Ohmmeter
V. Voltmeter
VI. Phase rotation meter

a) I, IV, and VI
b) I, V, and VI
c) I, II, and V
d) I, IV, and V

25. D.P.D.T. letters would be used to identify a _____.

a) raceway
b) switch
c) circuit breaker
d) conductor insulation

26. The current will lag the voltage when _____ is present in the circuit.

a) conductance
b) inductance
c) capacitance
d) resistance

27. On a multi-wire, three-wire branch circuit, the maximum unbalanced load on the neutral conductor at any time would be when _____.

a) both circuits are fully loaded
b) one hot leg is shut off
c) the neutral is disconnected
d) both circuits are off

28. To find the circular-mil area of a stranded conductor, the most accurate method is to _____.

a) use a dialometer
b) square the number of strands
c) use a micrometer
d) multiply the number of strands by the circular-mil area of one strand

29. A dc voltmeter can be used directly to measure which of the following?

a) Polarity
b) Power
c) Power factor
d) Hertz

30. The equipment used to measure and compare peak, average, and root-mean-square values on an ac voltage curve is called a(n) _____.

a) mandrel
b) thermal couple
c) oscilloscope
d) voltmeter

31. If the circuit voltage is increased, and everything else remains the same, only the _____ will change.

a) power

b) conductivity

c) current

d) resistance

32. To calculate the power used by an appliance (not motor driven), one needs to know _____.

I. current and resistance

II. voltage and resistance

III. current and voltage

a) I only

b) II only

c) III only

d) I, II, or III

33. The voltage drop in a line for an inductive load can be decreased by _____.

I. decreasing the load

II. increasing the voltage

III. increasing the wire size

a) III only

b) I or III

c) II or III

d) I, II, or III

34. In a series circuit the current returning to the generator is equal to the current leaving the generator less the losses due to the loads.

a) True

b) False

35. Power _____ voltage multiplied by current, whether capacitive reactance is in the circuit or not.

a) is always

b) is never

c) in a dc circuit is

d) in an ac circuit is

36. To fasten a box to a hollow tile wall, you would use _____.

a) rawl plugs

b) toggle bolts

c) lag bolts with backplates

d) wood plugs or bolts

37. Average voltage of an alternating current system is the same in dc.

a) True

b) False

38. Two 120-volt light bulbs connected in series across 240 volts will _____.
   a) burn out quickly
   b) flicker with the cycle
   c) burn at full brightness
   d) burn at half-brightness
39. When referring to an 8-32 machine bolt, the number 32 refers to the _____.
   a) length of bolt
   b) diameter
   c) strength
   d) threads per inch
40. One horsepower equals _____ watts.
   a) 500
   b) 707
   c) 746
   d) 1000
41. The life of insulation used in electrical installations is directly affected by heat. Of the following electrical insulations, which can least withstand heat?
   a) Rubber
   b) Mica
   c) Enamel
   d) Fiberglass
42. A transformer is more efficiently utilized when the load has a(n) _____ power factor.
   a) high
   b) medium
   c) low
   d) average
43. The greater number of free electrons, the better the _____ of a metal.
   a) conductivity
   b) insulation value
   c) voltage drop
   d) resistance
44. A type of transformer that has only a single winding is called a(n) _____ transformer.
   a) isolation
   b) delta
   c) auto
   d) wye
45. Batteries connected in series _____.
   a) have longer amp-hour capacity
   b) provide less voltage than each individual battery
   c) provide greater voltage than each individual battery
   d) require less charging

46. Three 9-ohm resistors connected in parallel have a total resistance of ______ ohms.

a) 1½
b) 3
c) 9
d) 27

Questions 47 through 50 refer to the drawing below.

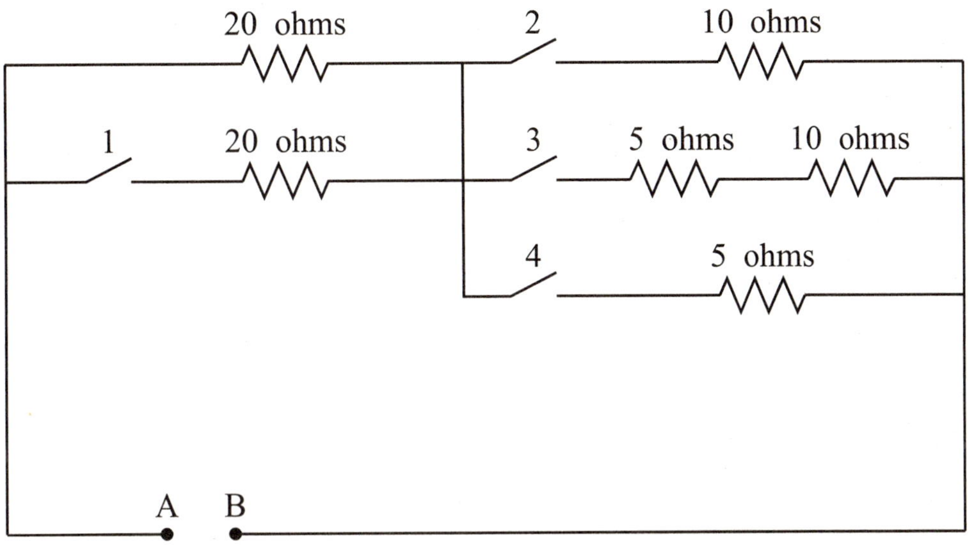

FIGURE 16-1

47. What is the voltage between A and B when switches 1, 3, and 4 are closed? The 5-ohm resistor is drawing 4 amperes.

a) 60 volts
b) 100 volts
c) 120 volts
d) 140 volts

48. If the current is 4 amperes when switch 3 is closed, what would be the applied voltage of the circuit?

a) 60 volts
b) 100 volts
c) 120 volts
d) 140 volts

49. If the voltage were 120 volts and switches 1 and 2 were closed, what would be the current in the circuit?

a) 12 amperes
b) 6 amperes
c) 4 amperes
d) 2 amperes

50. Which switches need to be closed to create the least amount of resistance?

a) 2 only
b) 1 and 2
c) 1, 2, and 3
d) 1, 2, 3, and 4

Exam Number 2

Time Limit: 1 Hour

1. What instrument is used to measure the RPM speed of a motor?
   a) An odometer
   b) A hydrometer
   c) A tachometer
   d) A megger
2. What would be the advantage of 240 volt rather than 120 volt on a load with the same wattage?
   a) Less voltage drop
   b) Less power used
   c) Both A and B
   d) Neither A nor B
3. Comparing incandescent lighting with fluorescent lighting at the same level of illumination, the cost of energy for fluorescent would be _____.
   a) 1.1 increase
   b) less
   c) greater
   d) the same
4. Electric motors are rated in terms of _____.
   a) kilowatts
   b) dynes
   c) foot-pounds
   d) horsepower
5. The liquid in a battery is called the _____.
   a) vermiculite
   b) electrolyte
   c) askarel
   d) hermetic
6. How many 4-way switches are required to switch a light from three places?
   a) Four
   b) Three
   c) Two
   d) One
7. The symbol for a delta connection is _____.
   a) Δ
   b) Ø
   c) Ω
   d) Y

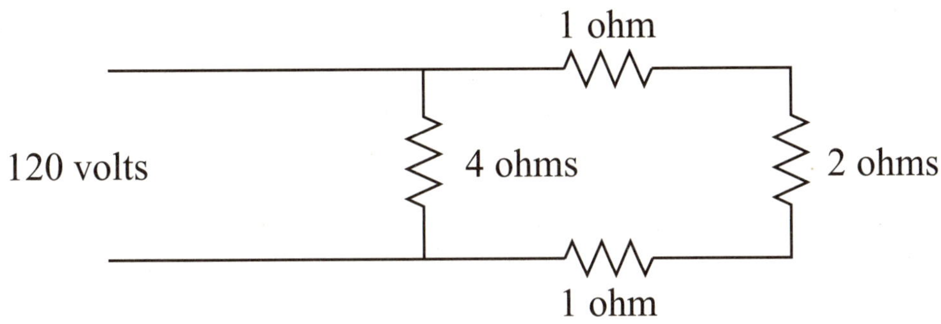

FIGURE 16-2

8. What is the current of the circuit in the figure above?
   a) 2 amperes
   b) 60 amperes
   c) 78 amperes
   d) 90 amperes
9. On a dc ammeter, to increase the range you would use a(n) _____.
   a) shunt
   b) condenser
   c) capacitor
   d) inductor
10. Root-mean-square is also called the _____ voltage.
   a) peak
   b) effective
   c) maximum
   d) average
11. The *National Electrical Code* is sponsored by the _____.
   a) Nationally Recognized Testing Laboratory
   b) Underwriters Laboratory
   c) National Fire Protection Association
   d) National Electrical Manufacturers Association
12. A drawing showing the floor arrangement of a building is referred to as a _____.
   a) plan
   b) perspective
   c) specification
   d) proposal
13. A _____ stores energy in much the same manner as a spring stores mechanical energy.
   a) resistor
   b) coil
   c) condenser
   d) none of these

14. Which of the following has the highest resistance?
   a) Paper
   b) Iron
   c) Brass
   d) Saltwater
15. With movable contacts, which of the following helps to prevent arcing?
   a) Condenser
   b) Capacitor
   c) Resistor
   d) All of these
16. A wattmeter is connected in _____ in the circuit.
   a) parallel
   b) series-parallel
   c) series
   d) none of these
17. The relationship of a transformer primary winding to the secondary winding is expressed in _____.
   a) turns-ratio
   b) kilowatt hours
   c) volt-amperes
   d) amperes
18. The difference between the thread on a bolt and the thread on a conduit is _____.
   a) no lubrication is required in threading conduit
   b) the conduit thread is tapered
   c) the conduit thread is deeper
   d) the conduit thread always has the same pitch for any diameter
19. For better illumination you would use _____.
   a) evenly spaced lights, higher ceilings
   b) random spacing of lights
   c) even spacing, numerous lights
   d) twice the number of lights
20. A clamp-on ammeter will measure _____.
   a) current when clamped on a multiconductor cable
   b) accurately only when clamped perpendicular to a conductor
   c) voltage when clamped on a single conductor
   d) accurately only when clamped parallel to a conductor
21. Which of the following motors does not have a commutator?
   a) Series
   b) Shunt
   c) Squirrel cage
   d) Synchronous

22. A fuse puller is used to replace _____ fuses.
    a) cartridge
    b) link
    c) current-limiting
    d) plug
23. An Erickson coupling is used to _____.
    a) connect EMT to flexible conduit
    b) connect two sections of rigid conduit when one section cannot be turned
    c) connect two sections of all thread together
    d) join sections of EMT together
24. Something that would affect the ampacity of a conductor would be _____.
    a) amperage
    b) length
    c) temperature
    d) voltage
25. Electromagnetic devices usually have a(n) _____ core.
    a) hard steel
    b) hard iron
    c) aluminum
    d) soft iron
26. For voltage and current to be in phase, _____.
    a) the circuit impedance has only resistance
    b) voltage and current appear at their zero and peak values at the same time
    c) both A and B
    d) neither A nor B
27. The most heat is created when current flows through which of the following?
    a) A 10-ohm resistor
    b) A 10-ohm condenser
    c) A 10-ohm inductance coil
    d) The heat would be equal in all of these.
28. The voltage drop in electrical circuits _____.
    a) is frequently a cause of burnouts of electrical heating elements
    b) may be eliminated by adjusting the generator field rheostat
    c) is determined by load, distance, and size of conductors
    d) is caused by poor regulation of the supply voltage
29. If the allowable current in a copper busbar is 1000 amperes per square inch of cross section, the width of a standard 1/4″ busbar designed to carry 1500 amperes would be _____.
    a) 8″
    b) 6″
    c) 4″
    d) 2″

30. XC, XL, and R each in its own branch circuit will flow in _____ direction(s).
   a) one
   b) two
   c) three
   d) none of these
31. Four heater coils with a given voltage will consume the most power when connected _____.
   a) two parallel pairs in series
   b) all in parallel
   c) two in parallel and two in series
   d) all in series
32. An electric bell outfit would be used to check for _____.
   a) current
   b) continuity
   c) voltage
   d) wattage
33. Which of the following is not a factor in calculating the feeder conductor size?
   a) Voltage drop
   b) Branch-circuit protection
   c) Demand factors
   d) Ambient temperature
34. A _____ is used for testing specific gravity.
   a) hydrometer
   b) calorimeter
   c) mandrel
   d) galvanometer
35. The load side is usually wired to the blades of a knife switch to _____.
   a) prevent arcing when opening the switch
   b) allow changing of fuses without opening the switch
   c) make the blades dead when the switch is open
   d) prevent blowing the fuse when opening the switch
36. New brushes for a generator should be fitted to the commutator by using a _____.
   a) strip of sandpaper
   b) polishing stone
   c) bastard file
   d) strip of emery cloth
37. To handle a three-phase unbalanced system, balance the system by making all loads equal to the _____ single-phase load.
   a) unbalanced
   b) largest
   c) average
   d) smallest

38. A light bulb usually contains _____.
   a) either a vacuum or gas
   b) air
   c) neon
   d) H_2O
39. How is a measurement of the ceiling taken?
   a) From the top of the slab on the first floor to the bottom of the slab on the second floor
   b) From the top of the slab on the first floor to the top of the slab on the second floor
   c) From the bottom of the slab on the first floor to the bottom of the slab on the second floor
   d) None of the above
40. Real power is expressed as _____.
   a) IR
   b) E/W
   c) $V \times C \times \cos\theta$
   d) none of these
41. Overloading a conductor continually is a poor practice because it will cause _____.
   a) damage to the raceway
   b) the conductor to weaken
   c) the conductor to melt
   d) the insulation to deteriorate
42. The voltage of a circuit is best defined as _____.
   a) the potential between two conductors
   b) the greatest difference of potential between two conductors
   c) the effective difference of potential between two conductors
   d) the average RMS difference of potential between any two conductors
43. With respect to a common light bulb, it is correct to state that the _____.
   a) filament is made of carbon
   b) lower-wattage bulb has the higher resistance
   c) circuit voltage has no effect on the life of the bulb
   d) base has a left-hand thread
44. What is the resistance of a 1000-watt heater connected to a source voltage of 120 volts?
   a) 8.3 ohms
   b) 14.4 ohms
   c) 8333 ohms
   d) 0.069 ohms
45. What pushes current through a line?
   a) Power
   b) Amperes
   c) EMF
   d) Ohms

46. A single-phase, 120/240-volt motor is connected to a source voltage of 120 volts. The windings are connected in _____.

a) series
b) parallel
c) series-parallel
d) none of these

47. Which of these conductors has the highest resistance?

a) Coated copper
b) Uncoated copper
c) Aluminum
d) Silver

48. A circuit that has four 150-watt, 120-volt heaters that are connected in series to a source voltage of 600 volts will have a total resistance of _____ and have a total current draw of _____.

a) 384 ohms; 0.4 amperes
b) 384 ohms; 1.56 amperes
c) 96 ohms; 1.25 amperes
d) 384 ohms; 1.25 amperes

49. The *National Electrical Code* requires that conduit be continuous from outlet to outlet, must be mechanically and electrically connected to all fittings, and must be suitably grounded. The reason for having the conduit electrically continuous and grounded is to _____.

a) provide a metallic return conductor
b) make it easy to test wiring connections
c) shield the wires inside the conduit from external magnetic fields
d) prevent electrical shock, which might otherwise result from contact with the conduit

50. A wye transformer has a phase current of 20 amperes. The line current is equal to _____.

a) the phase current
b) the phase current × 1.732
c) the phase current × 3
d) not enough information given

Exam Number 3

Time Limit: 1 Hour

1. Effective voltage and RMS voltage are both the same value in a(n) _____.
   I. ac circuit
   II. dc circuit
   III. generator circuit
   a) I only
   b) II only
   c) I and II
   d) III only
2. What amount of current is being drawn by four 150-watt, 120-volt heaters that are connected in series to a source voltage of 600 volts?
   a) 1 ampere
   b) 1.56 amperes
   c) 5 amperes
   d) 6.24 amperes
3. The advantage(s) of a 4-wire, 3-phase voltage source over a 3-wire, 3-phase voltage source is/are _____.
   I. a grounded neutral
   II. better speed control
   III. two voltage levels
   a) I only
   b) I and II
   c) I and III
   d) III only
4. The factor that will have the least effect on the voltage at the most distant point from the source of supply for a two-wire circuit is _____.
   a) the length of the circuit
   b) the amount of load on the circuit
   c) whether the supply is 25-cycle or 60-cycle ac
   d) the gauge of the circuit wires
5. Which of the following exists in a three-phase, 4-wire system, 208/120-volt system?
   a) 3Ø 208-volt, 3Ø 120-volt, and 1Ø 120-volt
   b) 3Ø 208-volt, 1Ø 120-volt, and 1Ø 208-volt
   c) 3Ø 208-volt, 3Ø 120-volt, and 1Ø 208-volt
   d) 1Ø 208-volt, 3Ø 208-volt, and 3Ø 120-volt

6. Mica is commonly used for _____.
   a) appliance insulation
   b) panelboards
   c) pole insulators
   d) commutator bar separators
7. The resistance in a circuit may be increased by _____.
   a) increasing the EMF
   b) increasing the current flow
   c) decreasing the current flow
   d) a loose connection
8. The *National Electrical Code* requires all circuits to have at least _____ and adequate grounding.
   a) two hot wires, a grounding wire,
   b) an ungrounded wire, a grounding wire,
   c) at least two wires
   d) a hot wire, a grounding wire,
9. In a highly inductive ac circuit, what device is used to correct the power factor towards unity?
   a) Inductor
   b) Resistor
   c) Capacitor
   d) Rectifier
10. _____ is/are the letter(s) used to indicate capacitive reactance in a circuit.
    a) XC
    b) XL
    c) HZ
    d) Z
11. The opposite of resistance would be _____.
    a) reactance
    b) capacitance
    c) reluctance
    d) conductance
12. The purpose of having a rheostat in the field of a dc shunt motor is to _____.
    a) reduce sparking at the brushes
    b) control the speed of the motor
    c) minimize the starting current
    d) limit the field current to a safe value
13. If you are using two wattmeters to measure the power, you are measuring power in a _____ system.
    a) double
    b) high-voltage
    c) three-phase
    d) single-phase

14. The advantage of ac over dc includes which of the following?
   a) Ease of voltage variations
   b) Impedance is greater
   c) Lower resistance at high currents
   d) Better speed control
15. Nichrome wire is mainly used for _____.
   a) generator interpoles
   b) heater coils
   c) motor windings
   d) light bulb filaments
16. A voltmeter will read which of the following on a 60-Hertz, ac system?
   a) Maximum value
   b) Average value
   c) RMS or effective value
   d) Instantaneous value
17. The primary purpose for grounding a raceway is to prevent the raceway from becoming _____.
   a) magnetized
   b) a path for eddy currents
   c) accidentally energized at a higher potential than ground
   d) a source of induction
18. Electrical equipment shall be installed _____.
   a) in accordance with the local code when more stringent than the *NEC*
   b) in accordance with the local code when less stringent than the *NEC*
   c) in accordance with the *NEC* regardless of the local code
   d) better than the minimum code allows
19. A(n) _____ is a device similar to a rectifier.
   a) commutator
   b) condenser
   c) inverter
   d) transformer
20. When two 1-ohm resistors are in parallel, the total resistance is _____ ohm(s).
   a) 2
   b) 1
   c) ½
   d) 0.05
21. A generator works on the basis of _____.
   a) an inductor moving inside a magnetic field
   b) a conductor moving inside a magnetic field
   c) a magnetic field moving around a conductor
   d) none of these

22. Ohm's law states _____.
    a) E is equal to I ÷ R
    b) E is equal to I × R
    c) E is equal to R ÷ I
    d) none of these
23. To measure ac cycles per second, you would use a(n) _____.
    a) power factor meter
    b) manometer
    c) ammeter
    d) frequency meter
24. When working on live 600-volt equipment where rubber gloves might be damaged, an electrician should _____.
    a) carry a spare pair
    b) work without gloves
    c) wear leather gloves over rubber ones
    d) reinforce the fingers with tape
25. In a 6-pole machine, 360 electrical degrees is equal to _____ mechanical degrees.
    a) 180
    b) 120
    c) 90
    d) 60
26. A megger is used to measure _____.
    a) specific gravity
    b) insulation resistance
    c) current
    d) reactance
27. The proper way to mount an instrument meter is to _____.
    a) drill from the back
    b) use the meter to mark your holes
    c) use a template
    d) drill from the front
28. The standard unit of electrical pressure is the _____.
    a) volt
    b) ampere
    c) watt-hour
    d) watt
29. What changes ac to dc?
    a) Transformer
    b) Rectifier
    c) Generator
    d) Cathode

30. The best way to cut Wiremold® is to _____.
   a) use a large pair of tin snips
   b) use an armored cable rotary cutter
   c) use a hacksaw and remove the burrs
   d) use a file
31. What is the total resistance of a 1-ohm resistor and a 2-ohm resistor that are connected in parallel?
   a) 1.5 ohms
   b) 1 ohm
   c) ⅔ ohm
   d) ½ ohm
32. It is good practice to use _____ in a long, straight, horizontal run of conduit.
   a) expansion joints
   b) universal joints
   c) insulation joints
   d) flexible joints
33. The least ballast noise comes from which type of ballast?
   a) A
   b) D
   c) F
   d) AB
34. A foreman should advise his crew, when working with low voltage, to _____.
   a) work when the load is zero
   b) never work on the circuit alone
   c) consider the circuit is always alive
   d) wait until the circuit has been killed
35. One million volts can be expressed as _____.
   a) one kilovolt
   b) one megavolt
   c) one millivolt
   d) one microvolt
36. When working near lead-acid storage batteries, extreme care should be taken to guard against sparks, because of _____.
   a) an explosion
   b) a short circuit
   c) an electrical shock
   d) overheating the electrolyte
37. A range is rated at 240 volts and draws 40 amperes. There is a splice that has a resistance of 0.1 ohm. How much power is being lost in the splice?
   a) 4 watts
   b) 40 watts
   c) 160 watts
   d) 400 watts

38. A fire alarm system is considered _____.
    a) a legally required standby system
    b) an optional standby system
    c) an interconnected system
    d) a progressive system
39. Electricity _____.
    a) cannot be stored
    b) can be stored in mechanical form
    c) can be stored in chemical form
    d) can be stored in solid form
40. An autotransformer is generally used rather than an isolation transformer _____.
    a) when safety is a factor
    b) when cost is a factor
    c) when there are several branch circuits
    d) where the ratio of transformation is low
41. A way of expressing the power factor is _____.
    a) θ
    b) R/Z
    c) sin θ
    d) XL/Z
42. A wattmeter indicates _____.
    a) power factor
    b) real power
    c) apparent power if PF is not in unity
    d) all of these
43. On a drawing, all edges that are invisible should be represented by lines that are _____.
    a) broken
    b) solid
    c) dotted
    d) curved
44. The horsepower rating of a motor _____.
    a) is the input to the motor
    b) is the output of the motor
    c) cannot be changed to watts
    d) is a measure of motor efficiency
45. A coil has 100 turns. A current of 10 amperes is passed through the coil. The coil develops _____.
    a) mutual inductance
    b) 1000 watts
    c) 1 kVA
    d) 1000 amp turns

46. A voltage of 2000 microvolts is the same as _____ volts.
   a) 0.020
   b) 0.002
   c) 0.200
   d) 2000
47. A hook on the end of a fish tape is not to _____.
   a) protect the end of the wire
   b) tie the wires to be pulled
   c) keep it from catching on joints and bends
   d) tie a swab to
48. Which of the following systems will furnish single-phase, three-phase, and two voltage levels?
   a) Single-phase, 4-wire
   b) Three-phase, 4-wire
   c) Three-phase, 3-wire
   d) Two-phase, 3-wire
49. If the clips on a cartridge fuse become hot, this is a good indication that _____.
   a) the fuse clips are too tight
   b) the fuse clips are too loose
   c) the voltage is too high
   d) the fuse rating is too high
50. In a three-phase circuit, the legs of the phase are _____ electrical degrees apart.
   a) 240
   b) 180
   c) 120
   d) 90

Exam Number 4

Time Limit: 1 Hour

1. A thermocouple will transform _____ into electricity.
   a) current
   b) heat
   c) power
   d) work
2. Batteries supply _____ current.
   a) negative
   b) positive
   c) direct
   d) alternating
3. The required diameter of an electrical conductor is determined by _____.
   I. volts
   II. length
   III. amperes
   a) I only
   b) III only
   c) I and III
   d) I, II, and III
4. In ac circuits, inductance is measured in units called the _____.
   a) ohm
   b) coulomb
   c) henry
   d) farad
5. Most electric power tools, such as electric drills, come with a third conductor in the power lead, which is used to connect the case of the tool to a grounded part of the electric outlet. The purpose for having this extra conductor is to _____.
   a) prevent accumulation of static charge on the case
   b) protect the user of the tool should the winding break down to the case
   c) eliminate sparking between the tool and the material being worked on
   d) provide for continued operation of the tool should the regular grounded line-wire open
6. Aluminum and copper-clad aluminum of the same circular mil and insulation have _____.
   a) different ampacities
   b) the same ampacity
   c) the same physical characteristics
   d) the same termination methods
7. _____ is current that flows through a circuit in the same direction at all times and with almost constant strength.
   a) Wattage
   b) AC
   c) Ampacity
   d) DC

8. If applied voltage to a heat strip is reduced, the current will _____.
   a) remain the same
   b) increase
   c) decrease
   d) double
9. A plug fuse contains a safety element composed of a piece of metal that has a _____.
   a) low resistance and low melting point
   b) low resistance
   c) low resistance and high melting point
   d) high melting point
10. Pure XL in a circuit will cause a _____ lag of the ampere curve to the voltage curve in an ac circuit.
    a) 90-degree
    b) 120-degree
    c) 180-degree
    d) 240-degree
11. The volt-ampere rating in an ac circuit is a way to indicate the _____ power.
    a) apparent
    b) peak
    c) true
    d) effective
12. DC voltage is reduced with a _____.
    a) condenser
    b) transformer
    c) capacitor
    d) resistor
13. Electrical capacity is measured in units called the _____.
    a) watt
    b) ohm
    c) farad
    d) henry
14. A megawatt is _____.
    a) 1000 watts
    b) 10,000 watts
    c) 100,000 watts
    d) 1,000,000 watts
15. A clamp-on ammeter will give the best reading at what angle to the conductor?
    a) Perpendicular
    b) Vertical
    c) Horizontal
    d) Diagonal

16. Which one of the following statements is not an advantage of using an autotransformer?
    a) Better voltage regulation
    b) Higher efficiency
    c) Less expensive
    d) Safe for stepping down higher voltages
17. If the voltage is doubled, the ampacity of a conductor _____.
    a) doubles
    b) is cut in half
    c) remains the same
    d) increases
18. If frequency is constant, the inductive reactance of a circuit will _____.
    a) remain constant regardless of voltage or current change
    b) vary directly with current
    c) not affect the impedance
    d) vary with voltage
19. What is the minimum number of wattmeters required to measure three-phase power?
    a) Four
    b) Three
    c) Two
    d) One
20. When rerouting conduit, it may be necessary to increase the wire size if the distance is considerably greater to _____.
    a) allow for possible resistance drop
    b) increase the mechanical strength
    c) account for current drop
    d) compensate for voltage drop
21. The disadvantage of ac compared to dc would be _____.
    a) reduced cost of transmission with high-voltage transformers
    b) that it cannot be used in electroplating
    c) that smaller conductors are needed to transmit electrical energy
    d) that less copper is required with ac at higher voltages
22. If a test lamp lights when placed in series with a condenser and a suitable source of dc, it is a good indication that a condenser is _____ if the light burns continuously.
    a) short circuited
    b) open circuited
    c) fully discharged
    d) fully charged

Questions 23 through 26 are to be answered using the drawing below.
The following delta load is being supplied from a three-phase, 208/120-volt transformer.

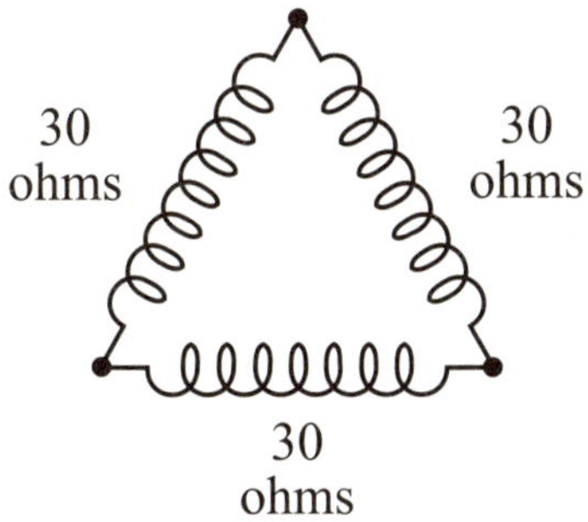

FIGURE 16-3

23. What is the approximate phase current for the load in the above drawing?

a) 4 amperes
b) 6.9 amperes
c) 12 amperes
d) 20.8 amperes

24. What is the approximate line current for the load in the above drawing?

a) 4 amperes
b) 6.9 amperes
c) 12 amperes
d) 20.8 amperes

25. What are the approximate volt-amps per phase for the load in the above drawing?

a) 1442 VA
b) 2497 VA
c) 4326 VA
d) 12,978 VA

26. What is the approximate load in volt-amperes for the above drawing?

a) 1442 VA
b) 2497 VA
c) 4326 VA
d) 12,978 VA

27. In multiple motor circuits there is feeder protection, branch-circuit protection, and motor protection. If the feeder protection trips, the fault would be expected to be in _____.

a) one of the starters
b) one of the branch circuits
c) one of the motors
d) the feeder

28. In the course of normal operation the instrument that will be least effective in indicating that a generator may overheat because it is overloaded is a(n) _____.
    a) ammeter
    b) wattmeter
    c) voltmeter
    d) stator thermocouple
29. A tungsten lamp will work on _____.
    a) ac
    b) dc
    c) both ac and dc
    d) neither ac nor dc
30. Which of the following breaks down rubber insulation?
    a) Water
    b) Acid
    c) Oil
    d) None of these
31. The definition of ambient temperature is _____.
    a) the insulation rating of the conductor
    b) the maximum operating temperature of the conductor
    c) the temperature of the conductor
    d) the temperature of the area surrounding the conductor
32. The weight of a fixture that can be supported by an outlet box shall not exceed _____ pounds.
    a) 50
    b) 40
    c) 30
    d) 25
33. An ammeter is connected _____ with the load.
    a) in parallel
    b) in series
    c) both in parallel and in series
    d) neither in parallel nor in series
34. Electroplating is an example of how electricity can cause _____.
    a) heat
    b) chemical activity
    c) pressure
    d) galvanic action

35. The feeder neutral conductor must be capable of carrying the maximum _____ load.
   a) demand
   b) balanced
   c) unbalanced
   d) connected
36. The current will lead the voltage when _____ is present in the circuit.
   a) reluctance
   b) inductance
   c) capacitance
   d) resistance
37. AC voltage of the system is not _____.
   a) RMS
   b) average
   c) EMF
   d) effective
38. Some materials give up electrons readily and others attract them; this is the basis for _____.
   a) photovoltaics
   b) piezoelectricity
   c) thermocouple
   d) electrochemistry
39. True power is always voltage multiplied by current _____.
   a) where frequency is constant
   b) in a dc circuit
   c) in an ac circuit
   d) regardless of whether capacitive reactance is in the circuit or not
40. Raceways on exterior surfaces of buildings shall be _____.
   a) rainproof and guarded
   b) weatherproof and covered
   c) raintight and arranged to drain
   d) watertight and arranged to drain
41. An ac ammeter is calibrated to read RMS values; this means the meter is reading the _____.
   a) effective value
   b) average value
   c) peak value
   d) maximum value
42. The current will lead the voltage when _____.
   a) resistance exceeds reactance in the circuit
   b) inductive reactance exceeds the capacitive reactance in the circuit
   c) capacitive reactance exceeds the inductive reactance in the circuit
   d) reactance exceeds the resistance in the circuit

43. If an ac sine wave reaches a peak voltage of 100, what is the effective root-mean-square voltage?
   a) 57.7 volts
   b) 70.7 volts
   c) 86.6 volts
   d) 141.4 volts
44. What is the function of a neon glow tester?
   I. Determines if circuit is ac or dc
   II. Determines polarity of dc circuits
   III. Determines if the circuit is alive
   a) III only
   b) I and II
   c) II and III
   d) I, II, and III
45. If a live conductor is contacted accidentally, the severity of the electrical shock is determined primarily by _____.
   a) the current in the conductor
   b) the size of the conductor
   c) the contact resistance
   d) whether the current is ac or dc
46. Heating elements in toasters have _____.
   a) high conductivity
   b) high resistance
   c) low resistance
   d) none of these
47. When an autotransformer is used to step up voltage, part of the single winding acts as the _____.
   a) primary
   b) secondary
   c) core
   d) none of these
48. The vector sum of the phase current is equal to _____ in a balanced, resistive three-phase system.
   a) 1.732 × phase current
   b) 0.707 × phase current
   c) phase current × power factor
   d) zero
49. _____ is the ability of a material to permit the flow of electrons.
   a) Resistance
   b) Current
   c) Voltage
   d) Conductance
50. An exciter may be used for more than one generator simultaneously.
   a) True
   b) False

Exam Number 5

Time Limit: 1 Hour

1. A signal circuit is used for _____.
   a) remote control
   b) low voltage
   c) annunciating
   d) low-energy power
2. A toaster will produce less heat on low voltage because _____.
   a) its total watt output increases
   b) its total watt output decreases
   c) the current increases
   d) the resistance has changed
3. Sixty-cycle frequency travels 270 degrees in what length of time?
   a) 1/80 sec.
   b) 1/120 sec.
   c) 1/180 sec.
   d) 1/240 sec.
4. An adjustable trip coil is rated 5 amperes on a 200-ampere switch. The trip coil should be set at _____ to trip at 120 amperes.
   a) 5 amperes
   b) 4 amperes
   c) 3 amperes
   d) 2 amperes
5. Power in a three-phase system may be measured with a minimum of _____.
   a) two voltmeters
   b) two ammeters
   c) one wattmeter
   d) none of these
6. An ammeter is disconnected from a current transformer (CT); the leads should be _____.
   a) taped
   b) insulated
   c) rerouted
   d) shorted
7. Which conductor insulation is approved for underground installations?
   a) TF
   b) TW
   c) SF
   d) THHN

8. How much power is being consumed by two 8-ohm resistors that are in parallel? The circuit is drawing 20 amperes.
   a) 6400 watts
   b) 3200 watts
   c) 1600 watts
   d) 320 watts
9. High-voltage cable that is to be installed in underground ducts is generally protected with a ______.
   a) steel wire armor
   b) copper outer jacket
   c) lead sheath
   d) tarred jute covering
10. When a load other than a motor is operating at a voltage 10% higher than the nameplate rating, it will ______.
   a) draw a lower current
   b) use more power
   c) have a longer life
   d) none of these
11. A 10-ohm resistance carrying 10 amperes of current uses ______ watts of power.
   a) 100
   b) 200
   c) 1000
   d) 10,000
12. A flat, circular piece of insulating material installed with individual lamps in a series circuit to provide circuit continuity when the lamp fails is a ______.
   a) fuse
   b) film-disc cutout
   c) pothead
   d) bond
13. A multimeter is a combination of ______.
   a) voltmeter, wattmeter, and ammeter
   b) voltmeter, ammeter, and megger
   c) ammeter, ohmmeter, and wattmeter
   d) voltmeter, ohmmeter, and ammeter
14. The current transformer ratio is 100 to 10 on an ammeter that has a reading of 3 amperes across the secondary. What is the actual current the circuit is drawing?
   a) 0.03 amperes
   b) 0.3 amperes
   c) 3 amperes
   d) 30 amperes

15. The _____ is the angle between the apparent power and the real power.
    a) reactive power
    b) power factor
    c) lag
    d) obtuse angle
16. Two 6-volt batteries in parallel _____.
    a) will draw lower currents
    b) will draw higher currents
    c) will give longer service
    d) will give 12 volts
17. On a 115-volt system, a 20-ampere fuse will blow when a load of _____ watts is connected to it.
    a) 1500
    b) 2000
    c) 2500
    d) 1920
18. Given: The voltage at the main service panel is 120 volts and the voltage at the remote panelboard is 116.5 volts. The load is fed from the remote panelboard. What is the minimum allowable voltage at the load? Do not exceed *NEC* recommendations.
    a) 113 volts
    b) 114 volts
    c) 115 volts
    d) 116 volts
19. A Western-Union splice is _____.
    a) for use on the West Coast only
    b) for the use of utility companies only
    c) for the purpose of strengthening a splice
    d) for the use of telephone companies only
20. The voltage will lead the current when _____.
    a) resistance exceeds reactance in the circuit
    b) capacitive reactance exceeds the inductive reactance in the circuit
    c) inductive reactance exceeds the capacitive reactance in the circuit
    d) reactance exceeds the resistance in the circuit
21. The diameter of a 4-inch conduit is measured from _____.
    a) the inside
    b) the outside
    c) the cross-sectional area
    d) none of these
22. The best conductor of electricity is _____.
    a) copper
    b) aluminum
    c) gold
    d) silver

23. A voltmeter measures _____.
   a) only voltage to ground
   b) voltage difference
   c) ac voltage only
   d) dc voltage only
24. The currents of a three-phase system are _____ degrees out of phase with each other.
   a) 1.732
   b) 90
   c) 120
   d) 270
25. Use _____ to lubricate the bushing on an armature.
   a) oil
   b) grease
   c) graphite
   d) none of these
26. When three light bulbs are wired in a single fixture, they are connected in _____.
   a) series-parallel
   b) series
   c) parallel
   d) none of these
27. A pendant fixture is a _____.
   a) bracket fixture
   b) recessed fixture
   c) hanging fixture
   d) none of these
28. Incandescent light bulbs, when operated at a voltage less than the rated voltage, will normally result in _____.
   a) more current
   b) longer bulb life
   c) shorter bulb life
   d) more light
29. A dc steady state condition is _____.
   a) ampacity reactance
   b) inductive reactance
   c) capacitive reactance
   d) resistance
30. To reverse the rotation of a three-phase motor you would _____.
   a) reverse any two of the three leads
   b) reverse all the leads
   c) reverse two of the four leads
   d) turn the motor around

31. A hydrometer is used to measure _____.
   a) revolutions per minute
   b) water pressure
   c) specific gravity
   d) water flow
32. A soldering iron should not be heated to excess because it will usually _____.
   I. anneal the copper tip
   II. loosen the copper tip on the iron rod
   III. ruin the tin on the surface of the tip
   a) I only
   b) II only
   c) III only
   d) I and III
33. If the maximum voltage of an ac circuit is 50 volts, the RMS value would be approximately _____ volts.
   a) 40
   b) 35
   c) 30
   d) 25
34. A standard lamp base is called _____.
   a) Edison
   b) standard
   c) candle
   d) mogul
35. The letter K used in the voltage drop formulas refers to the constant that represents the positive temperature coefficient of resistance.
   a) True
   b) False
36. When measuring to determine the size of a stranded conductor, you would place the wire gauge over _____.
   a) the insulation
   b) the outer covering
   c) all of the strands
   d) one strand of the conductor
37. A continuous load is where the _____ current is expected to continue for three hours or more.
   a) average
   b) maximum
   c) minimum
   d) none of these

38. How much power is being consumed by the circuit in the drawing below?

a) 90 amperes

b) 4800 watts

c) 7200 watts

d) 10,800 watts

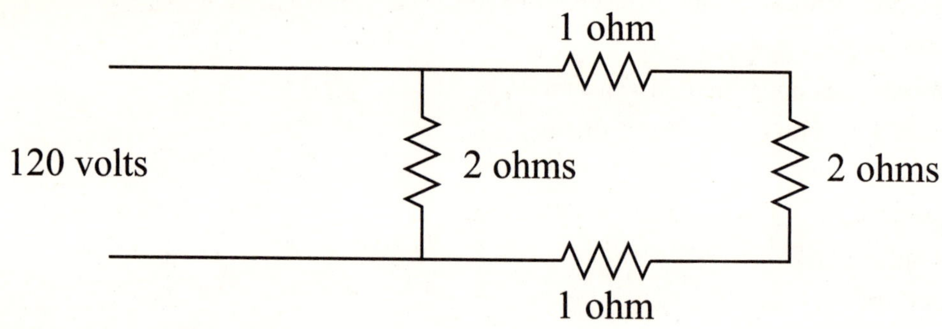

FIGURE 16-4

39. Ohm's law states that the current is directly proportional to the resistance in a circuit and inversely proportional to the voltage.

a) True

b) False

40. When the power factor in a given circuit is unity, the reactive power is ______.

a) a negative quantity

b) zero

c) at maximum

d) 1.1414 times the volt-amperes

41. Sixty-cycle frequency travels 90 degrees in what length of time?

a) 1/60 sec.

b) 1/90 sec.

c) 1/240 sec.

d) 1/3 sec.

42. Grounding the metallic cover of flexible metal conduit and armored cable is for protection against ______.

a) open field shorts

b) shock or injury

c) lightning

d) change of frequency

43. The current through a dc arc lamp can be controlled by a ______.

a) condenser

b) inductor

c) capacitor

d) shunt

44. A(n) _____ overcurrent device is not designated to interrupt short circuit current.
   a) thermal cutout
   b) Type S fuse
   c) inverse-time breaker
   d) dual-element fuse
45. What are incandescent light bulbs filled with?
   a) Freon
   b) Vacuum
   c) Air
   d) Gas
46. A substance that would be good as an electrical insulation is which of the following?
   a) Lead
   b) Iron
   c) Carbon
   d) Oil
47. A galvanometer is used to indicate _____.
   a) wattage
   b) current
   c) voltage
   d) resistance
48. To check the power factor in an electrical load, if a power factor meter is not available, you would use a(n) _____.
   a) ammeter and voltmeter
   b) kilowatt-hour meter
   c) ammeter, wattmeter, and voltmeter
   d) wattmeter
49. Impedance is measured in _____.
   a) henrys
   b) farads
   c) coulombs
   d) ohms
50. Which of the following would have the least amount of resistance?
   a) Resistor
   b) Semiconductor
   c) Conductor
   d) Insulator

Exam Number 6

Time Limit: 1 Hour

1. In a three-phase, 4-wire, wye system having 277 volts for lighting, the three-phase motors would be operating on a voltage of _____.
   a) 190 volts
   b) 277 volts
   c) 480 volts
   d) 208 volts
2. The least important reason for using a steel measuring tape around electrical equipment is _____.
   a) magnetic effect
   b) short circuit hazard
   c) danger of entanglement in rotating machinery
   d) shock hazard
3. The current through a dc arc lamp can be controlled by a _____.
   a) rheostat
   b) transformer
   c) half-wave rectifier
   d) full-wave rectifier
4. What winding of a current transformer will carry more current?
   a) Interwinding
   b) Secondary
   c) Phase
   d) Primary
5. For discharge lighting lamps to operate properly, they would require _____.
   a) some sort of ballast
   b) a resistive starter
   c) a cool ambient temperature
   d) a starter
6. Which of these is not a lamp base?
   a) Bayonet
   b) Admedium
   c) Optical
   d) Disk
7. Inductive reactance is measured in _____.
   a) ohms
   b) coulombs
   c) henrys
   d) farads

8. A motor with a wide speed range is a(n) _____ motor.
   a) synchronous
   b) induction
   c) dc
   d) ac
9. The term pothead as used in the electrical trade means _____.
   a) cable terminal
   b) protective device
   c) current-limiting fuse
   d) insulator on a line pole
10. Solders used for the connection of electrical conductors are alloys of _____.
   a) zinc and tin
   b) tin and lead
   c) copper, zinc, and tin
   d) zinc and lead
11. Which of the following is not the force that moves electrons?
   a) Current
   b) Potential
   c) Voltage
   d) EMF
12. A motor that operates the same in ac or dc is called a(n) _____ motor.
   a) squirrel cage
   b) induction
   c) universal
   d) synchronous
13. Magnetic lines generated from the same source cannot cross each other.
   a) True
   b) False
14. Two 20-ohm resistors in series across 240 volts will draw _____ from the line.
   a) 6 amperes
   b) 12 amperes
   c) 24 amperes
   d) 3 amperes
15. Which one of the following is considered the best insulator between live parts?
   a) Mica
   b) Carbon
   c) Porcelain
   d) Paper

16. A transformer that steps down 4800 volts to 240 volts can deliver 500 amperes at full load. The primary current would be _____ amperes.
   a) 10,000
   b) 250
   c) 25
   d) 20
17. Which of these cannot be connected to a branch circuit?
   a) Delta
   b) Auto
   c) Isolation
   d) High leg
18. Since fuses are rated by amperes and voltage, a fuse will work on _____.
   I. ac
   II. dc
   III. any voltage level
   a) I only
   b) II only
   c) III only
   d) either I or II
19. An ac voltage that is usually called 120 volts has a maximum value of approximately _____.
   a) 240 volts
   b) 208 volts
   c) 168 volts
   d) 120 volts
20. Luminaires hung by chains should be wired so that the _____.
   a) wires support the light
   b) chain is not grounded
   c) light is insulated from the chain
   d) wires do not support the light
21. In an incandescent lamp, the resistance of the filament is _____.
   a) lowest when the bulb has been on for more than 3 hours
   b) usually the same at all times
   c) highest when the bulb is on
   d) highest when the bulb is off
22. When relay contacts make and break frequently, a condenser is used across the relay contacts to _____.
   a) increase the efficiency
   b) delay the coil energizing
   c) energize the relay quicker
   d) reduce pitting of the contacts

23. Wound rotor and squirrel cage are two types of _____ motors.
   a) induction
   b) single-phase
   c) synchronous
   d) three-phase
24. The purpose of locknuts in making electrical connections on studs is to _____.
   a) avoid having to torque the studs
   b) connect multiple conductors on the stud
   c) make connections tamper proof
   d) prevent the connection from loosening under vibration
25. With respect to the safety value of insulation on electrical tools, it can be correctly stated that _____.
   a) the insulation should not be used as the only protective measure
   b) insulation is really not needed
   c) the insulation provides very little protection
   d) the insulation provides complete safety to the user
26. Voltage drop in a conductor is _____.
   a) the conductor resistance times the voltage
   b) a percentage of the applied voltage
   c) part of the load voltage
   d) dependant on the type of insulation
27. If the phase voltages are 120 degrees apart, the line voltages will be _____ degrees apart.
   a) 40
   b) 60
   c) 90
   d) 120
28. Three conductors (phase A, B, and neutral) running from a busway to a load center should not be installed in three separate raceways _____.
   a) to avoid using a gutter
   b) to have room for future fill
   c) because of inductance
   d) because it's easier to pull the conductors
29. The reason for rigid steel conduit to be galvanized or enameled is to _____.
   a) make painting easier
   b) make it easier to thread
   c) prevent corrosion
   d) provide better grounding
30. When three equal resistors are connected in parallel the total resistance is _____.
   a) one half of the sum of the three resistors
   b) less than any one resistor
   c) the sum of the three resistors
   d) one third of the sum of the three resistors

31. The current used for charging storage batteries is _____.
   a) alternating
   b) direct
   c) negative
   d) positive
32. Materials containing numerous free electrons are _____.
   a) good conductors
   b) copper conductors
   c) aluminum conductors
   d) good insulators
33. Electrical current is measured in terms of _____.
   a) watts
   b) electrons passing a point per second
   c) ohms
   d) electrical pressure
34. "The sum of the voltage drop around a circuit is equal to the source voltage" is known as _____.
   a) Watt's law
   b) Kirchoff's law
   c) Tesla's theory
   d) Ohm's law
35. Which of the following instruments would you use to indicate the phase relationship between the voltage and the current of an ac circuit?
   a) Growler
   b) Kilowatt hour meter
   c) Power factor meter
   d) Voltmeter
36. Voltaic reaction is _____.
   a) solar
   b) chemical
   c) both of these
   d) none of these
37. Lubrication would never be applied to a _____.
   a) controller drum
   b) knife switch
   c) bearing
   d) commutator
38. A motor works in the principles of _____.
   a) mechanical force
   b) residual force
   c) thermal action
   d) magnetism

39. The lead covering on conductors is used _____.
   a) for moisture proofing
   b) for easier soldering
   c) for grounding
   d) to prevent vibration
40. Which of the meters in the drawing below is/are hooked up correctly?
   a) 1 only
   b) 2 only
   c) 3 only
   d) 1, 2, and 3

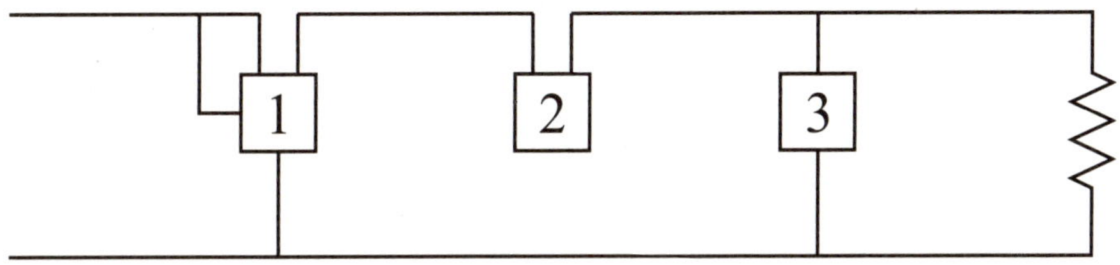

FIGURE 16-5

41. The heating of two different metals will cause _____.
   a) electron flow
   b) corrosion
   c) fusion
   d) galvanic action
42. Galvanized rigid conduit is made of _____.
   a) aluminum
   b) zinc
   c) iron
   d) nickel
43. The common fuse depends on the principle that the _____.
   a) current flow develops heat
   b) voltage develops heat
   c) over voltage will expand the link
   d) increase of resistance will occur
44. Ohm's law is _____.
   a) an equation for determining power
   b) the relationship among voltage, current, and power
   c) the relationship among voltage, current, and resistance
   d) a measurement of the IR losses
45. Piezoelectricity is caused by crystals or binding _____.
   a) static
   b) pressure
   c) heat
   d) chemical

46. Frequency is not a problem in most electrical work because _____.
   a) it never affects circuit values
   b) it concerns only dc
   c) it varies directly with voltage
   d) it is a constant value throughout the U.S. for distribution systems

47. A dc voltmeter may also be used to measure _____ with a dc voltage source.
   a) ohms
   b) power
   c) frequency
   d) amperes

48. What is the covering on a driven ground rod electrode?
   a) Zinc
   b) Enamel
   c) Aluminum
   d) Lead

49. When insulating a splice, _____.
   I. plastic tape has no dielectric strength
   II. a single layer of plastic tape is sufficient for 600 volts
   III. two layers of half lap tape should be used for mechanical protection
   a) I only
   b) II only
   c) III only
   d) I, II, and III

50. What is the voltage at the load in the drawing below?
   a) 115 volts
   b) 112.5 volts
   c) 110 volts
   d) 100 volts

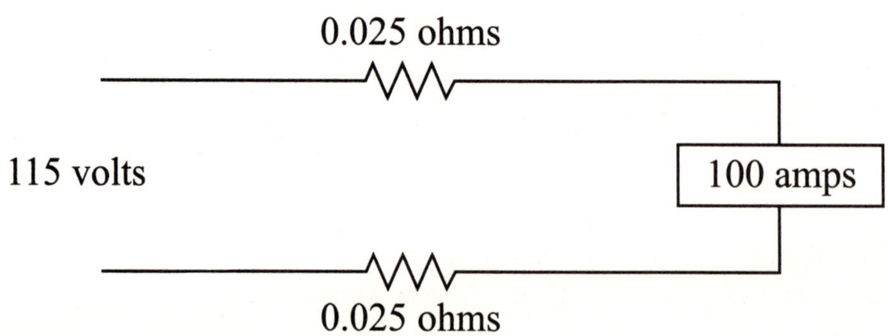

FIGURE 16-6

Exam Number 7

Time Limit: 1 hour

1. The best way to cut rigid metal conduit is to _____.
   a) use a pipe cutter
   b) use a hacksaw
   c) use a hacksaw and ream the ends
   d) use a cutting torch and ream the ends
2. What is the most common 3-phase, 480/208-volt connected transformer?
   a) Delta-wye
   b) Wye-delta
   c) Delta-delta
   d) Wye-wye
3. Sixty-cycle frequency travels 360 degrees in what length of time?
   a) $\frac{1}{60}$ sec.
   b) $\frac{1}{120}$ sec.
   c) $\frac{1}{180}$ sec.
   d) $\frac{1}{240}$ sec.
4. Resistance can be found directly from wattage and voltage measurements by the equation _____.
   a) $R = \frac{W}{V}$
   b) $R = \frac{W^2}{V}$
   c) $R = \frac{V^2}{W}$
   d) $R = \frac{V}{W}$
5. A high spot temperature in a corroded electrical connection is caused by _____.
   a) an increase in the voltage drop across the connection
   b) a decrease in the flow of current through the connection
   c) a decrease in the effective resistance of the connection
   d) a decrease in the voltage drop across the connection
6. The term *anode* refers to _____.
   a) inductor
   b) rectifier
   c) cathode
   d) capacitor
7. Although electrical tape is rated at 600 volts for one layer, a second layer should be used for mechanical protection.
   a) True
   b) False

8. A hickey is _____.
   a) part of a conduit
   b) part of a wall box
   c) a tool used to bend small sizes of rigid conduit
   d) not used in the electrical trade
9. A transformer has a voltage ratio of 10 to 1. If there are 6000 turns on the primary winding, there would be _____ turns on the secondary winding.
   a) 1000
   b) 600
   c) 60
   d) 60,000
10. A dual voltage motor will run more efficiently _____.
    a) at the higher voltage
    b) at the lower voltage
    c) the same at either voltage
    d) at 10% lower than the higher voltage
11. A box contains one grounded conductor and three ungrounded conductors. From one of the ungrounded conductors to the grounded conductor, 208 volts is measured. The other two ungrounded conductors measure 120 volts to the grounded conductors. The system is _____.
    a) wye, 4-wire
    b) wye, 3-wire
    c) open delta, 4-wire
    d) delta
12. One identifying feature of a squirrel-cage induction motor is that it has no _____.
    a) commutator or slip rings
    b) iron core in the rotating part
    c) windings on the stationary part
    d) air gap
13. The lubricant used for a motor sleeve bearing would be _____.
    a) oil
    b) grease
    c) graphite
    d) synthetic lubricant
14. What is the maximum size of rigid metal conduit that can be bent with a hand-held hickey bender?
    a) 3/4″
    b) 1″
    c) 1¼″
    d) 1½″

15. What is the power of two 0.62-ohm resistors that are in series with a voltage of 120 volts?
   a) 148.8 watts
   b) 11,612.9 watts
   c) 23,225.8 watts
   d) 46,451.6 watts
16. Which of the following is best to fight electrical fires?
   a) Soda-acid fire extinguisher
   b) Foam fire extinguisher
   c) CO_2 fire extinguisher
   d) Fine spray of water
17. A luminaire that can be controlled from five central points would require _____.
   a) five 4-ways
   b) two 2-ways and three 4-ways
   c) three 3-ways and two 4-ways
   d) two 3-ways and three 4-ways
18. Which of the following is the best conductor of electricity?
   a) Carbon
   b) Aluminum
   c) Iron
   d) Tungsten
19. A good insulator for very high temperature is _____.
   a) mica
   b) bakelite
   c) rubber
   d) plastic
20. When operated on a voltage 10% higher than nameplate rating, an appliance will _____.
   I. have a shorter life
   II. draw a higher current
   III. use more power
   a) I only
   b) II only
   c) III only
   d) I, II, and III
21. Which of these is not an incandescent lampholder?
   a) Mogul
   b) Standard
   c) Intermediate
   d) Medium

22. When referring to an 8-32 machine bolt, the number 8 refers to the _____.

a) length of bolt
b) diameter
c) strength
d) threads per inch

23. To determine directly whether all finished wire installations possess resistance between conductors and between conductors and ground, use _____.

a) clamps
b) set screws
c) a megger
d) shields

24. In a residence, receptacle outlets shall be installed so that no point along the floor line in any wall space is more than _____ feet between receptacles in that space.

a) 6
b) 6½
c) 10
d) 12

25. Mandatory *Code* rules are identified by the use of what word?

a) Should
b) Must
c) Could
d) Shall

26. If secondary voltage is lower than primary voltage, secondary current is _____ primary current.

a) lower than
b) larger than
c) equal to
d) variable to

27. Covered conductor ampacity is _____ a TW insulated conductor of the same size.

a) greater than
b) less than
c) equal to
d) exactly half of

28. What type of meter is represented by the drawing below?

a) Wattmeter
b) Ammeter
c) Voltmeter
d) Ohmmeter

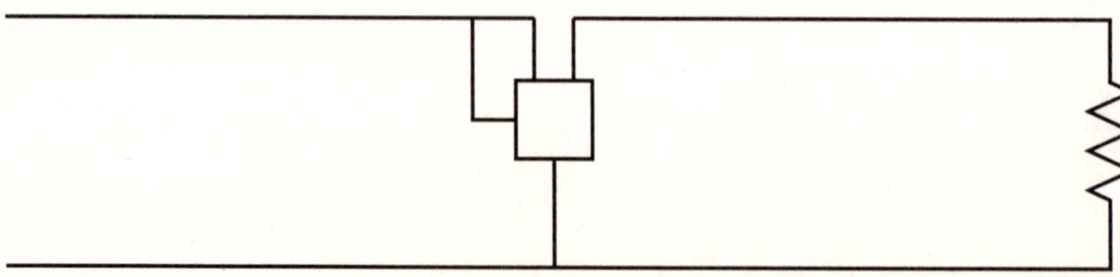

FIGURE 16-7

29. ALS stands for _____.
    a) American Line Switch
    b) American Luminaire Society
    c) Aluminum Sheathed Cable
    d) Aluminum Lighting Service
30. When capacitive reactance is in a circuit for both ac and dc, power can be determined by $W = V \times I$.
    a) True
    b) False
31. How are batteries charged?
    a) AC source
    b) Rectifier
    c) DC motor
    d) DC generator driven by a dc motor
32. The resistance of a copper conductor is _____.
    a) inversely proportional to its cross-sectional area
    b) directly proportional to its diameter
    c) inversely proportional to its length
    d) directly proportional to the square of its diameter
33. To fasten a box to a terra cotta wall you would use _____.
    a) expansion bolts
    b) rawl plugs
    c) toggle bolts
    d) lag bolts
34. In a series circuit with four unknown resistive loads, it is certain that the _____.
    a) current flowing through each load will be the same
    b) total current in the circuit is the sum of all four loads
    c) voltage drop across each load will be the same
    d) wattage consumed by each load will be the same
35. The main reason that water should not be used to extinguish fires involving electrical equipment is that water _____.
    a) will ruin the electrical equipment
    b) may short out the equipment
    c) may transmit shock to the user
    d) is ineffective in extinguishing electrical fires
36. Batteries supply _____ current.
    a) positive
    b) negative
    c) alternating
    d) direct

37. Polarity wiring is necessary to avoid _____.
   a) ground faults
   b) short circuits
   c) using color codes
   d) using ground-fault current interrupters
38. When using a 12-2 AWG with ground nonmetallic sheathed cable, the ground will carry current under normal operation.
   a) True
   b) False
39. What is the power of a circuit with 15 ohms of resistance that draws 20 amperes?
   a) 300 watts
   b) 3000 watts
   c) 4500 watts
   d) 6000 watts
40. Electron flow produced by means of applying a pressure to a material is called _____.
   a) piezoelectricity
   b) thermoelectricity
   c) photo conduction
   d) electrochemistry
41. The resistance of a circuit may vary due to _____.
   a) a change in voltage
   b) a change in current
   c) induction
   d) a loose connection
42. A stop switch is wired _____ in a motor control circuit.
   a) series-parallel
   b) parallel
   c) series
   d) none of these
43. Although silver is a better conductor than either copper or aluminum, it is seldom used in the manufacturing of electric conductors or bus. The reason for this is _____.
   a) it's difficult to terminate
   b) cost
   c) color
   d) tensile strength
44. A _____ is used to test the electrolyte of a battery.
   a) manometer
   b) hydrometer
   c) growler
   d) voltmeter

45. A conduit coupling is sometimes tightened by using a strap wrench rather than a slip-joint wrench; the strap wrench is used when it is important to avoid _____.

a) damaging the outside finish
b) crushing the conduit
c) bending the conduit
d) stripping the threads

46. The maximum size fuse to be used in a branch circuit containing no motors depends on the _____.

a) wire size
b) voltage
c) switch size
d) load

47. Connecting batteries in series will give greater _____.

a) amp-hours
b) voltage
c) amperage
d) all of these

48. The purpose of motor interpoles is to _____.

a) reduce armature current
b) maintain the motor speed constant
c) cancel the magnetic fields generated around the armature coils
d) cancel a portion of the magnetic field generated by the field winding

49. What is the maximum number of bonding conductors permitted under a bronze clamp that is bolted to a metal water pipe? The listing of the clamp does not indicate the number of conductors permitted.

a) 1
b) 2
c) 3
d) 4

50. How does a power supply convert dc voltage to ac and then back to dc voltage?

a) Resistors and capacitors
b) Transformers and transistors
c) Filters and resistors
d) Rectifiers and filters

PART III *NEC* Exam Questions

Exam Number 1

Time Limit: 2 Hours

1. A conductor used to connect equipment or the grounded circuit of a wiring system to a grounding electrode or electrodes is called a _____.
   a) grounded conductor
   b) ground
   c) common main
   d) grounding conductor
2. Each switchboard or panelboard, if used as service equipment, shall be provided with a _____.
   a) main bonding jumper
   b) battery charging panel
   c) power circuit
   d) 4-wire delta connected system
3. Fixed stage switchboards shall _____.
   I. be of the dead front type
   II. be recessed into the wall
   III. be provided with a metal hood
   a) I or II
   b) II or III
   c) I or III
   d) I, II, or III
4. Luminaires (fixtures) shall be wired with conductors having insulation suitable for the environmental conditions and _____ to which the conductors will be subjected.
   I. current
   II. voltage
   III. temperature
   a) I only
   b) II only
   c) III only
   d) I, II, and III
5. The minimum size conductors (for elevator traveling cables) operating control and signaling circuits are _____.
   a) 24 AWG
   b) 20 AWG
   c) 14 AWG
   d) 12 AWG
6. Receptacles rated 30 amperes that provide shore power for boats shall be _____.
   a) tamperproof
   b) GFCI
   c) of the locking and grounding type
   d) able to be locked in the off position

7. Signs and metal equipment of outline lighting systems shall have copper bonding conductors not smaller than _____.
   a) 14 AWG
   b) 12 AWG
   c) 10 AWG
   d) 8 AWG
8. What is the cross-sectional area of 2″ electrical metallic tubing?
   a) 1.342″
   b) 2.016″
   c) 3.205″
   d) 3.356″
9. Type MC cable shall not be used _____.
   a) for control circuits
   b) for signal circuits
   c) for outdoor circuits
   d) where exposed to destructive corrosive conditions, unless the metallic sheath is suitable for the conditions or is protected by material suitable for the conditions
10. Each _____ shall be controlled by an externally operable switch or circuit breaker that will open all ungrounded conductors.
    I. outline lighting system
    II. fixed sign
    III. portable sign
    a) I only
    b) II only
    c) I and II
    d) I, II, and III
11. Storable swimming pools are those that are constructed on or above the ground and are capable of holding water to a maximum depth of 42 inches and have a maximum dimension of _____.
    a) 5′
    b) 10′
    c) 18′
    d) the dimension does not matter
12. The proper switching sequence for series capacitors, over 600 volts, shall be ensured by use of _____.
    I. interlocks
    II. a switching procedure prominently displayed at the switching location
    III. mechanically sequenced isolating and bypass switches
    a) I or III
    b) II or III
    c) II only
    d) I, II, or III

13. The high leg of a three-phase, 4-wire, delta-connected system shall be identified _____.
    I. where connecting to a three-phase motor
    II. when terminating in an equipment panel
    III. when the grounded conductor is present
    a) II only
    b) III only
    c) II or III
    d) I, II, or III
14. Receptacles rated 20 amperes or less and designed for the direct connection of aluminum conductors shall be marked _____.
    a) CO
    b) AL
    c) CO/ALR
    d) none of these
15. All devices, excluding panelboards, provided with terminals for the attachment of conductors and intended for connection to more than one side of the circuit shall have _____ properly marked for identification.
    a) terminals
    b) sides
    c) conductors
    d) none of these
16. The grounding conductor of a coaxial cable shall be _____.
    a) solid copper
    b) stranded copper
    c) a corrosion-resistant conductor
    d) any of these
17. With open wire on insulators _____.
    I. surface-type snap switches do not need boxes
    II. conductor supports shall be within 6″ from a tap
    III. surface-type snap switches shall be mounted on insulating material
    a) I, II, and III
    b) I and III only
    c) II only
    d) I only
18. Electric ranges and clothes dryers used in mobile homes shall be _____.
    a) installed with the neutral bonded to the grounding conductor
    b) installed with a grounding conductor run in parallel with the neutral
    c) installed with the equipment enclosure ungrounded
    d) installed with the grounded circuit conductor insulated from the grounding conductor and equipment enclosures

19. A(n) _____ is a tank or vat in which electrochemical reactions are caused by applying electrical energy for the purpose of refining or producing usable materials.
   a) electrolytic cell
   b) duct heater
   c) heating boiler
   d) radiant heater
20. The minimum insulation level for neutral conductors of solidly grounded systems shall be _____ volts.
   a) 1000
   b) 600
   c) 480
   d) 250
21. Unguarded live parts above working spaces for a line voltage of 2300 volts shall be maintained at elevations not less than _____.
   a) 6′ 8″
   b) 8′ 6″
   c) 9′
   d) 9′ 6″
22. The point of attachment of service-drop conductors to a building shall not be less than _____ feet above finish grade.
   a) 8
   b) 10
   c) 12
   d) 14
23. In locations where walls are frequently washed, conduit should be mounted with a _____ air space between the wall and conduit.
   a) 1⁄8″
   b) 1⁄4″
   c) 3⁄8″
   d) 1⁄2″
24. In using multiple grounding electrodes, they shall be not less than _____ apart.
   a) 6 feet
   b) 10 feet
   c) 12 feet
   d) no separation is required
25. In calculating the load of a fluorescent luminaire (fixture), the calculated load shall be based on the _____ of the fixtures.
   a) watts of the lamp
   b) wattage of the ballast
   c) ampere rating of the ballast
   d) all of these

26. Plug fuses of the Edison-base type may be used in new installations.

a) True

b) False

27. Ferrous metal _____ shall be suitably protected against corrosion inside and outside (except threads at joints) by a coating of approved corrosion-resistant material.

I. raceways

II. boxes

III. support hardware

a) I only

b) II only

c) I or II

d) I, II, or III

28. Track lighting shall not be located within a zone measured _____ horizontally and _____ vertically from the top of the bathtub rim.

a) 8′; 3′

b) 5′; 8′

c) 3′; 8′

d) 6′; 8′

29. A multiwire branch circuit may supply _____.

a) 115/230 volts to only one utilization equipment

b) 115/230 volts where all ungrounded conductors are opened simultaneously

c) a and b

d) none of these

30. A single electrode consisting of a rod, pipe, or plate that does not have a resistance to ground of _____ shall be augmented by one additional electrode.

a) 25 ohms or less

b) 10 ohms or less

c) more than 25 ohms

d) that of a 3⁄4″ conduit

31. A 1″ EMT raceway (with over 2 conductors) is already filled with conductors totaling 20% of the raceway. What is the cross-sectional area of that 1″ raceway?

a) 1.049″

b) 0.173″

c) 0.346″

d) 0.864″

32. The connection between the grounded circuit conductor and the equipment grounding conductor at the service is called a(n) _____.

a) circuit bonding jumper

b) equipment bonding jumper

c) main bonding jumper

d) appliance bonding jumper

33. Plugmold shall not be permitted _____.

I. in hoistways

II. to be run within dry partitions

III. to be run through dry partitions

a) I only

b) II only

c) I and II

d) I, II, and III

34. Overhead conductors for festoon lighting shall not be smaller than _____.

a) 14 AWG

b) 12 AWG

c) 10 AWG

d) 8 AWG

35. The minimum spacing between bare metal parts of a 120-volt system, of opposite polarity, where mounted on the same surface is _____ inch(es).

a) ½

b) ¾

c) 1

d) 1¼

36. The length of a "panel thread" should not be less than _____.

a) the length of a standard pipe thread

b) twice the length of the coupling

c) the thickness of the panel plus the thickness of the locknut and bushing

d) the thickness of the panel plus the thickness of the locknut and bushing plus one half the length of the coupling

37. Hanging luminaires (fixtures) where located directly above any part of the bathtub shall be so installed that the fixture is not less than _____ feet above the top of the bathtub.

a) 2.5

b) 3

c) 8

d) 10

38. Where the number of current-carrying conductors contained in a sheet metal auxiliary gutter is _____ or less, the derating factors in 310.15(B)(2)(a) shall not apply.

a) 10

b) 20

c) 30

d) 40

39. How many conductors shall be reduced, in a junction box, for the following items: one receptacle, two internal cable clamps, and two grounding conductors?
    a) 3
    b) 4
    c) 5
    d) 6
40. Capacitors containing more than _____ gallons of flammable liquid shall be enclosed in vaults.
    a) 3
    b) 5
    c) 7
    d) 10
41. Each vented cell shall be equipped with a(n) _____ that is designed to prevent destruction of the cell.
    a) electrolyte
    b) flame arrester
    c) hydrometer
    d) gas arrester
42. For what kind of lighting loads does the *Code* state there shall be no reduction in the size of the neutral conductor?
    a) Motels
    b) Dwelling units
    c) Hospitals
    d) Nonlinear loads
43. In agricultural buildings, all cables shall be secured within _____ inches of a box.
    a) 6
    b) 8
    c) 10
    d) 12
44. Multispeed motors shall be marked with the code letter designating the locked-rotor kVA per hp for the highest speed at which the motor _____.
    a) needs to be rated
    b) can be stalled
    c) can be started
    d) can run safely

45. The minimum size service for a mobile home is _____ amperes.

a) 60

b) 100

c) 125

d) 200

46. A ceiling-suspended (paddle) fan outlet box shall be listed and shall not support a fan that weighs more than _____.

a) 30 pounds

b) 35 pounds

c) 50 pounds

d) 70 pounds

47. What is the maximum recommended percentage of voltage drop for a branch circuit?

a) 2 percent

b) 3 percent

c) 4 percent

d) 5 percent

48. What is the minimum width of working space in front of a panel that is 16 inches wide?

a) 16 inches

b) 24 inches

c) 30 inches

d) 36 inches

49. How many splices are permitted in the grounding electrode conductor with bolted connectors?

a) 0

b) 1

c) 2

d) 3

50. Class 2 circuit cables supplying wheelchair lift door interlocks shall be _____.

a) flame-retardant type

b) not less than 200° C

c) THN and rated 170° F

d) minimum size 14 AWG

Exam Number 2

Time Limit: 2 Hours

1. In communication circuits the bonding together of all separate electrodes ______.
   a) shall be permitted with a minimum size jumper 8-AWG copper
   b) shall be permitted with a minimum size jumper 6-AWG copper
   c) shall be permitted with a minimum size jumper 4-AWG copper
   d) shall not be permitted
2. Pull-type canopy switches shall not be located more than ______ inches from the center of the canopy.
   a) 1½
   b) 2
   c) 3
   d) 3½
3. Signs and outline lighting systems with incandescent lamp holders shall be marked to indicate the maximum allowable wattage of lamps. The markings shall be permanently installed, in letters at least ______ inch(es) high, and shall be located where visible during relamping.
   a) ¼
   b) ½
   c) 1
   d) 6
4. A transverse metal raceway for electrical conductors, providing access to predetermined cells of a precast cellular concrete floor, thereby permitting the installation of electric conductors from a distribution center to the floor cells, is called a(n) ______.
   a) cellular raceway
   b) header
   c) concrete raceway
   d) under-floor raceway
5. Tap conductors supplying household electric cooking equipment from a 50-ampere branch circuit shall have an ampacity of not less than ______ amperes.
   a) 20
   b) 30
   c) 40
   d) 50
6. A disconnecting means serving a hermetic refrigerant motor-compressor shall be selected on the basis of the nameplate rated-load current or branch-circuit selection current. The ampere rating shall be at least ______ of the nameplate rated-load current or branch-circuit selection current, whichever is greater.
   a) 80%
   b) 100%
   c) 115%
   d) 125%

7. Electrical installations in hollow spaces, vertical shafts, and ventilation or air-handling ducts shall be made so that the possible spread of fire or products of combustion will not be _____.
   a) exposed
   b) under rated
   c) substantially increased
   d) allowed
8. Tap devices used in Type FC assemblies shall be rated at not less than _____ or more than 300 volts to ground.
   a) 15 amperes
   b) 20 amperes
   c) 30 amperes
   d) 40 amperes
9. A residence has a front entrance on the south side of the house along with an attached garage that has an 8′ wide door, and there is also a back entrance door on the north side of the house. How many wall-switch-controlled lighting outlets are required for these outdoor entrances?
   a) One
   b) Two
   c) Three
   d) None of these
10. An attachment plug and receptacle shall be permitted to serve as the disconnecting means for a single-phase room air conditioner rated 250 volts or less if _____.
    a) it is protected by an inverse time circuit breaker
    b) it is rated 30 amps or less
    c) the power supply cord is 6′ or less
    d) the manual controls are readily accessible and located within 6′ of the floor
11. Secondary circuits of instrument transformers and instrument cases shall be grounded by using _____.
    I. copper
    II. aluminum
    III. metal
    a) I only
    b) II only
    c) I or II
    d) I, II, or III
12. Electric-discharge luminaires (fixtures) of less than 1000 volts that have _____ shall be installed so that there will not be any contact with combustible material.
    a) transformers
    b) exposed ballasts
    c) exposed ballasts or transformers
    d) none of these

13. Which of the following appliances, other than built-in, are considered portable in recreational vehicles?

    I. Refrigerators

    II. Gas ranges

    III. Clothes washers

    a) I only

    b) I, II, and III

    c) I and III only

    d) None of these

14. How would you connect the grounded circuit of a wiring system to the grounding electrode?

    a) Bonding jumper

    b) Grounded conductor

    c) Grounding conductor

    d) Bonding jumper main

15. An isolating switch is one that is _____.

    a) primarily used with an isolation transformer

    b) required to have a padlock

    c) intended for isolating an electrical circuit from its source of power

    d) used only for heavy motor overloads

16. Luminaires (fixtures) shall be constructed or installed so that adjacent combustible material will not be subjected to temperature in excess of _____ °C.

    a) 75

    b) 90

    c) 150

    d) 194

17. Bored holes in wood members for cable- or raceway-type wiring shall be bored so that the edge of the hole is not less than _____ inch(es) from the nearest edge of the wood member.

    a) 1

    b) 1⅛

    c) 1¼

    d) 1½

18. Each patient bed location in general care areas shall be provided with a minimum of _____ receptacles.

    a) six

    b) four

    c) three

    d) two

19. The minimum size conductor permitted in parallel for elevator lighting is _____ copper provided the ampacity is equivalent to at least that of 14-AWG copper.
   a) 20-AWG
   b) 16-AWG
   c) 18-AWG
   d) 24-AWG
20. The source of power for electrically operated pipe organs shall be a transformer-type rectifier, the dc potential of which shall not exceed _____ dc.
   a) 6 volts
   b) 12 volts
   c) 24 volts
   d) 30 volts
21. Field bends or modifications shall be so made that the _____ of the cable tray system and support for the cables is maintained.
   a) strength
   b) temperature
   c) electrical continuity
   d) rigidity
22. The spacing between wood or other combustible materials and an incandescent or HID lamp or lampholder shall not be less than _____ inches in electric signs.
   a) 2
   b) 4
   c) 6
   d) 8
23. A transformer equipped with coordinated thermal overload protection by the manufacturer and arranged to interrupt the primary current shall be permitted to have primary overcurrent protection rated or set at a current value that is not more than _____ times the rated current of the transformer for transformers having not more than 6 percent impedance.
   a) 1.25
   b) 2
   c) 2.5
   d) 6
24. Where internal combustion engines are used as the prime mover of a generator to supply emergency power, how much on-premises fuel supply is required?
   a) 2 hours' full-demand operation of the system
   b) 3 hours' full-demand operation of the system
   c) ½ hour's full-demand operation of the system
   d) 1 hour's full-demand operation of the system
25. Fire alarm circuits shall be grounded; the only exception is direct-current fire alarm circuits having a maximum current of _____ ampere(s) as specified in Article 760, Part III.
   a) 0.300
   b) 0.030
   c) 2.34
   d) 1.5

26. Elevators shall have a single means for disconnecting all ungrounded main power supply conductors for each unit.
    a) This does not include the emergency power service.
    b) No elevators are to operate on emergency power systems.
    c) This does include the emergency power service.
    d) This does not include the emergency power service if the system is automatic.
27. Conductors to the hoistway door interlocks from the hoistway riser shall be _____.
    a) rated 75° C minimum
    b) flame retardant
    c) rated 200° C
    d) b and c
28. In judging equipment, considerations such as _____ shall be evaluated.
    I. mechanical strength
    II. electrical insulation
    III. demand factor
    a) I only
    b) I and II only
    c) II only
    d) I, II, and III
29. For industrial establishments only, omission of overcurrent protection shall be permitted at points where busways are reduced in ampacity, provided that the length of the busway having the smaller ampacity does not exceed _____ and has an ampacity at least equal to one third the rating or setting of the overcurrent device next back on the line, and provided that such busway is free from contact with combustible material.
    a) 15 feet
    b) 50 feet
    c) 100 feet
    d) 200 feet
30. Auxiliary equipment for electric-discharge lamps shall be enclosed in noncombustible cases and treated as sources of heat.
    a) True
    b) False
31. Where the grounding electrode conductor is connected to a concrete-encased electrode as permitted in 250.52(A)(3), that portion of the conductor that is the sole connection to the grounding electrode shall not be required to be larger than _____ copper wire.
    a) 8-AWG
    b) 6-AWG
    c) 4-AWG
    d) 2-AWG
32. The terminals of an electric-discharge lamp shall be considered as energized where any lamp terminal is connected to a circuit of over _____ volts.
    a) 50
    b) 150
    c) 300
    d) 600

33. Electrically heated smoothing irons shall not be equipped with an identified temperature-limiting means.

a) True

b) False

34. Where the service-entrance phase conductors are larger than 1100 kcmil copper, the bonding jumper shall have an area that is not less than _____ percent of the area of the largest phase conductor.

a) 25

b) 12½

c) 80

d) 112

35. Receptacles located within the patient care areas of pediatric wards, rooms, or areas shall be _____.

a) GFCI protected

b) specification grade

c) listed tamper resistant

d) isolated

36. Bonding conductors of outline lighting systems shall be copper and not smaller than _____.

a) 14 AWG

b) 12 AWG

c) 10 AWG

d) 8 AWG

37. In patient care areas of hospitals, the equipment grounding terminal buses of the normal and essential branch-circuit panelboards serving the same individual patient vicinity shall be bonded together with an insulated continuous copper conductor not smaller than _____.

a) 12 AWG

b) 10 AWG

c) 8 AWG

d) 6 AWG

38. Each _____ carrying a number of infrared lampholders shall be considered an appliance.

I. section

II. panel

III. strip

a) I only

b) II only

c) III only

d) I, II, or III

39. Emergency system equipment for sources of power, where located within assembly occupancies greater than 1000 persons or in buildings above _____ feet in height with any of the following occupancy classes—assembly, educational, residential, detention and correctional, business, and mercantile—shall be installed either in spaces fully protected by approved automatic fire suppression systems or in spaces with a 1-hour fire rating.
    a) 23
    b) 50
    c) 75
    d) 100
40. Overcurrent devices shall not be located in the vicinity of easily ignitable material, such as in _____.
    a) clothes closets
    b) bedrooms
    c) garages
    d) kitchens
41. Cable trays shall include fittings or other suitable means for _____.
    I. changes in direction
    II. elevation of runs
    a) I only
    b) II only
    c) I and II
    d) none of these
42. The heating of conductors depends on _____ values, which, with generator field control, are reflected by the nameplate current rating of the motor-generator driving motor rather than by the rating of the elevator motor, which represents actual but short-time and intermittent full-load current values.
    a) voltage
    b) ozone
    c) root-mean-square current
    d) ambient temperature
43. For services exceeding 600 volts, where oil switches or air, oil, vacuum, or sulfur hexafluoride circuit breakers constitute the service disconnecting means, an isolating switch with _____ contacts shall be installed on the supply side of the disconnecting means and all associated service equipment.
    a) knife blade
    b) visible break
    c) double-throw
    d) closed
44. Every electric sign of any type, fixed or portable, shall be _____.
    a) electrically isolated
    b) listed
    c) approved
    d) permanently wired

45. Which of the following is the correct feeder load for a 4.5-kVA clothes dryer?
   a) 4.0 kVA
   b) 4.5 kVA
   c) 5 kVA
   d) None of these

46. Threaded entries into Class I, Division 1 explosionproof equipment shall be made up with at least _____ threads fully engaged.
   a) 2
   b) 3
   c) 4
   d) 5

47. Automatic transfer switches for emergency systems shall be _____.
   a) electrically operated and electrically held
   b) mechanically operated and mechanically held
   c) electrically operated and mechanically held
   d) mechanically operated and electrically held

48. What is the minimum bending radius for a 7⁄8-inch MI cable?
   a) 6½ inches
   b) 7 inches
   c) 8 inches
   d) 8¾ inches

49. Conductors emerging from the ground shall be protected from physical damage up to a point that is _____ above finished grade.
   a) 2 feet
   b) 4 feet
   c) 6 feet
   d) 8 feet

50. Low-voltage plenum cable installed in ducts containing dust shall be installed under what condition?
   a) If the cable contains no splices
   b) If the cable is shielded
   c) If the exhaust air exceeds 2000 CFM
   d) Never

Exam Number 3

Time Limit: 2 Hours

1. Motion picture projectors are _____.
   a) covered under theater and similar locations in the *Code*
   b) not covered in the *Code*
   c) covered in their own section of the *Code*
   d) part of the section in the *Code* on motion picture studios
2. Where the number of conductors in a raceway is seven, the individual ampacity of each conductor shall be reduced _____.
   a) to 70% due to number of conductors
   b) to 80% if they are continuous loads
   c) to both a and b if both conditions exist
   d) Neither a nor b applies if ambient temperature is below 30° C.
3. Loop wiring for underfloor raceways shall not be considered _____.
   I. a splice
   II. a tap
   a) I only
   b) II only
   c) neither I nor II
   d) both I and II
4. When rerouting conduit, it may be necessary to increase the wire size, if the distance is considerably greater, in order to _____.
   a) account for current drop
   b) compensate for voltage drop
   c) increase the mechanical strength
   d) allow for possible resistance drop
5. When changing wires in an existing circuit _____.
   a) it is not permissible to use a larger size wire
   b) it is permissible to use one size smaller wire
   c) the type of insulation is important
   d) no more wires may be pulled in than were originally pulled in
6. How many grounding terminals shall be provided in junction boxes, transformer enclosures, and ground-fault circuit-interrupter enclosures connected to a conduit that extends directly to the forming shell of a swimming pool?
   a) 2
   b) 3
   c) 4
   d) At least one more than the number of conduit entries

7. In the supply circuit of X-ray equipment, a disconnecting means of adequate capacity shall be _____.
   a) at least 50 percent of the input required for the momentary rating
   b) at least 100 percent of the input required for the momentary rating
   c) at least 50 percent of the input required for the long-time rating
   d) at least 125 percent of the input required for the long-time rating
8. Copper-clad aluminum conductors have an ampacity _____.
   a) equal to aluminum
   b) equal to copper
   c) rating of their own
   d) lower than copper but higher than aluminum
9. A secondary tie is a circuit operating at _____ volts, nominal, or less between phases that connects two power sources or power supply points, such as the secondaries of two transformers.
   a) 250
   b) 600
   c) 1000
   d) 5000
10. In a commercial garage, all 125-volt, single-phase, 15- and 20-ampere receptacles shall have ground-fault circuit-interrupter protection for _____.
    a) electrical diagnostic equipment
    b) electrical hand tools
    c) portable lighting equipment
    d) all of these
11. Floor-mounted Type FCC cable shall be covered with _____.
    I. vinyl flooring
    II. carpet squares
    III. commercial carpet
    a) I only
    b) II only
    c) II or III
    d) I, II, or III
12. The wiring of an outdoor sign that is portable or mobile and is readily accessible shall be _____.
    a) provided with factory-installed ground-fault circuit-interrupter protection
    b) protected by a 20-ampere overcurrent protection device
    c) protected by a ground-fault circuit-interrupter
    d) none of the above

13. An overcurrent protective device for a branch circuit over 600 volts shall be capable of detecting and interrupting all values of current that can occur at their location _____.
    I. in excess of their trip setting
    II. within 50% of their melting point
    III. in excess of their melting point
    a) I only
    b) II only
    c) I or II
    d) I or III
14. What is the maximum distance between supports for a straight run of 1¼″ rigid metal conduit made up with threaded couplings?
    a) 10′
    b) 12′
    c) 14′
    d) 16′
15. Boxes shall not be required for _____ when installed with open wiring on insulators.
    I. surface-type snap switch devices
    II. receptacle devices
    III. tap devices
    a) I only
    b) I and II
    c) II and III
    d) I, II, and III
16. Sign and outline lighting system enclosures _____.
    I. shall be constructed of metal
    II. shall be listed
    a) I only
    b) II only
    c) either I or II
    d) neither I nor II
17. _____ may be made in conduit bodies.
    I. Splices
    II. Taps
    a) I only
    b) II only
    c) neither I nor II
    d) either I or II

18. Alternating-current circuits of less than 50 volts shall be grounded under which of the following conditions?
    I. Where supplied by transformers if the transformer supply system is less than 150 volts to ground
    II. Where installed as overhead conductors outside of buildings
    III. Where supplied by transformers if the transformer supply system is grounded
    a) I only
    b) II only
    c) III only
    d) I, II, and III
19. The sum of cross-sectional areas of all contained conductors at any cross section of a metal wireway shall not exceed _____ percent of the interior cross-sectional area of the wireway.
    a) 20
    b) 30
    c) 40
    d) 60
20. The type of disconnecting means for a motor shall be _____.
    I. a listed motor-circuit switch rated in horsepower
    II. a listed molded case circuit breaker
    III. a listed molded case switch
    a) I only
    b) II only
    c) III only
    d) I, II, or III
21. Receptacles mounted on portable generators need not be connected to a grounding electrode.
    a) True
    b) False
22. Motors smaller than a ⅛ hp shall be marked with the rated horsepower.
    a) True
    b) False
23. Voltage shall not exceed 600 volts between conductors on branch circuits supplying only ballasts for electric-discharge lamps in tunnels with a height of not less than _____ feet.
    a) 22
    b) 18
    c) 15
    d) 12
24. General power and lighting in a health care facility _____.
    a) are connected to the alternate source generator
    b) are part of the continuous power system
    c) are part of the equipment system
    d) are not part of the emergency system

25. If completely enclosed in the motor control enclosure, 12-AWG aluminum conductors shall be permitted to be protected with an overcurrent device that has a maximum rating of _____ amperes.
   a) 15
   b) 20
   c) 30
   d) 100

26. The letter suffix shall indicate the following: _____ for two insulated conductors laid parallel within an outer nonmetallic covering.
   a) D
   b) M
   c) C
   d) A

27. A(n) _____ circuit is a circuit in which any spark or thermal effect is incapable of causing ignition of a mixture of flammable or combustible material in air under prescribed test conditions.
   I. sealed
   II. intrinsically safe
   III. low voltage
   a) I only
   b) II only
   c) III only
   d) I, II, or III

28. There shall be an air space of at least _____ between the walls, back, gutter partition if of metal, or door of any cabinet and the nearest exposed current-carrying part of devices mounted within the cabinet where the voltage is 480.
   a) ¼″
   b) ½″
   c) 1″
   d) 1½″

29. A generator set using internal combustion engines for legally required standby systems shall _____.
   I. have a means provided for automatic transfer
   II. have a means provided for automatically starting the prime mover
   III. be provided with at least 2 hours of on-site fuel
   a) III only
   b) I and II
   c) II and III
   d) I, II, and III

30. The approximate square-inch area of a 6-AWG bare conductor is _____.
   a) 0.0726
   b) 0.059
   c) 0.184
   d) 0.027

31. A type of surface, flush, or freestanding raceway designed to hold conductors and receptacles, assembled in the field or at the factory, is called _____.

a) underfloor raceway

b) recessed receptacles

c) multioutlet assembly

d) cellular metal floor raceway

32. Unused cable or raceway openings in boxes shall be effectively closed to afford protection substantially equivalent to _____.

a) a plug at least 1⁄16 inch

b) the wall of the equipment

c) the voltage of the conductors

d) the ampacity of the conductors

33. Type CMX limited-use communications cable shall be listed as being _____.

I. resistant to flame spread

II. suitable for use in dwellings

III. suitable for use in raceway

a) I only

b) II only

c) II and III

d) I, II, and III

34. If festoon lighting exceeds _____ feet the conductors shall be supported by messenger wire.

a) 50

b) 40

c) 12

d) 10

35. The feeder conductor ampacity shall not be less than that of the service conductors where the feeder conductors carry the total load supplied by service conductors with an ampacity of _____ or less.

a) 50 amperes

b) 55 amperes

c) 60 amperes

d) 100 amperes

36. The branch-circuit overcurrent devices in emergency circuits shall be _____.

a) fuses only

b) a slow-blow type

c) of the reset type only

d) accessible to authorized personnel only

37. Electrically heated smoothing irons intended for use in a residence shall be _____.

a) provided with a signal

b) equipped with an approved stand, which is a separate piece of equipment

c) equipped with an identified temperature-limiting means

d) equipped with an approved stand, which is part of the appliance

38. What is the approximate square inch area of a 14-AWG RHH conductor with an outer covering?
   a) 0.0293
   b) 0.163
   c) 0.193
   d) 0.0209

39. Which of these are examples of resistive heaters?
   I. Heating blankets
   II. Heating tape
   III. Heating barrels
   a) I only
   b) II only
   c) I and II
   d) I, II, and III

40. Energized parts of service equipment that are not enclosed shall be installed on a _____ and guarded in accordance with 110.18 and 110.27.
   I. switchboard
   II. panelboard
   III. control board
   a) I only
   b) I or III
   c) II or III
   d) I, II, or III

41. The rating of the surge arrester shall be _____ the maximum continuous phase-to-ground power frequency voltage available at the point of application.
   I. equal to
   II. greater than
   III. less than
   a) I only
   b) I or II
   c) I or III
   d) III only

42. For a feeder supplying household electric ranges and electric dryers, the maximum unbalanced load shall be considered as _____ percent of the load on the ungrounded conductors.
   a) 70
   b) 40
   c) 80
   d) 50

43. Flexible cords shall be secured to the underside of showcases so that _____.
   I. the wiring is not exposed to mechanical damage
   II. the free lead at the end of a group of showcases has a female fitting not extending beyond the case
   III. the separation between cases is not in excess of 2″, or more than 12″ between the first case and the supply receptacle
   a) I and II
   b) I and III
   c) II and III
   d) I, II, and III

44. Because aluminum is not a magnetic metal, there will be _____.
   a) no heating due to hysteresis
   b) no heating due to voltage
   c) no induced currents
   d) none of these

45. Outdoor antennas shall not be attached to which of the following?
   a) Electric service mast
   b) Poles carrying open electric light wires of over 250 volts between conductors
   c) Poles carrying open trolley wires of over 250 volts between conductors
   d) All of the above

46. Derating factors shall not apply to conductors in nipples having a length not exceeding _____ inches.
   a) 24
   b) 30
   c) 36
   d) 48

47. All 120-volt, single-phase, 15- and 20-ampere receptacles in bedrooms of dwellings shall be _____.
   a) GFCI protected
   b) arc-fault protected
   c) listed
   d) GFCI and arc-fault protected

48. What is the minimum headroom of working space about service equipment?
   a) 56 inches
   b) 66 inches
   c) 78 inches
   d) 96 inches

49. Certain wiring methods shall be permitted in ducts used _____.
   a) for environmental air
   b) to transport dust
   c) to transport loose stock
   d) for vapor removal

50. What is the maximum ampere receptacle that can be installed on a 15-ampere branch-circuit?
   a) 15
   b) 20
   c) 25
   d) 30

Exam Number 4

Time Limit: 2 Hours

1. What is the maximum number of 12-AWG conductors allowed in a 4″ square metal box that is 1¼″ deep?
   a) 6
   b) 8
   c) 9
   d) 10
2. A luminaire (fixture) shall not be attached to or suspended from its outlet box.
   a) True
   b) False
3. A hoistway is a _____ in which an elevator or dumbwaiter is designed to operate.
   a) shaftway
   b) hatchway
   c) well hole
   d) all of these
4. Five ranges rated at 12 kW shall be rated at _____ kW minimum demand.
   a) 17
   b) 20
   c) 21
   d) 27
5. AC voltage of the system is not _____.
   a) EMF
   b) effective
   c) average
   d) RMS
6. When necessary, a shallow outlet box not less than _____ deep may be installed.
   a) ½″
   b) ¾″
   c) 1″
   d) 1¼″
7. The branch circuit load for a 6-kW wall-mounted oven shall be _____.
   a) 4.8 kW
   b) 6 kW
   c) 6.4 kW
   d) 8 kW

8. Overhead service drop conductors shall not be installed within an area extending _____ feet horizontally from the inside of the wall of a pool.
   a) 10
   b) 14
   c) 18
   d) 20
9. On hermetic refrigerant motor-compressors, the value of branch-circuit selection current will always be _____ the marked rated-load current.
   I. less than
   II. equal to
   III. greater than
   a) I only
   b) I or II
   c) II or III
   d) III only
10. Where practicable, a separation of at least _____ feet shall be maintained between any coaxial cable and lightning conductors.
   a) 3
   b) 6
   c) 8
   d) 10
11. Electric-discharge lighting systems having an open-circuit voltage exceeding _____ volts shall not be installed in dwelling occupancies unless such equipment is designed so that there will be no exposed live parts when lamps are being inserted, are in place, or are being removed.
   a) 300
   b) 600
   c) 1000
   d) 5000
12. Cord-connected appliances of recreational vehicles, such as _____, shall be grounded by means of an approved cord with equipment grounding conductor and grounding-type attachment plug.
   I. cloths dryers
   II. refrigerators
   III. the electrical system of gas ranges
   a) I only
   b) III only
   c) I and II
   d) I, II, and III
13. What is the maximum number of 18-AWG TFFN conductors that are allowed in ½″ electrical nonmetallic tubing?
   a) 18
   b) 20
   c) 22
   d) 23

14. Where portions of a raceway system are known to be subjected to different temperatures and where condensation is known to be a problem, as in cold-storage areas of buildings, the raceway shall be filled with an approved material to prevent the circulation of _____ air to a _____ section of the raceway.

    a) cold; warmer
    b) warm; colder
    d) both a and b
    d) none of these

15. Which of these is not required to be grounded?

    a) A 2-wire dc system supplying premises wiring
    b) A dc system operating at 50 volts or less between conductors
    c) A dc system operating at 100 volts or less between conductors
    d) A dc system operating at more than 50 volts between conductors

16. Which of these are acceptable methods of preventing pull on flexible cords from being transmitted to joints or terminals?

    I. Tying a knot in the cord
    II. Winding tape on the cord
    III. Fittings designed for the purpose

    a) III only
    b) II and III
    c) I and III
    d) I, II, and III

17. The load for the required branch circuit installed for the supply of an exterior sign shall be computed at a minimum of _____ volt-amperes.

    a) 20
    b) 180
    c) 1200
    d) 1500

18. Steel cable trays shall not be used as equipment grounding conductors for circuits _____.

    a) with ground-fault protection above 600 amperes
    b) without ground-fault protection above 600 amperes
    c) with ground-fault protection below 600 amperes
    d) without ground-fault protection below 600 amperes

19. An oil switch can be installed on a circuit having a rating that _____ or 100 amperes.

    a) exceeds 600 volts
    b) does not exceed 250 volts
    c) does not exceed 600 volts
    d) does not exceed 1000 volts

20. A hermetic refrigerant motor-compressor's branch-circuit short-circuit and ground-fault protective device shall be capable of carrying _____.
   a) the starting current of the motor
   b) 110% of the starting current of the motor
   c) 125% of the starting current of the motor
   d) 150% of the starting current of the motor
21. The type of grounding receptacle or attachment plug required in a hazardous (classified) location of an anesthetizing location shall be listed as _____.
   a) Class II, Division 2 type receptacle
   b) hazardous (classified) receptacle
   c) Class I, Division 1 type receptacle
   d) Class I, Group C type receptacle
22. The short-circuit and ground-fault protective device protecting the branch circuit shall have sufficient _____ to permit the motor-compressor and other motors to start and accelerate their loads.
   a) capacity
   b) voltage
   c) current
   d) time delay
23. A boiler employing resistance-type heating elements rated more than 48 amperes and not contained in an ASME rated and stamped vessel shall have the heating elements subdivided into loads not exceeding _____ and protected at not more than _____ amperes.
   a) 60; 48
   b) 48; 80
   c) 48; 60
   d) 24; 48
24. A building or other structure served shall be supplied by _____.
   a) one service
   b) two services
   c) six services
   d) none of these
25. Sealing compound is employed with Type MI cable termination fittings in a Class I location for the purpose of _____.
   a) excluding moisture
   b) limiting a possible explosion
   c) preventing a possible explosion
   d) preventing passage of gas or vapor
26. The maximum length of an equipment bonding jumper on the outside of a raceway is _____.
   a) 25 feet
   b) 6 feet
   c) 3 feet
   d) 1.8 feet

27. Control circuit overcurrent protection for tapped conductors, 14 AWG and larger, not with communication circuits and other than motor control devices, shall be protected by overcurrent devices rated _____.
    a) at not more than 125% of the ampacity of the conductors
    b) at not more than the ampacity of the remote control conductors
    c) in accordance with the conductor ampacity
    d) at not more than 300% of the ampacity of the remote control conductors
28. Non-current-carrying conductive materials enclosing electrical conductors shall be connected to earth so as to limit the voltage _____.
    I. imposed by lightning
    II. imposed by line surges
    III. to ground on these materials
    a) I only
    b) III only
    c) I and II
    d) I, II, and III
29. The grounded circuit conductor shall not be used to ground equipment on the supply side of the ac service-disconnecting means.
    a) True
    b) False
30. When a three-phase motor is to be installed in an isolated, unattended location and supplied by a three-wire, three-phase, ungrounded system, the number and location of running overcurrent protection units required are _____.
    a) two, one per phase in any ungrounded phase wire
    b) two, in any two conductors
    c) three, one in each phase wire
    d) three, one of which may be nonautomatic motor controller
31. The secondary circuits of wound-rotor ac motors, including _____, shall be permitted to be protected against overload by the motor-overload device.
    I. conductors
    II. controllers
    III. resistors
    a) I only
    b) I and II
    c) II and III
    d) I, II, and III
32. Direct buried conductors and cables emerging from the ground shall be protected, but in no case shall the protection be required to exceed _____ below finished grade.
    a) 6 inches
    b) 12 inches
    c) 18 inches
    d) 24 inches

33. Uninsulated conductors shall have a clearance from signs of not less than _____.

a) 3′
b) 5′
c) 8′
d) 10′

34. For capacitors (over 600 volts), group-operated switches shall be used for capacitor switching and shall be capable of _____.

I. carrying continuously not less than 135 percent of the rated current of the capacitor installation
II. interrupting the maximum continuous load current of each capacitor, capacitor bank, or capacitor installation that will be switched as a unit
III. carrying currents due to faults on capacitor side of switch

a) I only
b) II only
c) III only
d) I, II, and III

35. The disconnecting means for motor circuits rated 600 volts, nominal, or less, shall have an ampere rating of at least _____ percent of the full-load current rating of the motor.

a) 100
b) 115
c) 125
d) 150

36. Amplifiers with output circuits carrying audio program signals shall be permitted to employ _____ wiring where the amplifier is listed and marked for use with the specific class of wiring method.

I. Class 1
II. Class 2
III. Class 3

a) II only
b) III only
c) II or III
d) I, II, or III

37. The horsepower rating for an air-conditioner disconnect shall be selected from Tables 430.248, 430.249, or 430.250 corresponding to the _____.

I. rated-load current
II. branch-circuit selection current
III. loads

a) I only
b) III only
c) I or II
d) I, II, or III

38. The connections from headers to cabinets, in cellular concrete floor raceways, shall be made by means of _____.

I. listed metal raceways

II. listed nonmetallic raceways

III. listed fittings

a) I only

b) II only

c) I and III

d) I, II, and III

39. The forming shell of a wet-niche lighting fixture shall be grounded with a(n) _____ copper conductor where installed in rigid nonmetallic conduit.

a) 8-AWG bare

b) 8-AWG insulated

c) 12-AWG bare

d) 12-AWG insulated

40. Power feed, grounding connection, and shield system connection between the FCC system and other wiring systems shall be accomplished in a _____ identified for this use.

a) trench

b) raceway

c) transition assembly

d) metal shield connection

41. Insulated conductors shall have a clearance from signs of not less than _____.

a) 3′

b) 5′

c) 8′

d) 10′

42. The entire floor area in a commercial garage where Class I liquids are transferred, up to a level of _____ inches above the floor, shall be considered a Class I, Division 2 location.

a) 10

b) 12

c) 18

d) 24

43. Temporary electrical power and lighting installations shall be permitted _____.

I. for test

II. for experiments

III. during emergencies

a) I and II

b) I and III

c) II and III

d) I, II, and III

44. What is the minimum size equipment grounding conductor required for a branch circuit load of 75 amperes?
    a) 4 AWG
    b) 6 AWG
    c) 8 AWG
    d) 10 AWG
45. What standard size inverse time circuit breaker is required for three parallel runs of 500-kcmil THW copper conductors?
    a) 1000
    b) 1100
    c) 1200
    d) None of these
46. What is the maximum recommended percentage of voltage drop for a feeder circuit?
    a) 2 percent
    b) 3 percent
    c) 5 percent
    d) 10 percent
47. What is the maximum length for a power-supply cord feeding a mobile home?
    a) 20 feet
    b) 25½ feet
    c) 32 feet
    d) 36½ feet
48. What is the minimum current rating for a sign control switch with a transformer having a rating of 10 amperes?
    a) 15 amperes
    b) 20 amperes
    c) 30 amperes
    d) 40 amperes
49. Which of the following wiring methods shall not be permitted in ducts or plenums used for environmental air?
    a) MI cable
    b) MC cable
    c) NM cable
    d) Electrical metallic tubing
50. A 500-foot run of conductors has been sized to compensate for voltage drop. What must be done to the grounding conductor?
    a) Nothing
    b) It must be sized proportionately according to circular mil area of the ungrounded conductors.
    c) Double the size of the grounding conductor.
    d) Increase the grounding conductor one size.

Exam Number 5

Time Limit: 2 Hours

1. Thermoplastic insulation on fixture wires may become deformed at _____ under normal conditions.

   I. 0°F

   II. 72°F

   III. 80°F

   a) I only

   b) I and III

   c) I and II

   d) II and III

2. The use of nonmetallic underground conduit with conductors and fittings shall be permitted _____.

   I. in cinder fill

   II. in exposed locations

   III. embedded in concrete

   a) I only

   b) I and II

   c) I and III

   d) I, II, and III

3. Appliances that are fastened in place and connected to branch circuits supplying lighting loads shall not exceed _____ of the branch-circuit rating.

   a) 50%

   b) 80%

   c) 100%

   d) 125%

4. Rigid nonmetallic conduit, when used in a motor fuel dispensing facility, shall _____.

   a) be permitted where buried under not less than 1 foot of cover

   b) be permitted where buried under not less than 2 feet of cover

   c) be permitted where buried under not less than 4 feet of cover

   d) not be permitted to be buried

5. Circuits for lighting shall not be connected to any system containing trolley wires with a ground return.

   a) True

   b) False

6. Individual dry-type transformers of more than _____ kVA rating shall be installed in a transformer room of fire-resistant construction.

   a) 50

   b) 100

   c) 112½

   d) 125

7. Type MI cable conductors shall be of _____ with a resistance corresponding to standard AWG and kcmil sizes.
   I. nickel
   II. nickel-coated copper
   III. solid copper
   a) I only
   b) I or II
   c) II or III
   d) I, II, or III
8. Type AC cable shall be permitted to be unsupported where the cable is not more than _____ in length at terminals where flexibility is necessary.
   a) 2′
   b) 28″
   c) 30″
   d) 3′
9. What is the classification of a grease pit that is not ventilated?
   a) Class I, Division 1
   b) Class I, Division 2
   c) Class II, Division 1
   d) Class II, Division 2
10. Cable tray systems shall not be used in _____ or where subject to damage.
    a) hazardous locations
    b) hoistways
    c) 600-volt systems
    d) tunnels
11. Where conduit is threaded in the field, a standard cutting die with a _____ shall be used.
    a) 1″ taper per foot
    b) ¼″ taper per foot
    c) ½″ taper per foot
    d) ¾″ taper per foot
12. A _____ may consist of one or more sensing elements integral with the motor and an external control device.
    a) thermal protector
    b) vented power fuse
    c) motor-circuit switch
    d) cutout
13. Switches, flashers, and similar devices controlling transformers and electronic power supplies shall be rated for controlling inductive loads or have a current rating not less than _____ of the current rating of the transformer.
    a) 100%
    b) 125%
    c) 200%
    d) 300%

14. A pool panelboard, not part of the service equipment, shall have an equipment grounding conductor installed between its grounding terminal and _____.
    - a) a separate ground
    - b) a ground rod
    - c) the grounding terminal of the service equipment
    - d) the bonding grid
15. Conductors may not be spliced by means of brazing.
    - a) True
    - b) False
16. A thermal barrier shall be required if the space between the resistors and reactors and any combustible material is less than _____ inches.
    - a) 12
    - b) 6
    - c) 36
    - d) 24
17. An electrically operated pipe organ shall have both the generator and motor frames grounded unless _____.
    - a) effectively insulated from ground
    - b) effectively insulated from ground and the motor driving it
    - c) both the generator and motor have double insulation
    - d) effectively insulated from ground and from each other
18. The 4-AWG solid copper conductor connecting the grounding electrode to the service has been accidentally cut. Which of the following could be acceptable methods for splicing this conductor?
    - I. A split bolt connector
    - II. An irreversible compression-type connector listed for the purpose
    - III. Exothermic welding
    - a) I only
    - b) I or II
    - c) II or III
    - d) I, II, or III
19. A public address system _____.
    - a) is covered in the *Code* under Article 640
    - b) is not covered in the *Code*
    - c) has its own *Code* section
    - d) none of these
20. The radius of a 3″ conduit containing conductors without lead sheath shall not be less than _____ inches.
    - a) 36
    - b) 15
    - c) 21
    - d) 18

21. In general, the voltage limitation between conductors in surface metal raceway is _____.
   a) 300 volts
   b) 600 volts
   c) 900 volts
   d) 1000 volts
22. Which of the following is not a standard classification for a branch circuit supplying several loads?
   a) 50 amperes
   b) 30 amperes
   c) 25 amperes
   d) 20 amperes
23. When an outlet from an underfloor raceway is discontinued, the sections of circuit conductors supplying the outlet _____.
   a) may be handled like abandoned outlets on loop wiring
   b) shall be removed from the raceway
   c) may be spliced
   d) may be reinsulated
24. Motor branch-circuit short-circuit and ground-fault protection and motor overload protection shall be permitted to be combined in a single protective device where the rating or setting of the device provides the _____ protection specified in 430.32.
   a) combined overload
   b) overload
   c) overcurrent
   d) branch circuit
25. Infrared lamp commercial and industrial heating appliances shall have overcurrent protection not exceeding _____ amperes.
   a) 30
   b) 40
   c) 50
   d) 60
26. Branch-circuit conductors within 3 inches of a ballast shall have an insulation temperature rating not lower than _____ unless supplying a luminaire (fixture) listed and marked as suitable for a different insulation temperature.
   a) 60° C
   b) 75° C
   c) 85° C
   d) 90° C
27. Conductors of light and power systems of all voltages may occupy the same enclosure or raceway _____.
   a) if the power system is over 600 volts and the light system is under 600 volts
   b) if less than 600 volts and if insulated for at least the maximum circuit voltage applied to any conductor within the enclosure, cable, or raceway
   c) in most instances without qualification
   d) if the power system is over 600 volts and individual circuits are ac

28. Steel raceways installed in tunnels shall not be used as an equipment grounding conductor over _____ volts.
    a) 150
    b) 300
    c) 600
    d) 1000
29. The maximum operating temperature of Type SFF-1 silicone-insulated fixture wire is _____.
    a) 302° C
    b) 200° C
    c) 150° C
    d) 90° C
30. The grounding conductor in each cord, cable, or raceway of an electrically driven irrigation machine shall be _____.
    a) 8-AWG copper
    b) 10-AWG copper
    c) 12-AWG copper
    d) sized not less than the largest supply conductor
31. In an industrial commercial (loft) building, where the actual number of general-purpose receptacle outlets is unknown, an additional load of _____ volt-amp(s) per square foot shall be included in the load calculation.
    a) 0
    b) ½
    c) 1
    d) 1½
32. Ferrous metal enclosures for grounding electrode conductors shall be _____.
    a) bonded
    b) rigid conduit only
    c) not less than ¾″ in diameter
    d) electrically continuous
33. For systems over 1000 volts, the neutral of a solidly grounded neutral system shall be permitted to be grounded at more than one point for _____.
    I. transformers supplying conductors to a building or other structure
    II. underground circuits where the neutral is exposed
    III. overhead circuits installed outdoors
    a) I only
    b) II only
    c) III only
    d) I, II, and III
34. When bare grounded conductors are used with insulated conductors, their allowable ampacities shall be limited to _____.
    a) 60° C
    b) 75° C
    c) 90° C
    d) those permitted for the adjacent insulated conductors

35. Heaters installed within _____ feet of the outlet of an air-moving device, a heat pump, an air conditioner, elbows, baffle plates, or other obstructions in ductwork may require turning vanes, pressure plates, or other devices on the inlet side of the duct heater to ensure an even distribution of air over the face of the heater.

a) 6

b) 3

c) 4

d) 2

36. A (luminaire) fixture that weighs more than 6 pounds or exceeds _____ inches in any dimension shall not be supported by the screw shell of a lampholder.

a) 24

b) 18

c) 16

d) 12

37. A green wire with yellow stripes, used in branch circuits, may be used as a(n) _____.

I. grounded conductor

II. grounding conductor

III. ungrounded conductor

a) I only

b) II only

c) I or II

d) I, II, or III

38. Flexible metal conduit shall be secured at which of the following?

a) Where fished

b) At intervals not exceeding 4½ feet

c) Within 12 inches of each box where fished

d) Lengths not exceeding 3 feet at motor terminals for 1-inch conduit

39. A 10-ohm resistor in an ac circuit creates _____ heat than (as) the same resistor in a dc circuit.

a) 30 percent less

b) the same

c) more

d) half as much

40. Rigid conduit buried in an area subject to heavy vehicular traffic shall have a minimum cover of _____ inches.

a) 24

b) 18

c) 12

d) 6

41. Every circuit breaker having an interrupting rating other than _____ amperes shall have its interrupting rating shown on the circuit breaker.
    a) 1000
    b) 4000
    c) 5000
    d) 10,000
42. Transformers for outline lighting installations shall have a secondary-circuit current rating of not more than _____ milliamperes.
    a) 30
    b) 60
    c) 100
    d) 300
43. FC assemblies can be installed _____.
    I. where installed for exposed work
    II. in locations where they will not be subjected to physical damage
    III. in surface metal raceways identified for the use
    a) I and II
    b) II and III
    c) I and III
    d) I, II, and III
44. Conductors of dissimilar metals _____ be intermixed in a terminal or splicing connector where physical contact occurs between dissimilar conductors.
    a) shall
    b) may
    c) shall not
    d) will
45. In an information technology equipment room, a means shall be provided to disconnect _____.
    I. power to all electronic equipment
    II. power to all required fire/smoke dampers
    III. power to all dedicated HVAC systems serving the room
    a) I and II
    b) II and III
    c) I and III
    d) I, II, and III
46. For a cord-and-plug-connected motor not to have a horsepower-rated attachment plug, the motor must be rated _____ or less.
    a) ¼ hp
    b) ⅓ hp
    c) ½ hp
    d) 1 hp

47. In calculating the load of fluorescent lights with ballasts, the load must be based on _____.

a) the total wattage of the lamps
b) the total ampere rating of each fixture
c) the total VA of the lamps
d) the total current of the lamps

48. Which of the following conductors is moisture and heat resistant?

a) TBS
b) TFE
c) TW
d) RHH

49. A panelboard feeding swimming pool pumps at a motel is below grade and behind a wall inside a room. The panelboard must be _____.

a) weatherproof
b) GFCI protected
c) within sight of the pool equipment
d) vapor tight

50. For a building over 75 feet in height with one floor of dwelling occupancies, equipment for emergency feeder circuits shall be located either in spaces fully protected by approved automatic fire suppression systems or in spaces with a _____-hour fire resistance rating.

a) 1
b) 2
c) 3
d) 4

Exam Number 6

Time Limit: 2 Hours

1. Knob and tube wiring splices shall be _____ unless approved splicing devices are used.
   a) bolted
   b) soldered
   c) taped
   d) clamped
2. The equipment grounding conductor in type NM cable for 15-, 20-, and 30-ampere branch circuits _____.
   a) is required only with aluminum cable
   b) must be the same size as the insulated circuit conductors
   c) may be at least one size smaller than the insulated circuit conductor
   d) none of these
3. Generators operating at _____ volts or less and driven by individual motors shall be considered as protected by the overcurrent device protecting the motor if these devices will operate when the generators are delivering not more than 150 percent of their full-load rated current.
   a) 65
   b) 70
   c) 75
   d) 85
4. Temporary electrical power and lighting installations shall be permitted for a period not to exceed _____ days for holiday decorative lighting and similar purposes.
   a) 15
   b) 30
   c) 60
   d) 90
5. Knife switches rated at over 1200 amperes at 250 volts or less _____.
   a) may be opened under load
   b) should be connected in parallel
   c) shall be used only as isolating switches
   d) should be placed so that gravity tends to close them
6. The size of the system bonding jumper for a system that supplies a Class 1, Class 2, or Class 3 circuit, and is derived from a transformer rated not more than 1000 volt-amperes, shall not be smaller than the derived phase conductors and shall not be smaller than _____.
   a) 14-AWG aluminum
   b) 12-AWG copper
   c) 10-AWG copper
   d) 14-AWG copper

7. The screw shell of all sign lampholders in grounded circuits shall be connected to _____.
   I. the grounded conductor
   II. the neutral
   III. the white or natural gray conductor
   a) I only
   b) II only
   c) III only
   d) I, II, or III
8. DC conductors used for electroplating shall be protected from overcurrent by _____.
   I. a current-sensing device that operates a disconnecting means
   II. fuses or circuit breakers
   III. other approved means
   a) I or II
   b) II only
   c) I, II, or III
   d) II or III
9. In general, busways shall be securely supported at intervals not exceeding _____ feet.
   a) 10
   b) 5
   c) 6
   d) 3
10. The *National Electrical Code* scope includes _____.
   a) conductors and equipment on all buildings or premises where electrical power is employed
   b) conductors and equipment on certain public or private premises
   c) conductors and equipment wherever and whenever electricity is employed
   d) conductors and equipment on all buildings, premises, and conveyances where electrical power is employed
11. A lighting and appliance branch circuit panelboard contains six 3-pole circuit breakers and eight 2-pole circuit breakers. The maximum allowable number of single pole breakers permitted to be added is _____.
   a) 28
   b) 12
   c) 8
   d) 6
12. A threaded enclosure of 80 cubic inches that contains a device shall be considered adequately supported if two conduits are threaded wrenchtight into the same side of the enclosure, and if each conduit is supported within _____ inches of the enclosure.
   a) 24
   b) 36
   c) 30
   d) 18

13. A copper bus bar is 4″ wide by ½″ thick. What is the ampacity?
    a) 1500 amperes
    b) 2000 amperes
    c) 1000 amperes
    d) 500 amperes
14. Where multiconductor portable cables are used to connect mobile equipment and machinery above 2000 volts, the conductors shall be _____.
    a) shielded
    b) MK braid
    c) unshielded
    d) without ground
15. Circuit breakers rated at _____ amperes or less and 600 volts or less shall have the ampere rating molded, stamped, etched, or similarly marked into their handles or escutcheon area.
    a) 600
    b) 200
    c) 100
    d) 50
16. Compliance with the provisions of the *Code* will result in _____.
    a) good electrical service
    b) freedom from hazard
    c) an efficient system
    d) all of these
17. Open conductors on insulators must be covered when they are within _____ feet of a building.
    a) 10
    b) 12
    c) 15
    d) 25
18. Conductors of ac circuits rated 600 volts, nominal, or less may occupy the same raceway with conductors of dc circuits that are less than 600 volts.
    a) True
    b) False
19. Emergency power for hospitals shall have a transition time, from instant of failure to normal service to the emergency generator source, not to exceed _____.
    a) 30 seconds
    b) 10 seconds
    c) 5 seconds
    d) 1 second
20. Tap conductor raceways for recessed luminaire (fixture) connections shall be limited to _____ feet in length.
    a) 2
    b) 4
    c) 6
    d) 10

21. Optional standby systems are typically installed to provide an alternate source of electric power to serve loads such as _____.
    I. communication systems
    II. data processing systems
    III. refrigeration systems
    a) II and III
    b) I, II, and III
    c) I and III
    d) III only
22. Metallic enclosures of reactors and adjacent metal parts shall be installed so that the temperature rise _____ is not hazardous to personnel or does not constitute a fire hazard.
    a) from circulating currents
    b) of 40° C
    c) from more than three current-carrying conductors
    d) from induced circulating currents
23. An auto-transformer fault sensing system can be accomplished by the use of two subtractive-connected donut-type current transformers installed to sense and signal when an unbalance occurs in the line current to the auto-transformer of _____ or more of rated current.
    a) 25%
    b) 33%
    c) 50%
    d) 66%
24. Where it is impracticable to locate the service head above the point of attachment, the service head location shall be permitted not further than _____ from the point of attachment.
    a) 1′
    b) 2′
    c) 4′
    d) 6′
25. The grounded conductor of a mineral-insulated, metal-sheathed cable shall be identified at the time of installation by _____.
    a) distinctive marking at its terminations
    b) continuous markings of white or natural gray
    c) continuous markings of green
    d) noncontinuous markings of white or natural gray
26. Cases or frames of current transformers, the primaries of which are not over _____ volts to ground and that are used exclusively to supply current to meters, are not required to be grounded.
    a) 150
    b) 300
    c) 600
    d) 1000

27. All heating cables, the splice between the heating cable and nonheating leads, and _____ minimum of the nonheating lead at the splice shall be embedded in plaster or dry board in the same manner as the heating cable.
   a) 1-inch
   b) 3-inch
   c) 6-inch
   d) 12-inch
28. If the supply conductors for a welder are protected by an overcurrent device rated or set at not more than _____ percent of I_{1max} or rated primary current of the welder, a separate overcurrent device shall not be required.
   a) 25
   b) 50
   c) 100
   d) 200
29. Individual branch circuits of Type FCC cable shall have ratings not exceeding _____ amperes.
   a) 15
   b) 20
   c) 30
   d) 40
30. On electrode-type boilers, the size of branch-circuit conductors and overcurrent protective devices shall be calculated on the basis of _____ of the total load.
   a) 50%
   b) 80%
   c) 100%
   d) 125%
31. An insert installed in a cellular concrete floor raceway shall be grounded to the raceway by _____.
   a) a connection on the header
   b) a grounding device
   c) the insert
   d) a grounding fitting
32. What is the maximum length for a feeder without overcurrent protection?
   a) 6′
   b) 10′
   c) 50′
   d) 100′
33. The ampacity of the conductors from the generator terminals to the first distribution device(s) containing overcurrent protection shall not be less than _____ percent of the nameplate current rating of the generator.
   a) 80
   b) 100
   c) 115
   d) 125

34. Where grounding means does not exist in the receptacle enclosure, a non-grounding-type receptacle shall be permitted to be replaced with _____.

I. another non-grounding-type receptacle
II. a ground-fault circuit-interrupter type receptacle
III. a grounding-type receptacle where supplied through a ground-fault circuit-interrupter

a) I only
b) I or II
c) II or III
d) I, II, or III

35. General power and lighting in a health care facility _____.

a) are part of the continuous power system
b) are part of the equipment system
c) are not part of the emergency system
d) are connected to the alternate source generator

36. The rating or setting of the overcurrent devices for capacitors shall be _____.

a) as low as practicable
b) not more than 80% of the capacitor
c) not more than 125% of the capacitor
d) not more than 135% of the capacitor

37. A recessed incandescent luminaire (fixture) with a completely enclosed lamp, when installed in a closet, must _____.

a) be mounted on the ceiling, provided there is a minimum clearance of 12″ between the luminaire (fixture) and the nearest point of a storage space
b) be mounted on the wall, provided there is a minimum clearance of 12″ between the luminaire (fixture) and the nearest point of a storage space
c) both a and b
d) none of these

38. For capacitors over 600 volts, group-operated switches shall be used for capacitor switching and shall be capable of _____.

I. carrying continuously not less than 135 percent of the rated current of the capacitor installation
II. withstanding the maximum inrush voltage, including contributions from adjacent capacitor installations
III. interrupting the maximum continuous load current of each capacitor, capacitor bank, or capacitor installation that will be switched as a unit

a) I only
b) I and III
c) I and II
d) I, II, and III

39. Switch enclosures may be used as _____ as long as adequate space is provided.

I. junction boxes
II. auxiliary gutters
III. raceways

a) I only
b) I and III
c) I, II, and III
d) none of these

40. All heating elements that are replaceable in the field and are a part of an electric heater shall be legibly marked with the rating in _____.

a) volts and watts
b) volts and amperes
c) either a or b
d) both a and b

41. All 125-volt, single-phase, 15- and 20-ampere receptacles installed _____ in hotels and motels shall have ground-fault circuit-interrupter protection for personnel.

a) in guest rooms
b) in hallways
c) in bathrooms of guest rooms
d) all of these

42. The resistance of a 6-AWG THW copper conductor is 0.491 ohms; the resistance of a 6-AWG THHN copper conductor will be _____.

a) higher
b) lower
c) the same
d) cannot determine

43. A sign shall be placed at the _____ indicating type and location of on-site legally required standby power sources.

a) main disconnect
b) service entrance
c) front door
d) alarm panel

44. The ampacity of capacitor-circuit conductors shall not be less than _____ percent of the rated current of the capacitor.

a) 50
b) 80
c) 100
d) 135

45. Neon secondary-circuit conductors over 1000 volts shall be _____.

I. insulated

II. listed as gas tube sign and ignition cable type GTO

III. not smaller than 18 AWG

a) I and III

b) I and II

c) II and III

d) I, II, and III

46. Outlet boxes shall be permitted to support lighting fixtures weighing _____ or less, unless supported independently of the outlet box.

a) 70 pounds

b) 50 pounds

c) 35 pounds

d) 25 pounds

47. A parallel set of 250-kcmil THHN copper conductors terminate in a 500-ampere overcurrent protective device located in a subpanel. What size ground is required in each raceway?

a) 4 AWG

b) 2 AWG

c) 1 AWG

d) 1/0 AWG

48. In aircraft hangars, the area within 5 feet horizontally from aircraft power plants or aircraft fuel tanks shall be classified as a Class I, Division 2 location that shall extend upward from the floor to a level _____ feet above the upper surface of wings and engine enclosures.

a) 5

b) 8

c) 10

d) 12

49. Power-limited fire alarm cable installed in spaces used for environmental air _____.

a) must be Type NPLFP

b) must be THHN

c) must be Type FPLP

d) cannot be installed

50. In outdoor pool areas, lighting fixtures must be installed at least _____ above the maximum water level of the pool.

a) 6 feet

b) 8 feet

c) 10 feet

d) 12 feet

Exam Number 7

Time Limit: 2 Hours

1. Where a double-throw knife switch has a vertical throw, _____ shall be provided to hold the blades in the open position when so set.
   a) a tamperproof device
   b) a locking device
   c) an automatic device
   d) integral mechanical means
2. Auxiliary contacts of a renewable type shall be provided on all knife switches rated 600 volts and designed for use in breaking current over _____ amperes.
   a) 200
   b) 150
   c) 100
   d) 50
3. A Class II hazardous location is where _____ is (are) present.
   a) fibers and flyings
   b) radioactive material
   c) gases and vapors
   d) combustible dust
4. The words "thermally protected" appearing on the nameplate of a motor or motor-compressor indicate that the motor is provided with a _____.
   a) thermal protector
   b) breaker
   c) switch
   d) fuse
5. Insulated conductors shall be marked or tagged with which of the following?
   I. Type letters
   II. Manufacturer identification
   III. Maximum rated voltage
   a) I only
   b) II only
   c) I and III
   d) I, II, and III
6. Sheet metal auxiliary gutters shall be supported throughout their entire length at intervals not exceeding _____.
   a) 10′
   b) 6′
   c) 5′
   d) 3′

7. In pull boxes or junction boxes having any dimension over _____ feet, all conductors shall be cabled or racked up in an approved manner.
   a) 6
   b) 4
   c) 2
   d) 1.8
8. Service-drop conductors over residential property and driveways, and those commercial areas not subject to truck traffic where the voltage does not exceed 300 volts to ground, shall have a minimum clearance of _____ feet.
   a) 18
   b) 15
   c) 12
   d) 10
9. The space measured horizontally above a show window at its maximum width must have at least one receptacle for each _____ linear feet.
   a) 6
   b) 12
   c) 8
   d) 10
10. Locations where combustible dust is normally in heavy concentrations are designated as _____.
    a) Class II, Division 1
    b) Class III, Division 1
    c) Class I, Division 2
    d) Class II, Division 2
11. All cord-and-plug-connected smoothing irons and electrically heated appliances that are rated at more than _____ watts and product temperatures in excess of 121° C on surfaces with which the cord is likely to be in contact shall be provided with one of the types of approved heater cords listed in Table 400.4.
    a) 50
    b) 100
    c) 300
    d) 500
12. All theater fixed-stage switchboards that are not completely enclosed _____ shall be provided with a metal hood extending the full length of the board to protect all equipment on the board from falling objects.
    I. dead front
    II. dead rear
    III. recessed into a wall
    a) I only
    b) I and II
    c) III only
    d) I and II, or III

13. The minimum size of a receiving station outdoor antenna conductor, where the span is 75 feet, shall be at least _____ if a copper-clad steel conductor is installed.
    a) 14 AWG
    b) 17 AWG
    c) 10 AWG
    d) 12 AWG
14. An enclosure that is designed for either surface mounting or flush mounting and is provided with a frame, mat, or trim in which a swinging door or doors are or can be hung is a _____.
    a) cabinet
    b) panelboard
    c) switchboard
    d) wireway
15. A crane rail, when used as a conductor, shall be _____.
    I. grounded by the fittings used for the suspension of the rail to a building
    II. grounded by a bonding conductor to the grounding electrode system
    III. effectively grounded at the transformer
    a) I only
    b) I and III
    c) I and II
    d) I, II, and III
16. Which of the following is not a true statement concerning an equipment grounding conductor?
    a) Under certain conditions, equipment grounding conductors may be required to be larger than circuit conductors.
    b) One equipment grounding conductor may serve multiple circuits.
    c) Under certain conditions, equipment grounding conductors may be run in parallel.
    d) The size of equipment grounding conductors shall be increased to compensate for voltage drop.
17. Conductors that supply one or more arc welders shall be protected by an overcurrent device rated or set at not more than _____ of the conductor rating.
    a) 80%
    b) 125%
    c) 200%
    d) 300%
18. Unless part of listed decorative lighting assemblies, pendant conductors shall not be smaller than _____ for mogul-base screw-shell lampholders.
    a) 12 AWG
    b) 14 AWG
    c) 16 AWG
    d) 18 AWG

19. In parking garages used for parking or storage and where no repair work is done, open flame, welding, or the use of volatile flammable liquids are _____.

a) not classified

b) Class III

c) Class II

d) Class I

20. Festoon lighting is a string of outdoor lights suspended between two points _____.

a) less than 15 feet apart

b) less than 25 feet apart

c) more than 15 feet apart

d) regardless of distance apart

21. Where feeder conductors have an ampacity greater than required by 430.24, the rating or setting of the feeder overcurrent protective device shall be permitted to be based on _____.

a) the connected load

b) the largest motor × 125%

c) the ampacity of the feeder conductors

d) the FLC × demand factor

22. Grounding-type receptacles shall be installed only on circuits of the _____ for which they are rated, except as provided in Tables 210.21(B)(2) and (B)(3).

I. voltage class

II. wattage

III. current

a) I only

b) I and III

c) I and II

d) I, II, and III

23. One or more nonmetallic surface extensions shall be permitted to be run in any direction from an existing outlet, but not on the floor or within _____ from the floor.

a) 2″

b) 6″

c) 8″

d) 12″

24. Where required, drawings for feeder installations must be submitted before _____.

a) using the feeders

b) completing the installation

c) the installation of the feeders

d) the building is occupied

25. For permanently connected appliances rated at not over 300 volt-amperes or _____ horsepower, the branch-circuit overcurrent device shall be permitted to serve as the disconnecting means.
   a) 1/8
   b) 1/4
   c) 1/2
   d) 1

26. What is the maximum operating temperature of rubber-covered, Type RFH-1 heat-resistant fixture wire?
   a) 224° F
   b) 194° F
   c) 167° F
   d) 124° F

27. A controller serving more than one air-conditioner motor-compressor shall have _____.
   a) a continuous-duty full-load current rating
   b) a locked-rotor current rating not less than the combined load as determined in accordance with 440.12(B)
   c) either a or b
   d) both a and b

28. Portable electric lamps in Class I locations _____.
   a) shall be used in a spray area during operation of the finishing process
   b) shall not be used in a spray area during spray operations
   c) shall be used in a spray area during cleaning or repairing operations
   d) shall be used in a spray area during spray operations if grounded

29. Audio system equipment supplied by branch-circuit power shall not be placed laterally within _____ feet of the inside wall of a pool, spa, hot tub, or fountain.
   a) 5
   b) 10
   c) 15
   d) 20

30. Each unit of an information technology system supplied by a branch circuit shall be provided with a manufacturer's nameplate, which shall also include the input power requirements for _____.
   I. voltage
   II. frequency
   III. maximum rated load in amperes
   a) I only
   b) II only
   c) I and III
   d) I, II, and III

31. When calculating the maximum number of conductors permitted in a conduit or tubing, all of the same size (total cross-sectional area including insulation), the next higher whole number shall be used to determine the maximum number of conductors permitted when the calculation results in a decimal of _____ or larger.

a) 0.008

b) 0.8

c) 0.5

d) 0.08

32. Luminaires (lighting fixtures) mounted in walls shall be installed with the top of the luminaire lens not less than _____ below the normal water level of the pool.

a) 6″

b) 8″

c) 18″

d) 24″

33. Tubing used for arms and stems on luminaires (lighting fixtures) where provided with cut threads shall not be less than _____ inches in thickness.

a) 0.025

b) 0.015

c) 0.020

d) 0.040

34. Unless specified otherwise, live parts of electrical equipment operating at _____ volts or more shall be guarded against accidental contact.

a) 120

b) 50

c) 24

d) 12

35. Dry-type transformers exceeding 112½ kVA shall not be located within _____ of combustible materials of buildings.

a) 30″

b) 24″

c) 16″

d) 12″

36. For straight pulls the length of the box shall be not less than _____ the outside diameter, over sheath, of the largest shielded conductor or cable entering the box on systems over 600 volts.

a) 36 times

b) 8 times

c) 48 times

d) 6 times

37. Underground copper service-lateral conductors shall not be smaller than _____.

a) 6 AWG

b) 8 AWG

c) 4 AWG

d) 3 AWG

38. Current supply shall be such that, in the event of failure of the normal supply to (or within) the building or group of buildings concerned, emergency lighting, emergency power, or both shall be available within the time required for the application but not to exceed _____ seconds.

a) 60

b) 30

c) 10

d) 5

39. A raceway contains 45 current-carrying conductors. The ampacity of each conductor shall be reduced _____ percent.

a) 35

b) 40

c) 50

d) 60

40. A(n) _____ is a protective device for limiting surge voltages by discharging or bypassing surge current, and it also prevents continued flow of follow current while remaining capable of repeating these functions.

a) circuit breaker

b) surge arrester

c) auto fuse

d) fuse

41. If two 1-ohm resistors are in parallel, the total resistance is _____ ohm(s).

a) 2

b) ½

c) 1

d) cannot be calculated without voltage

42. Color braid of flexible cords used to identify the grounded conductor shall be finished to show a _____ color, and the braid on the other conductor or conductors finished to show a readily distinguishable solid color or colors.

I. gray

II. green

III. white

a) I or III

b) II or III

c) I or II

d) III only

43. Sign transformers and power supplies provided with an integral enclosure, including a primary and secondary circuit splice enclosure, shall _____.

a) be enclosed in metal

b) be enclosed in a lead sheath

c) be enclosed in noncombustible material

d) not require an additional enclosure

44. Where a flexible cord is used to supply a 120-volt room air conditioner, the length of such cord shall not exceed _____.

a) 12′

b) 10′

c) 8′

d) 6′

45. Service-entrance cables mounted in contact with a building shall be supported at intervals not exceeding _____.

a) 2½′

b) 4½′

c) 12′

d) 30′

46. Transformers supplied from power-supply conductors of power-limited fire alarm circuits shall be protected by an overcurrent device rated not over _____ amperes.

a) 15

b) 20

c) 30

d) 40

47. As a general rule, low-voltage cables shall be separated by at least _____ inches from high-voltage conductors.

a) 2

b) 4

c) 6

d) 8

48. What is the minimum horizontal clearance for outside branch-circuit wiring that is not on a building?

a) 1 foot

b) 3 feet

c) 8 feet

d) 10 feet

49. What is the maximum rating of overcurrent protection percentage for a 600-volt, 9-ampere transformer where there is no secondary protection?

a) 115%

b) 125%

c) 150%

d) 167%

50. How many motor overload protection units are required on a three-phase ac motor?

a) 1

b) 2

c) 3

d) 4

Exam Number 8

Time Limit: 2 Hours

1. A surge arrester is a protective device for limiting surge voltages by _____ or bypassing surge current, and it also prevents continued flow of follow current while remaining capable of repeating these functions.
   a) decreasing
   b) discharging
   c) limiting
   d) derating
2. Two-wire dc circuits in an integrated electrical system shall be permitted to be _____.
   a) ungrounded
   b) uninsulated
   c) over 600 volts
   d) grounded
3. Power supply cords for audio equipment shall be suitable for the use and shall be permitted to be used where the _____ of such equipment is facilitated through the use of a power supply cord.
   I. maintenance
   II. repair
   III. interchange
   a) I only
   b) I or II
   c) I or III
   d) I, II, or III
4. Open motors having commutators shall be located so that sparks cannot reach adjacent combustible material, but this _____.
   a) shall not prohibit these motors from being classified as Class I locations
   b) shall not prohibit the installation of these motors on wooden floors
   c) is required only for motors over 600 volts
   d) none of these
5. Branch circuits supplying two or more outlets for fixed electric space-heating equipment in a dwelling shall be rated _____ amperes.
   a) 15-20-30-40
   b) 15-20-25-30
   c) 20-30-40
   d) 15-20-30

6. For general wiring in Class I, Division I locations it is permissible to use _____.
   I. rigid metal conduit
   II. EMT
   III. flexible metal conduit
   a) I only
   b) I or II
   c) I or III
   d) I, II, or III
7. The rating of the branch circuit for flat cable assemblies shall not exceed _____ amperes.
   a) 50
   b) 30
   c) 20
   d) 15
8. The voltage to ground, in electric signs, of any output terminals of the secondary circuit shall not exceed _____ volts, under any load conditions.
   a) 1000
   b) 5000
   c) 7500
   d) 15,000
9. Each plate electrode shall expose not less than _____ square foot (feet) of surface to exterior soil.
   a) 1
   b) 2
   c) 3
   d) 4
10. Exposed live parts within porcelain luminaires (fixtures) shall be suitably recessed and located so as to make it improbable that wires will come in contact with them. There shall be a spacing of at least _____ between live parts and the mounting plane of the luminaire (fixture).
   a) 3⁄4″
   b) 1⁄2″
   c) 1⁄4″
   d) 1⁄8″
11. A unit or assembly of units or sections and associated fittings forming a rigid structural system used to securely fasten or support cables and raceways is defined as a _____.
   a) multioutlet assembly
   b) flat cable assembly
   c) cable tray system
   d) wireway

12. Bare conductors in auxiliary gutters shall be securely and rigidly supported so that the minimum clearance between bare current-carrying metal parts of different potential that are held free in the air will not be less than _____.
    a) ¼″
    b) ½″
    c) 1″
    d) 2″
13. Recessed portions of luminaire (fixture) enclosures, other than at the points of support, shall be spaced at least _____ from combustible materials.
    a) ¼″
    b) ½″
    c) 1″
    d) 3″
14. Shore power for boats shall be provided by single receptacles rated not less than _____ and shall be of the locking and grounding type.
    a) 15 amperes
    b) 20 amperes
    c) 30 amperes
    d) 50 amperes
15. All Type FCC cable connections shall use connectors identified for their use, installed such that _____ is provided.
    I. insulation
    II. electrical continuity
    III. sealing against dampness and liquid spillage
    a) I only
    b) II only
    c) III only
    d) I, II, and III
16. In community antenna television and radio distribution systems, the grounding conductor shall not _____.
    a) be larger than 12 AWG
    b) be smaller than 12 AWG
    c) be larger than 14 AWG
    d) be smaller than 14 AWG
17. _____ shall be permitted to be embedded or encased in concrete or in direct contact with the earth where provided with supplementary corrosion protection.
    I. Boxes
    II. Nonferrous metal raceways
    III. Cable sheathing
    a) I or III
    b) I or II
    c) II or III
    d) I, II, or III

18. The minimum spacing between the busbars of a 240-volt system, where the opposite polarities are held in free air, is _____.

a) ½ inch
b) ¾ inch
c) 1¼ inches
d) 2 inches

19. The installations of communication wires and cables in _____ shall be so made that the possible spread of fire will not be substantially increased.

I. hollow spaces
II. vertical shafts
III. air-handling ducts

a) I only
b) II only
c) I or II
d) I, II, or III

20. Which of these do not need to be inspected at the time of installation?

I. Subcontractor wiring under 12 volts
II. Factory-installed internal wiring
III. Warranted wiring

a) I only
b) II only
c) II and III
d) I, II, and III

21. Equipment having an open-circuit voltage exceeding _____ shall not be installed in dwelling occupancies unless such equipment is designed so that there will be no exposed live parts when lamps are being inserted, are in place, or are being removed.

a) 250 volts
b) 300 volts
c) 600 volts
d) 1000 volts

22. Enclosures, in Class I, Division 2 locations, for isolating switches for transformer or capacitor banks that are not intended to interrupt current in the normal performance of the function for which they are installed, shall _____.

a) have interlocking devices
b) not have doors in hazardous areas
c) be permitted to be installed in general-purpose enclosures
d) be explosionproof enclosures

23. For cord-and-attachment-plug-connected motor-compressors and equipment on 15- or 20-ampere branch circuits, the rating of the attachment plug and receptacle shall not exceed 15 amperes at _____ volts or 20 amperes at _____ volts.

a) 250; 125
b) 250; 250
c) 125; 250
d) 125; 125

24. Secondary-circuit voltage in electric signs shall not exceed _____, nominal, under any load condition.

a) 1000 volts

b) 5000 volts

c) 7500 volts

d) 15,000 volts

25. Floor-mounted Type FCC cable, cable connectors, and insulating ends shall be covered with carpet squares not larger than _____ inches square.

a) 18

b) 24

c) 36

d) 48

26. Solid dielectric insulated conductors operated above 2000 volts in permanent installations shall have _____ insulation and shall be shielded.

a) hi-temperature

b) ozone-resistant

c) asbestos

d) perfluoro-alkoxy

27. Which applications are covered in the article that covers the construction and installation of induction and dielectric heating equipment?

I. Industrial and scientific applications

II. Medical or dental applications

III. Line frequency pipelines and vessels heating

a) I and II

b) I and III

c) I only

d) I, II, and III

28. Any room or location in which flammable anesthetics or volatile flammable disinfecting agents are stored shall be considered to be a _____ location from floor to ceiling.

a) Class II, Division 1

b) Class I, Division 1

c) Class I, Division 2

d) nonhazardous

29. Each (luminaire) fixture or each secondary circuit of tubing in electric-discharge lighting systems that have an open-circuit voltage of 1500 volts shall have a clearly legible marking in letters not less than 1/4 inch high, reading _____.

I. "Caution 1500 Volts"

II. "Caution High Voltage"

III. "Danger High Voltage"

a) I only

b) I or II

c) I or III

d) I, II, or III

30. Receptacle outlets in floors shall not be counted as part of the required number of receptacle outlets unless located within _____ of the wall.

a) 1½′

b) 5½′

c) 6′

d) 18′

31. Emergency system equipment for sources of power, where located within assembly occupancies greater than _____ persons or in buildings above 75 feet in height, with any of the following occupancy classes—assembly, educational, residential, detention and correctional, business and mercantile—shall be installed either in spaces fully protected by approved automatic fire suppression systems or in spaces with a 1-hour fire rating.

a) 100

b) 200

c) 500

d) 1000

32. Heating panels or heating panel sets shall not exceed _____ per square foot of heated area.

a) 355 watts

b) 15 watts

c) 16½ watts

d) 33 watts

33. Landscape lighting installed with Type UF cable and connected to a 24-volt transformer shall be buried with a minimum cover of _____.

a) 6″

b) 12″

c) 18″

d) 24″

34. The maximum length permitted for conduit nipples is _____ when filling raceways to 60 percent of their total cross-sectional area.

a) 12″

b) 23″

c) 24″

d) 48″

35. A locknut is considered a(n) _____.

a) device

b) fitting

c) outlet

d) conduit body

36. Luminaires (fixtures) shall be wired with conductors having insulation suitable for the _____ to which the conductors will be subjected.

I. environmental conditions

II. current and voltage

III. temperature

a) I and II

b) II and III

c) II only

d) I, II, and III

37. Action requiring personal intervention for its control is defined as _____.

a) manual

b) non-self-acting

c) automatic

d) nonautomatic

38. Each commercial building and each commercial occupancy accessible to pedestrians shall be provided with at least one _____ in an accessible location at each entrance to each tenant space for sign or outline lighting system use.

a) receptacle

b) duplex receptacle

c) outlet

d) disconnect

39. Each conductor of a lead-in from an outdoor antenna at a receiving station shall be provided with a listed _____.

a) antenna discharge unit

b) self-supporting antenna

c) antenna

d) lead-in unit

40. Conduit bodies shall not contain splices, taps, or devices _____.

a) at all

b) unless they are marked by the manufacturer with their volume

c) unless they have been approved by the authority having jurisdiction

d) unless they are durably and legibly marked by the manufacturer with their volume

41. Where used with the connection of portable lamps or appliances, each flexible cord shall be equipped with a(n) _____ and shall be energized from a receptacle outlet.

a) cord cap

b) attachment plug

c) plug

d) device

42. Service conductors are the conductors from the service point to the _____.

a) service lateral

b) service entrance

c) service disconnecting means

d) service drop

43. Outdoor antennas and lead-in conductors shall not be attached to poles or similar structures carrying open electric light or power wires or trolley wires of over _____ volts between conductors.

a) 150

b) 250

c) 600

d) 1500

44. A separate branch circuit shall supply the car _____ on each elevator car.

I. ventilation

II. air-conditioning

III. lights and receptacle(s)

a) III only

b) I and II

c) I and III

d) I, II, and III

45. Receptacles located within the rooms, bathrooms, playrooms, activity rooms, and patient care areas of pediatric wards shall _____.

I. be listed tamper resistant

II. be a listed locking and grounding type

III. employ a listed tamper-resistant cover

a) I only

b) I or III

c) I or II

d) I, II, or III

46. What is the smallest size grounding conductor required for a 250-VA transformer?

a) 14 AWG

b) 12 AWG

c) 10 AWG

d) 8 AWG

47. Interior metal water piping located more than _____ feet from the point of entrance to the building shall not be used as a part of the grounding electrode system or as a conductor to interconnect electrodes that are part of the grounding electrode system.

a) 3

b) 5

c) 6

d) 10

48. Feeder conductors installed in a rigid metal conduit are feeding a subpanel in the garage. The conduit will be installed underground near the swimming pool. This conduit must not be within _____ from the inside wall of the pool.

   a) 3 feet
   b) 4 feet
   c) 5 feet
   d) 10 feet

49. What is the minimum size multiconductor power-limited fire alarm cable permitted to be installed in buildings?

   a) 26 AWG
   b) 24 AWG
   c) 18 AWG
   d) 16 AWG

50. A ceiling-suspended (paddle) fan that weighs _____ must be supported independently of the box.

   a) 35 pounds
   b) 50 pounds
   c) 70 pounds
   d) 75 pounds

Exam Number 9

Time Limit: 2 Hours

1. Cable tray systems shall not be used in _____ or where subject to severe physical damage.
   I. hazardous locations
   II. hoistways
   III. corrosive locations
   a) II only
   b) III only
   c) I and II
   d) I, II, or III
2. What is the maximum number of 12-AWG THW copper conductors allowed in a 3″ × 2″ × 1½″ device box?
   a) 2
   b) 3
   c) 4
   d) 5
3. An autotransformer starter shall provide _____.
   I. an "off" position
   II. a running position
   III. at least one starting position
   a) I only
   b) I and II
   c) I and III
   d) I, II, and III
4. For a lead acid battery that requires water additions, in an emergency system, the container shall _____.
   I. not be required to be transparent
   II. be transparent
   III. be translucent
   a) I only
   b) II or III
   c) II only
   d) I or III
5. An isolated winding type swimming pool transformer shall have _____ between the primary and secondary windings.
   a) an isolation space
   b) a nonmagnetic barrier
   c) a grounded metal barrier
   d) none of these

6. A _____ rated duplex receptacle may be installed on a 20-ampere branch circuit.
   I. 15-ampere
   II. 20-ampere
   III. 30-ampere
   a) I only
   b) II only
   c) I and II
   d) I, II, and III
7. What is the minimum size branch-circuit neutral conductor for an 8¾-kW electric range?
   a) 12 AWG
   b) 10 AWG
   c) 8 AWG
   d) 6 AWG
8. A motel conference room is designed for the assembly of 100 or more persons. The room is fire rated construction. Which of the following wiring methods shall be required?
   I. rigid nonmetallic conduit
   II. Type MI cable
   III. Type AC cable
   a) I only
   b) II only
   c) II or III
   d) I, II, or III
9. An alternating current general-use snap switch is suitable for use on ac circuits for controlling _____.
   I. tungsten-filament lamp loads not exceeding the ampere rating of the switch at 120 volts
   II. resistive and inductive loads, including electric-discharge lamps, not exceeding the ampere rating of the switch at the voltage involved
   III. motor loads not exceeding 125 percent of the ampere rating of the switch at its rated voltage
   a) I and II
   b) I and III
   c) II and III
   d) I, II, and III
10. The _____ in (into) motor vehicles of alcohol-based windshield washer fluid in areas used for the service and repair operations of the vehicles shall not cause such areas to be classified as hazardous (classified) locations.
    I. storage
    II. handling
    III. dispensing
    a) I only
    b) III only
    c) II or III
    d) I, II, or III

11. Phase current is equal to line current divided by the square root of three in a wye transformer.
    a) True
    b) False
12. A(n) _____ or larger copper or aluminum grounding electrode conductor shall be protected where exposed to physical damage.
    a) 8 AWG
    b) 6 AWG
    c) 4 AWG
    d) 2 AWG
13. Splices and taps shall be permitted within a wireway, provided they are accessible. The conductors, including splices and taps, shall not fill the wireway to more than _____ of its area at that point.
    a) 20%
    b) 30%
    c) 75%
    d) 80%
14. The disconnecting means of a sign shall be within _____ of the sign that it controls.
    a) sight
    b) 50′
    c) 75′
    d) none of these
15. When portable lamps are used with Edison-base lampholders, the grounded conductor _____ and attached to the screw shell and the identified blade of the attachment plug.
    a) shall be colored green
    b) shall be identified
    c) both a and b
    d) neither a nor b
16. Ground clamps shall be approved for general use without protection or shall be protected from physical damage by enclosing in _____.
    I. wood
    II. metal
    III. an equivalent protective covering
    a) II only
    b) I or II
    c) II or III
    d) I, II, or III
17. The bottom of a sign or outline lighting system equipment shall not be less than _____ above areas accessible to vehicles unless protected from physical damage.
    a) 10′
    b) 14′
    c) 16′
    d) 18′

18. The grounding conductor for secondary circuits of instrument transformer and for instrument cases shall not be smaller than _____.

a) 10-AWG copper

b) 12-AWG aluminum

c) 12-AWG copper

d) 14-AWG copper

19. Each electric-discharge luminaire (fixture) or secondary circuit of tubing having an open-circuit voltage of over 1000 volts shall have a clearly legible marking in letters not less than _____ high.

a) ¼″

b) ½″

c) 1″

d) 6″

20. All heating elements that are rated over 1 ampere, replaceable in the field, and a part of an appliance shall be legibly marked with _____.

I. the ratings in volts and amperes

II. the ratings in volts and watts

III. the manufacturer's part number

a) I only

b) I or II

c) III only

d) I, II, or III

21. The residual voltage of a capacitor shall be reduced to _____ volts, nominal, or less, within 1 minute after the capacitor is disconnected from the source of supply.

a) 12

b) 24

c) 50

d) 120

22. The maximum ampacity for a 500-kcmil Type IGS cable shall not exceed _____ amperes.

a) 119

b) 168

c) 380

d) 430

23. Optional standby systems are those systems intended to supply power to _____.

I. property where life safety does not depend on the performance of the system

II. public facilities

III. private facilities

a) I only

b) I or III

c) II or III

d) I, II, or III

24. Aluminum fittings shall be permitted to be used with _____.

I. IMC
II. steel RMC
III. aluminum RMC

a) I only
b) III only
c) II or III
d) I, II, or III

25. Upon completion of an emergency system, the authority having jurisdiction shall _____ a test on the complete system upon installation and periodically afterward.

a) conduct
b) witness
c) conduct or witness
d) none of these

26. Conductors within enclosures of switches or overcurrent devices shall not fill the wiring space at any cross section to more than _____ percent of the cross-sectional area of the space.

a) 30
b) 40
c) 60
d) 75

27. For circuits supplying lighting units that have autotransformers, the calculated load shall be based on _____.

a) the total watts of the lamps
b) the total ampere rating of such units
c) the total volt-amperes of the lamps
d) none of these

28. A mechanically integrated assembly of modules or panels with a support structure and foundation, tracker, and other components, as required, to form a direct-current power-producing unit is known as a(n) _____.

a) blocking diode
b) stand-alone system
c) photovoltaic output circuit
d) array

29. The *National Electrical Code* is _____.

a) intended as a design specification
b) an instruction manual for untrained persons
c) neither a nor b
d) both a and b

30. Live parts of electrical equipment operating at 50 volts or more shall be guarded against accidental contact by approved enclosures or by which of the following?

I. By location in a room that is accessible only to qualified persons

II. By location on a suitable balcony elevated and arranged so as to exclude unqualified persons

III. By elevations of 8 feet or more above the floor or other working surface

a) I or II

b) I or III

c) II or III

d) I, II, or III

31. Type THW insulation has a _____ C rating within fluorescent luminaires.

a) 60°

b) 75°

c) 90°

d) 194°

32. The _____ is that portion of a wiring system prior to the final overcurrent protective device protecting the circuit.

a) branch circuit

b) feeder

c) service

d) power

33. Which of these are not covered in the *National Electrical Code?*

a) Electronic organs

b) Distribution of sound

c) Speech input systems

d) Video systems

34. Bonding is the permanent joining of _____ parts to form an electrically _____ path, ensuring electrical continuity and the capacity to conduct safely any current likely to be imposed.

a) bonded; continuous

b) metal; continuous

c) metallic; conductive

d) none of these

35. Type _____ cable is a single or multiconductor solid dielectric insulated cable rated 2001 volts or higher.

a) MC

b) MI

c) MV

d) none of these

36. Circuits supplying electrically operated pipe organs shall be so arranged that 26-AWG and 28-AWG conductors shall be protected by an overcurrent device rated at not more than _____ amperes.

a) 5

b) 6

c) 15

d) 20

37. Where the voltage exceeds 600 volts, nominal, permanent and conspicuous warning signs shall be provided, reading as follows: "Danger—High Voltage—Keep Out."

a) True

b) False

38. All splices and joints and the free ends of conductors shall be covered with an insulation equivalent to that of the conductor.

a) True

b) False

39. An isolating switch is one that is _____.

a) required to have a padlock

b) used only for heavy motor overloads

c) primarily used with an isolation transformer

d) intended for cutting off an electrical circuit from its source of power

40. Readily accessible is _____.

a) requiring to move only small obstacles

b) requiring only a 6-foot ladder

c) capable of being reached quickly for operation

d) within 50 feet

41. The neutral conductor _____.

a) carries the balanced current

b) carries the unbalanced current

c) is always the "white" grounded conductor

d) is always the "green" conductor

42. The letters CATV refer to _____.

a) TV

b) video

c) communications

d) antenna systems

43. _____ means so constructed or protected that exposure to the weather will not interfere with successful operation.

a) Weatherproof

b) Weather-resistant

c) Weather-tight

d) Weather-sealed

44. The definition of a continuous load is _____.
    a) operating at a substantially constant load for an indefinitely long time
    b) intermittent operation in which the load conditions are regularly recurrent
    c) a load where the maximum current is expected to continue for 1 hour or more
    d) a load where the maximum current is expected to continue for 3 hours or more
45. In other than residential calculations, an ordinary outlet shall be calculated at _____.
    a) 2 amperes
    b) 746 watts
    c) 180 VA
    d) 660 VA
46. Which of the following wiring methods are not permitted in wheelchair lifts?
    a) AC
    b) FMC
    c) IMC
    d) NMC
47. The conductor between a 1-kV surge arrester and the grounding connection shall not be smaller than _____ copper.
    a) 6-AWG
    b) 8-AWG
    c) 12-AWG
    d) 14-AWG
48. Double locknuts are required on rigid metal conduit when the voltage exceeds _____ to ground.
    a) 120 volts
    b) 150 volts
    c) 240 volts
    d) 250 volts
49. An ac or dc general-use snap switch feeding an inductive load shall not exceed _____ of the ampere rating of the switch at the applied voltage.
    a) 50 percent
    b) 80 percent
    c) 100 percent
    d) 125 percent
50. What is the minimum percentage factor for a separate overload device installed on a 1½-hp motor with a 55° C temperature rise?
    a) 110 percent
    b) 115 percent
    c) 125 percent
    d) 140 percent

Exam Number 10

Time Limit: 2 Hours

1. Where extensive metal in or on buildings may become energized and is subject to personal contact, _____ will provide additional safety.
   a) adequate bonding
   b) grounding
   c) bonding
   d) adequate bonding and grounding
2. The area of a 12-AWG conductor is _____ circular mils.
   a) 1.93
   b) 4.10
   c) 6530
   d) 1650
3. The grounded conductor must not be broken at receptacles _____.
   a) not having a grounded conductor
   b) in multiwire branch circuits
   c) on 2-wire circuits
   d) all of the above
4. A single receptacle installed on an individual branch circuit shall have a rating not less than _____ of the rating of the branch circuit.
   a) 50%
   b) 80%
   c) 100%
   d) 125%
5. Type _____ cable is a factory assembly of one or more conductors insulated with a highly compressed refractory mineral insulation and enclosed in a liquidtight and gastight continuous copper or alloy steel sheath.
   a) MTW
   b) MI
   c) IGS
   d) FCC
6. Where all loads are connected at the transformer supply points at each end of the tie and overcurrent protection is not provided in accordance with Parts I, II, and VIII of Article 240, the rated ampacity of the tie shall not be less than _____ of the rated secondary current of the highest rated transformer supplying the secondary tie system.
   a) 67%
   b) 75%
   c) 100%
   d) 125%

7. Noninsulated busbars shall have a minimum space of _____ inches between the bottom of the enclosure and the busbar.
   a) 6
   b) 8
   c) 10
   d) 12
8. Insulated conductors on resistors, used for connections between resistance elements and controllers, shall be suitable for an operating temperature of not less than _____.
   a) 60° C
   b) 75° C
   c) 90° C
   d) 100° C
9. _____ on switchboards having exposed live parts on the front of panels shall not have their cases grounded.
   I. Instruments
   II. Meters
   III. Relays
   a) I only
   b) II only
   c) III only
   d) I, II, and III
10. Branch circuits that supply neon tubing installations shall not be rated in excess of _____ amperes.
   a) 15
   b) 20
   c) 30
   d) 40
11. Which of the following equipment shall be grounded around a swimming pool?
   I. All electric equipment located within 5′ of the inside wall of the pool
   II. Ground-fault circuit interrupters
   III. Transformer enclosures
   a) I only
   b) III only
   c) I and III
   d) I, II, and III
12. Askarel-insulated transformers installed indoors, rated over 25 kVA, and installed in a poorly ventilated place shall be furnished with a _____.
   I. means for absorbing any gasses generated by arcing inside the case
   II. pressure relief vent that is connected to a chimney
   III. flue that will carry such gases outside the building
   a) I only
   b) II only
   c) III only
   d) I, II, or III

13. Where the dc system consists of a 3-wire balancer set or a balancer winding with overcurrent protection as provided in 445.12(D), the grounding electrode conductor shall _____.
   a) not be smaller than the neutral conductor
   b) be sized from Table 250.66
   c) not be smaller than 6-AWG copper
   d) not be smaller than any system conductor
14. Duty on escalator driving machine motors shall be rated as _____.
   a) intermittent
   b) continuous
   c) periodic
   d) varying
15. Power-supply cords for mobile homes shall be rated _____ amperes.
   a) 40
   b) 50
   c) 60
   d) 100
16. As a general rule, branch circuits shall not be supplied by autotransformers.
   a) True
   b) False
17. Power-limited fire alarm circuit conductors may be installed in the same raceway as _____.
   I. Class 1 conductors
   II. electric power conductors
   III. non-power-limited fire alarm conductors
   a) I only
   b) II only
   c) III only
   d) none of these
18. Where motors are provided with terminal housing, the housing shall be of _____ and of substantial construction.
   a) metal
   b) plastic
   c) a or b
   d) neither a nor b
19. If a protective device rating is marked on an appliance, the rating of the branch-circuit overcurrent device shall not exceed _____ the protective device rating that is marked on the appliance.
   a) 80%
   b) 125%
   c) 50%
   d) at all

20. The maximum number of 12-AWG RHH conductors that may be installed in a 1½″ × 4″ octagonal box containing a fixture stud is _____.

a) 5

b) 6

c) 7

d) 9

21. How many 12-AWG XHHW conductors can be installed in a ⅜″ flexible metal conduit with the use of inside fittings?

a) 0

b) 1

c) 2

d) 3

22. The paralleling efficiency of ground rods longer than _____ feet is improved by spacing greater than 6 feet.

a) 2.5

b) 6

c) 8

d) 10

23. Signs and outline lighting systems shall be installed so that adjacent combustible materials are not subjected to temperatures in excess of _____.

a) 75° C

b) 90° F

c) 90° C

d) 194° C

24. All bonded parts of a swimming pool shall be connected to an equipotential bonding grid with a solid copper conductor, _____, not smaller than 8 AWG.

I. insulated

II. covered

III. bare

a) I only

b) III only

c) I or III

d) I, II, or III

25. Every transformer shall be provided with a nameplate giving the _____.

I. amount and kind of insulating liquid where used

II. impedance of transformers 25 kVA and larger

III. frequency

a) I and II

b) I and III

c) II and III

d) I, II, and III

26. Each transformer having currents of 9 amperes, or more, rated 600 volts, nominal, or less, (with only primary protection) shall be protected by an individual overcurrent device, rated or set at not more than _____ of the rated primary current of the transformer.

a) 167%

b) 250%

c) 125%

d) 300%

27. A receptacle that provides power for a water-pump motor shall be permitted not less than _____ feet from the inside wall of the pool.

a) 15

b) 10

c) 8

d) 5

28. Type _____ cable, is a mineral-insulated, metal-sheathed cable that is a factory assembly of one or more conductors insulated with a highly compressed refractory mineral insulation and enclosed in a liquid tight and gastight continuous copper or alloy steel sheath.

a) MC

b) MI

c) MV

d) none of these

29. _____ means self-acting, operating by its own mechanism when actuated by some impersonal influence.

a) Self-operating

b) Non-automatic

c) Remote controlled

d) Automatic

30. Intermittent duty is a requirement of service that demands operation for alternate intervals of _____.

I. load and no load

II. load and rest

III. load, no load, and rest

a) I only

b) II only

c) III only

d) I, II, or III

31. Every circuit breaker having an interrupting rating other than _____ amperes shall have its interrupting rating shown on the circuit breaker.

I. 1000

II. 5000

III. 10,000

a) I only

b) II only

c) III only

d) I, II, and III

32. _____ duty is a type of service where both the load and the time intervals may have wide variations.
   a) Varying
   b) Continuous
   c) Periodic
   d) Intermittent
33. To open a single-throw switch, the switch should be mounted so that the switch blade must move _____.
   a) downward
   b) twice
   c) upward
   d) slowly
34. What is the highest voltage allowed in a dwelling occupancy?
   a) 480 volts
   b) 1000 volts
   c) 1200 volts
   d) 1320 volts
35. Which of these is considered an overload?
   a) A short between line A and line B
   b) A short between line A and ground
   c) Both a and b
   d) Neither a nor b
36. Which of these communication cables cannot be run vertically?
   a) CMX
   b) CMP
   c) CMR
   d) All of these
37. Which of these shall not be installed in direct contact with the earth or in areas subject to severe corrosive influences unless protected by an approved corrosion-resistant coating?
   a) PVC schedule 40
   b) Rigid metal conduit
   c) Orangeburg
   d) Ceramic
38. The scope of the *National Electrical Code* includes _____.
   a) all buildings or premises where electrical power is employed
   b) conductors and equipment wherever and whenever electricity is employed
   c) conductors and equipment on certain public or private premises
   d) all of these

39. A swimming pool panelboard not part of the service equipment shall have an equipment grounding conductor installed with the feeder conductors between the grounding terminal of the pool equipment panelboard and _____.

a) the equipotential bonding grid
b) a separate ground
c) the grounding electrode
d) the grounding terminal of the applicable service equipment

40. Identified as used in the *NEC*, in reference to a conductor or its terminals, means that such conductor or terminal is to be recognized as _____.

a) grounded
b) bonded
c) colored
d) marked

41. A(n) _____ conductor is a conductor encased within material of composition or thickness that is not recognized by the *NEC* as electrical insulation.

a) bare
b) insulated
c) wrapped
d) covered

42. A plastic bushing is considered a(n) _____.

a) mounting device
b) fitting
c) electric device
d) insulated device

43. The definition of automatic is self-acting, operating by its own mechanism when actuated by some impersonal influence, such as a change in _____.

I. pressure
II. current strength
III. temperature

a) II only
b) III only
c) II and III
d) I, II, and III

44. Ground-fault protection that functions to open the service disconnect _____ protect(s) service conductors or the service disconnect.

a) will
b) will not
c) adequately
d) totally

45. What is the ampacity of a round copper conductor that has a 2″ diameter?

a) 1000

b) 2000

c) 3142

d) 4000

46. A swimming pool has 230-volt insulated service conductors installed overhead. What is the minimum height of the conductors above the pool deck?

a) 12 feet

b) 14.5 feet

c) 18 feet

d) 22.5 feet

47. Receptacles and attachment plugs on a 30-ampere branch circuit feeding electric-discharge luminaires (lighting fixtures) that are provided with mogul-base, screw-shell lampholders shall not be less than _____ of the luminaire (fixture) full-load current.

a) 130 percent

b) 125 percent

c) 115 percent

d) 100 percent

48. Which of the following support methods is permissible for wiring that is located within the cavity of a fire-rated floor—ceiling or roof—ceiling assembly?

a) The wiring can be secured to the fire-rated assembly.

b) No wiring can be attached to the fire-rated assembly.

c) An independent means of secure support must be provided.

d) The wiring can be laid directly on the fire-rated assembly.

49. When an underfloor raceway outlet is abandoned, discontinued, or removed, the sections of circuit conductors supplying the outlet shall be _____.

a) removed from the raceway

b) capped

c) capped with Wire Nuts®

d) insulated

50. Pendant luminaires (lighting fixtures) in a Class 1, Division 2 location suspended by threaded rigid metal conduit stems longer than _____ inches shall be provided with a permanent and effective bracing against lateral displacement.

a) 12

b) 16

c) 18

d) 24

Exam Number 11

Time Limit: 2 Hours

1. The maximum rating of an Edison-base type plug fuse is _____.
   a) 20 amperes
   b) 30 amperes
   c) 40 amperes
   d) 50 amperes
2. Service conductors run above the top level of a window shall be _____.
   a) 8 feet above the window
   b) accessible
   c) 3 feet above the window
   d) considered out of reach
3. For circuits operating at 1000 volts or more, the conductor between the surge arrester and the line shall not be smaller than _____ copper.
   a) 12-AWG
   b) 10-AWG
   c) 8-AWG
   d) 6-AWG
4. In an emergency wiring system, current supply shall be such that, in the event of failure of the normal supply to, or within, the building or group of buildings concerned, _____ shall be available within the time required for the application but not to exceed 10 seconds.
   I. emergency lighting
   II. emergency power
   III. normal lighting
   a) I or III
   b) II or III
   c) I or II
   d) I, II, or III
5. Which of the following is required for branch circuit temporary wiring?
   a) Conductors shall be protected from overcurrent as provided in 240.4, 240.5, and 240.100.
   b) All branch circuits shall originate in an approved power outlet or panelboard.
   c) Conductors shall be permitted within multiconductor cord of a type identified in Table 400.4 for hard usage or extra-hard usage.
   d) All of these
6. Rigid nonmetallic conduit shall be permitted to be used under the chassis of a mobile home where _____.
   I. closely routed against frames
   II. closely routed against equipment enclosures
   III. exposed to physical damage
   a) I or II
   b) I or III
   c) I, II, or III
   d) none of these

7. Type FCC cable shall be clearly and durably marked on both sides at intervals of not more than 24 inches with the information required by 310.11(a) and with which of the following additional information?
   I. Material of conductors
   II. Maximum temperature rating
   III. Ampacity
   a) I only
   b) III only
   c) I and III
   d) I, II, and III
8. The *National Electrical Code* is _____.
   a) meant to be used as an instruction manual for untrained persons
   b) intended to protect persons and property
   c) published by the Bureau of Standards
   d) intended to be a design manual
9. What size equipment grounding conductor is required for a remote pool panelboard that is protected with a 30-ampere breaker?
   a) 14-AWG Cu
   b) 12-AWG Cu
   c) 10-AWG Cu
   d) 10-AWG Al
10. Bonding jumpers shall be used around concentric or eccentric knockouts to ensure continuity at service equipment and service raceways.
    a) True
    b) False
11. Where practicable, dissimilar metals in contact anywhere in the system shall be avoided to eliminate the possibility of _____.
    a) resistance
    b) galvanic action
    c) hysteresis
    d) specific gravity
12. Where raceway-type service masts are used, all raceway fittings shall be _____ for use with service masts.
    a) heavy-duty
    b) identified
    c) approved
    d) none of these

13. The discharge circuit of a capacitor bank shall not be connected in which way(s)?
    I. Permanently connected to the terminals
    II. Automatic means of connecting it to the terminals
    III. Manual means of connecting the discharge circuit
    a) I only
    b) II only
    c) III only
    d) I or II
14. Appliance receptacle outlets installed in a dwelling unit for specific appliances, such as laundry equipment, shall be installed within _____ of the intended location of the appliance.
    a) 4′
    b) 6′
    c) 8′
    d) 10′
15. Splices and taps shall not be located within luminaire (fixture) _____.
    a) pancake boxes
    b) splice boxes
    c) arms or stems
    d) junction boxes
16. _____ 12-AWG THHN and one bare equipment grounding conductor may be installed in ⅜″ liquidtight flexible metal conduit (fittings outside of conduit).
    a) One
    b) Two
    c) Three
    d) Four
17. Means shall be provided to disconnect the _____ of all fixed electric space-heating equipment from all ungrounded conductors.
    I. heater
    II. motor
    III. motor controller(s)
    IV. supplementary overcurrent protective device(s)
    a) I and II
    b) II and IV
    c) I, II, III, and IV
    d) I, III, and IV
18. A listed sheet metal strap-type ground clamp (having a rigid metal base that seats on the electrode and having a strap of such material and dimensions that it is not likely to stretch during or after installation) is not permitted for indoor telecommunication purposes.
    a) True
    b) False

19. Face plates for flush-mounted snap switches are not required to seat against the finished wall surface.
    a) True
    b) False
20. The maximum ampacity of twenty 12-AWG THW conductors pulled through a 1″ conduit nipple that is 18″ long is _____ amperes each.
    a) 10
    b) 12.5
    c) 20
    d) 25
21. Receptacles located more than _____ above the floor shall not be counted in the required number of receptacles along the wall.
    a) 5′
    b) 4′
    c) 5½′
    d) 6½′
22. Constant wattage heating cables, in poured concrete, shall not exceed _____ watts per linear foot of cable.
    a) 16½
    b) 25
    c) 33
    d) 54
23. What is the minimum size fixture wire?
    a) 22 AWG
    b) 18 AWG
    c) 20 AWG
    d) 16 AWG
24. Underground service conductors carried up a pole must be protected from mechanical injury to a height of at least _____ feet above finished grade.
    a) 8
    b) 10
    c) 12
    d) 15
25. Nonheating leads of heating cables operating in 240-volt systems shall have a _____ identification color.
    a) yellow
    b) blue
    c) red
    d) brown

26. When calculating the maximum number of conductors in a box, two 18-AWG fixture wires are _____.

a) counted as one conductor
b) counted as two conductors
c) not counted
d) none of these

27. Nominal battery voltage shall be calculated on the basis of _____.

a) 2.0 volts per cell for the lead-acid type and 1.5 volts per cell for the alkali type
b) 2.0 volts per cell for the lead-acid type and 1.2 volts per cell for the alkali type
c) 1.2 volts per cell for the lead-acid type and 2.0 volts per cell for the alkali type
d) 1.5 volts per cell for the lead-acid type and 2.0 volts per cell for the alkali type

28. The minimum size service for a mobile home is _____.

a) 60 amperes
b) 80 amperes
c) 100 amperes
d) 200 amperes

29. The minimum headroom of working spaces about motor control centers shall be _____.

a) 6′ ¼″
b) 6′ 3″
c) 6′ 4″
d) 6′ 6″

30. Which of these is covered in the *National Electrical Code?*

I. Optical fiber cable
II. Floating buildings
III. Carnivals

a) I only
b) I and III
c) II and III
d) I, II, and III

31. Partially protected locations under canopies, marquees, roofed open porches, and similar locations are considered _____.

a) dry locations
b) damp locations
c) wet locations
d) weatherproof locations

32. _____ is a generic term for a group of nonflammable synthetic chlorinated hydrocarbons used as electrical insulating media.

a) Hermetic
b) Prynol
c) Askarel
d) Electrolyte

33. _____ conductors shall be used for wiring on luminaire chains and on other movable or flexible parts.

a) Solid

b) Stranded

c) Covered

d) Insulated

34. The connection between two or more portions of the equipment grounding conductor is called a(n) _____.

a) bonding jumper main

b) equipment bonding jumper

c) grounding jumper

d) grounding conductor

35. A structure designed to support a wet-niche luminaire (lighting fixture) assembly and intended for mounting in a pool or fountain structure is a _____.

a) wet-niche shell

b) forming shell

c) dry-niche shell

d) none of these

36. The part of an electrical system that performs a mechanical rather than an electrical function is called a _____.

a) device

b) receptacle

c) parameter

d) fitting

37. What is the largest hole that is allowed to be bored in a 2″ × 4″ stud in a house?

a) ¾″

b) 1″

c) 1½″

d) None of these

38. Which of these wiring methods would not be an approved method to wire an ac receptacle that has to be routed under the chassis of a mobile home?

a) PVC schedule 80

b) UF cable

c) Rigid metal conduit

d) Electrical metallic tubing

39. What is the minimum size copper equipment grounding conductor required for a circuit that is being fed from a 100-ampere breaker?

a) 4 AWG

b) 6 AWG

c) 8 AWG

d) 10 AWG

40. The *National Electrical Code* is intended to be suitable for mandatory application by governmental bodies exercising legal jurisdiction _____.

I. over electrical installations

II. for use by insurance inspectors

a) I only

b) II only

c) neither I nor II

d) both I and II

41. A _____ is an overcurrent protective device with a circuit-opening fusible part that is heated and severed by the passage of overcurrent through it.

a) fuse

b) thermal protector

c) breaker

d) service drop

42. _____ is constructed so that moisture will not enter the enclosure under specified test conditions.

a) Waterproof

b) Watertight

c) Weatherproof

d) Raintight

43. All metal parts that are within _____ feet horizontally of the inside walls of a swimming pool shall be bonded.

a) 5

b) 10

c) 15

d) 20

44. _____ is any current in excess of the rated current of equipment or the ampacity of a conductor.

a) Ampacity

b) Overcurrent

c) Overload

d) Short

45. Fittings, devices, appliances, and fixtures are defined as _____.

a) equipment

b) appliances

c) connectors

d) appurtenance

46. What is the minimum bending radius for ¾-inch flexible metal conduit where it may be infrequently flexed in service after installation?

a) 10 inches

b) 12 inches

c) 17½ inches

d) 20 inches

47. In Class I circuits, overcurrent protection shall not exceed _____ for 16-AWG conductors.

a) 10 amperes

b) 15 amperes

c) 20 amperes

d) 30 amperes

48. Spanish moss is _____.

a) easily ignitable

b) explosive

c) combustible

d) noncombustible

49. Where the rated input current of the autotransformer is less than 9 amperes, an overcurrent device rated or set at not more than _____ of the input current shall be permitted.

a) 100 percent

b) 125 percent

c) 150 percent

d) 167 percent

50. In a wheelchair lift, liquidtight flexible metal conduit shall not exceed _____ feet.

a) 1.8

d) 2

c) 6

d) 10

Exam Number 12

Time Limit: 2 Hours

1. On circuits of 600 volts or less, overhead spans longer than 50 feet in length shall have open individual conductors not smaller than _____-AWG copper unless supported by a messenger wire.
   a) 10
   b) 8
   c) 6
   d) none of these
2. Conductors supplying outlets for arc projectors of the professional type shall not be smaller than _____.
   a) 14 AWG
   b) 12 AWG
   c) 10 AWG
   d) 8 AWG
3. A combination consisting of a compressor and motor, both of which are enclosed in the same housing, with no external shaft or seals, with the motor operating in the refrigerant, is the definition of a(n) _____.
   a) motor-compressor
   b) air conditioner
   c) hermetic refrigerant motor-compressor
   d) none of these
4. Household-type appliances with surface heating elements having a maximum demand of more than _____ amperes calculated in accordance with Table 220.55 shall have its power supply subdivided into two or more circuits, each of which shall be provided with overcurrent protection rated at not over 50 amperes.
   a) 100
   b) 60
   c) 50
   d) 40
5. Receptacles and cord connectors shall be _____.
   a) rated at not less than 30 amperes, 250 volts
   b) rated at not less than 20 amperes, 125 volts
   c) rated at not less than 15 amperes, 125 volts
   d) of a type suitable for use as lampholders
6. Conduit encased in 2 inches of concrete on ground floor shall be considered a _____ location.
   a) moist
   b) wet
   c) damp
   d) dry

7. Type MTW insulation would be used for _____.
   a) feeders only
   b) machine tool wiring
   c) switchboards only
   d) fixtures
8. Heating cables shall be furnished complete with factory-assembled nonheating leads at least _____ feet in length.
   a) 3
   b) 6
   c) 7
   d) 10
9. Accessible wire is _____.
   I. admitting close approach
   II. not permanently closed in by the structure
   III. not guarded by locked doors, elevation, or other methods
   a) I only
   b) II only
   c) I or III
   d) I, II, or III
10. Workmanship of the job is beyond the control of the inspector.
    a) True
    b) False
11. A branch circuit that supplies a number of outlets for lighting and appliances is known as a(n) _____ branch circuit.
    a) appliance
    b) multiwire
    c) multipurpose
    d) general-purpose
12. When a controller is not within sight from the motor location, the disconnect shall be capable of being _____ in the open position.
    a) shut off
    b) locked
    c) up
    d) down
13. Flexible cords shall not be used as a substitute for _____ wiring.
    a) concealed
    b) temporary
    c) fixed
    d) none of these

14. A ground ring encircling the building shall be buried at a depth below the earth's surface of not less than _____ feet.

a) 30

b) 3

c) 2½

d) 2

15. Wires run above heated ceilings and within thermal insulation shall be derated on the basis of _____.

a) 50° C

b) 86° C

c) 20° C

d) 30° C

16. Suitable covers shall be installed on all boxes, fittings, and similar enclosures to prevent accidental contact with _____ parts or physical damage to parts or insulation (over 600 volts, nominal).

a) energized

b) electrical

c) mechanical

d) none of these

17. Circuits for electrical cranes operating over combustible fibers in Class III locations shall not be _____.

a) THWN conductors

b) grounded

c) pulled in raceways

d) spliced

18. Types TPT and TST shall be permitted in lengths not exceeding _____ where attached directly, or by means of a special type of plug, to a portable appliance rated at 50 watts or less and of such nature that extreme flexibility of the cord is essential.

a) 4′

b) 6′

c) 8′

d) 10′

19. What is the minimum spacing between line terminals of a 480-volt motor where the terminal housings enclose rigidly mounted motor terminals?

a) ¼″

b) ⅜″

c) ½″

d) ⅝″

20. General-purpose receptacles shall be located at least _____ feet from the inside walls of a swimming pool.

a) 20

b) 15

c) 10

d) 5

21. The walls and roofs of transformer vaults shall be constructed of materials that have adequate structural strength for the condition with a minimum fire resistance of _____.

a) 3 hours

b) 2 hours

c) 1 hour

d) none of these

22. Feeders containing a common neutral shall be permitted to supply _____.

I. two sets of 3-wire feeders

II. three sets of 3-wire feeders

III. two sets of 4-wire feeders

a) II only

b) I or III

c) II or III

d) I, II, or III

23. Multi-outlet assembly may be used _____.

a) in hoistways

b) in storage battery rooms

c) in dry locations

d) where concealed

24. Electrical systems that are grounded shall be connected to earth in a manner that will _____.

I. limit the voltage imposed by lightning

II. limit the voltage imposed by line surges

III. aid the overcurrent device for shorts and ground faults

a) I only

b) III only

c) I and II

d) I, II, and III

25. A motor control circuit _____.

a) carries the electric signals directing the performance of the controller and carries the main power current

b) carries the electric signals directing the performance of the controller but does not carry the main power current

c) does not carry the electric signals directing the performance of the controller but carries the main power current

d) is any switch or device that is normally used to start and stop a motor by making and breaking the motor circuit current.

26. Multiwire branch circuits shall supply _____.

I. line to line loads

II. line to neutral loads

III. single phase voltages

a) I only

b) II only

c) III only

d) I and II

27. What is the operation of equipment in excess of normal, full-load rating, or of a conductor in excess of rated ampacity that, when it persists for a sufficient length of time, would cause damage or dangerous overheating?

a) Short

b) Continuous

c) Overcurrent

d) Overload

28. The *NEC* is designed for safety regardless of _____.

I. time

II. maintenance

III. efficiency

IV. cost

V. future expansion

a) I and II

b) III and IV

c) I through IV

d) I through V

29. Conductors such as nonmetallic sheath cable, when run in the space between studs, are defined as _____.

a) enclosed

b) concealed

c) inaccessible

d) buried

30. A dry location is classified as a location that _____.

I. is not normally subject to dampness

II. may be temporarily subject to dampness

III. may be temporarily subject to wetness

a) I only

b) I and II

c) I and III

d) I, II, and III

31. Luminaires (fixtures) shall be securely fastened to the suspended ceiling framing member by mechanical means, such as _____.

   I. bolts or screws

   II. rivets

   III. listed clips identified for the use

   a) I only
   b) I and II
   c) I and III
   d) I, II, and III

32. The weight of a fixture that can be supported by an outlet box shall not exceed _____ pounds.

   a) 50
   b) 40
   c) 30
   d) 25

33. What is the thickness required for the sealing compound in an explosionproof fitting that is fed by a ¾″ conduit?

   a) ½″
   b) ⅝″
   c) ¾″
   d) The code does not specify.

34. What is the classification of a motor fuel dispensing facility?

   a) Class I, Division 2
   b) Class I, Division 1
   c) Class II, Division 2
   d) Class II, Division 1

35. Equipment intended to interrupt current at fault levels shall have an _____ rating sufficient for the nominal circuit voltage and the current that is available at the line terminals of the equipment.

   a) ampacity
   b) equipment
   c) interrupting
   d) operating

36. Concrete, brick, and tile walls shall be considered as _____.

   a) wet
   b) dry
   c) grounded
   d) isolated

37. What is the minimum size for fixture wires?
   a) 22 AWG
   b) 20 AWG
   c) 18 AWG
   d) 16 AWG
38. A panelboard contains eight three-pole circuit breakers and four two-pole circuit breakers. The maximum allowable number of single-pole circuit breakers permitted to be added in this panelboard is _____.
   a) ten
   b) twelve
   c) sixteen
   d) eighteen
39. The conductor used to connect the grounded circuit of a wiring system to the grounding electrode is the _____.
   a) main bonding jumper
   b) grounded conductor
   c) grounding conductor
   d) bonding jumper
40. If a protective device rating is marked on an appliance, the branch-circuit overcurrent device rating shall not exceed _____ the protective device rating marked on the appliance.
   a) 125%
   b) 80%
   c) 50%
   d) at all
41. _____ is a general term including material, fittings, devices, appliances, luminaires (fixtures), apparatus, and the like used as part of or in connection with an electrical installation.
   a) Appurtenance
   b) Appliance
   c) Connectors
   d) Equipment
42. Is it permissible to disconnect a grounded circuit conductor through a switch or circuit breaker?
   a) No, the grounded conductor can never be switched.
   b) Yes, if it is disconnected before the ungrounded conductors.
   c) Yes, but only with an isolating switch.
   d) Yes, if all circuit conductors are disconnected simultaneously.
43. In an area that requires explosionproof wiring, raceways entering a box in this area (which contains equipment that may produce sparks) shall be provided with _____.
   a) triple locknuts
   b) two grounding conductors
   c) an approved sealing compound
   d) insulated bushings

44. When one electric circuit controls another electric circuit through a relay, the circuit is a _____ circuit.
   a) remote-control
   b) low-energy power
   c) transfer
   d) pilot
45. Utilization equipment uses _____.
   a) energy for electrical purposes
   b) chemical or mechanical energy
   c) electrical energy only
   d) mechanical energy only
46. Raceways on the outside of buildings shall be _____.
   a) rainproof and arranged to drain
   b) weatherproof and arranged to drain
   c) watertight and arranged to drain
   d) raintight and arranged to drain
47. Something constructed, protected, or treated so as to prevent rain from interfering with the successful operation of the apparatus under specific test conditions is called _____.
   a) raintight
   b) rainproof
   c) watertight
   d) weatherproof
48. Service equipment with eccentric or concentric knockouts shall be bonded by _____.
   a) locknuts
   b) bonding jumpers
   c) approved devices
   d) bushings
49. For garages and outbuildings on residential property, a snap switch or a set of _____ snap switches shall be permitted as the disconnecting means.
   I. 2-way
   II. 3-way
   III. 4-way
   a) I and III
   b) II and III
   c) II only
   d) I, II, and III
50. The *NEC* does not cover _____.
   a) public address systems
   b) optical fiber cable
   c) telephones
   d) premises wiring

Exam Number 13

Time Limit: 2 Hours

1. Liquidtight flexible nonmetallic conduit shall not be permitted in lengths longer than _____ feet.
   a) 4
   b) 6
   c) 8
   d) 10
2. A room air conditioner shall have its branch-circuit requirements determined by which of the following?
   a) It is cord-and-attachment-plug-connected with a rating of not more than 40 amperes and 250 volts, single phase.
   b) Total rated-load current is shown on the room air conditioner nameplate.
   c) The rating of the branch-circuit short-circuit and ground-fault protective device does not exceed the ampacity of the branch-circuit conductors or the rating of the receptacle, whichever is less.
   d) All of the above
3. Where magnesium, aluminum, or aluminum-bronze powders may be present, transformers _____.
   a) must be approved for Class II, Division 1
   b) may be pipe ventilated
   c) are not permitted
   d) must be dusttight
4. Solid dielectric insulated conductors operated above 2000 volts in permanent installations shall have ozone-resistant insulation and shall be _____.
   a) shielded
   b) surface mounted
   c) protected
   d) covered
5. What is the minimum thickness of a sheet metal box measuring 6″ × 4″ × 3½″?
   a) 0.0625″
   b) 15 MSG
   c) 0.0747″
   d) 16 MSG
6. Cells in jars of conductive material shall be installed in trays of nonconductive material with not more than _____ cells in the series circuit in any one tray.
   a) 24
   b) 20
   c) 18
   d) 16

7. Service entrance cables, where subject to physical damage, shall be protected by which of the following?
   I. EMT
   II. IMC
   III. Rigid metal conduit
   a) II or III
   b) III only
   c) I or III
   d) I, II, or III
8. How often must support be provided for vertical runs of 250-kcmil copper?
   a) 60′
   b) 80′
   c) 135′
   d) 180′
9. Wet-niche lighting fixtures shall be connected to an equipment grounding conductor not smaller than _____.
   a) 6 AWG
   b) 8 AWG
   c) 10 AWG
   d) 12 AWG
10. Signs or outline lighting systems operated by electronic or electromechanical controllers located external to the sign or outline lighting system shall be permitted to have a disconnecting means located _____.
    a) only in the controller
    b) only external to the controller
    c) within sight of the controller or in the same enclosure with the controller
    d) within sight of the sign
11. The *National Electrical Code* does not cover standby power generation systems.
    a) True
    b) False
12. Electrical nonmetallic tubing is permitted to be used in sizes up to _____.
    a) ½″
    b) 1″
    c) 2″
    d) 4″
13. At what angle does a header attach to a floor duct?
    a) Right angle
    b) Parallel
    c) Straight
    d) None of these

14. Where raceways containing ungrounded conductors _____ or larger enter a cabinet, box enclosure, or raceway, the conductors shall be protected by a substantial fitting providing a smoothly rounded insulating surface, unless the conductors are separated from the fitting or raceway by substantial insulating material that is securely fastened in place.
    a) 1/0 AWG
    b) 4 AWG
    c) 6 AWG
    d) 2 AWG
15. Where portions of a raceway are subjected to different temperatures and where condensation is known to be a problem, as in cold storage areas of buildings or where passing from the interior to the exterior of a building, the raceway shall be filled with an approved material to prevent the circulation of _____ section of the raceway.
    a) cold air to a warmer
    b) warm air to a colder
    c) any
    d) none of these
16. Battery voltage is calculated on the basis of _____ volts per cell for the lead-acid type and _____ volts per cell for the alkali type.
    a) 2.0; 1.5
    b) 2.0; 1.2
    c) 1.2; 2.0
    d) 1.5; 2.0
17. All electric equipment, including power-supply cords, used with storable pools shall be protected by _____.
    a) GFCIs
    b) double-insulation
    c) circuit breakers
    d) fuses
18. Conductors _____ and larger shall be stranded when installed in raceways.
    a) 10 AWG
    b) 12 AWG
    c) 8 AWG
    d) none of these
19. Silicone rubber insulated fixture wire should be limited to use where the voltage does not exceed _____ volts.
    a) 250
    b) 300
    c) 600
    d) 1000

20. Cord-and-plug-connected appliances connected to branch circuits with other lighting loads shall not exceed _____ of the branch circuit rating.

a) 50%

b) 80%

c) 100%

d) 125%

21. When overcurrent protection is provided as part of an industrial machine, the nameplate of the machine shall be marked _____.

I. with the maximum ampere rating of the short-circuit and ground-fault protective device

II. with the total volt-amps

III. "overcurrent protection provided at machine supply terminals"

a) II only

b) I and II

c) I and III

d) I, II, and III

22. When threading rigid conduit, the standard die should provide a _____-inch taper per foot.

a) ¼

b) ½

c) ¾

d) 1

23. _____ boxes may be weatherproof.

I. Watertight

II. Raintight

III. Rainproof

a) I only

b) II only

c) III only

d) I, II, and III

24. At least _____ inches of free conductor shall be left at each outlet, junction, and switch point for splices or the connection of luminaires (fixtures) or devices.

a) 12

b) 10

c) 8

d) 6

25. Busways rated over 600 volts shall have connection hardware accessible for _____.

I. maintenance

II. connection

III. installation

a) I only

b) II only

c) III only

d) I, II, and III

26. The rated primary current of an electric resistance welder is _____.
    a) determined by the percentage of time during which the welder is loaded
    b) the rated kVA multiplied by 1000 and divided by the rated primary voltage, using values given on the name plate
    c) the current drawn from the supply circuit during each welder's operation at the particular heat tap and control setting used
    d) all of these
27. Electrical equipment can be defined as _____.
    I. fittings
    II. devices
    III. appliances
    IV. luminaires (fixtures)
    a) II only
    b) II, III, and IV
    c) I and II
    d) I, II, III, and IV
28. All _____ installed in theater dressing rooms shall be controlled by wall switches installed in the dressing rooms.
    I. lights
    II. receptacles
    III. receptacles adjacent to the mirror(s) and above the dressing table counter(s)
    a) I only
    b) II only
    c) both I and III
    d) I, II, and III
29. A solidly grounded service disconnect (227/480 volt, 3-phase) rated 1000 amps containing 800-amp fuses is required to have ground fault circuit protection.
    a) True
    b) False
30. The definition of identified is _____.
    a) acceptable to the authority having jurisdiction
    b) equipment to which has been attached a label, symbol, or other identifying mark of an organization acceptable to the authority having jurisdiction
    c) materials included in a list published by an organization acceptable to the authority having jurisdiction and concerned with product evaluation
    d) recognizable as suitable for the specific purpose, function, use, environment, application, and so forth, where described in a particular *Code* requirement
31. An area subjected to quantities of hazardous dust is classified as _____.
    a) Class I, Division 1
    b) Class II, Division 1
    c) Class I, Division 2
    d) Class II, Division 2

32. Places of assembly are designed or intended for the assembly of _____ or more persons.
   a) 50
   b) 100
   c) 200
   d) 400
33. _____ may be connected ahead of the main service disconnect.
   a) Motors
   b) Luminaires
   c) Arresters
   d) Nothing
34. The conductor insulation with the highest temperature rating would be _____.
   a) THWN
   b) RUW
   c) THW
   d) RHH
35. Storable pools, hot tubs, and therapeutic pools are included in Article 680 with swimming pools.
   a) True
   b) False
36. A single receptacle is a _____ contact device with no other contact device on the same yoke.
   a) single
   b) grounded
   c) 2-wire
   d) none of these
37. A _____ is a device or group of devices that serves to govern, in some predetermined manner, the electric power delivered to the apparatus to which it is connected.
   a) branch circuit
   b) controller
   c) switch
   d) control circuit
38. The work space about electric equipment over 600 volts, nominal, where energized parts are exposed, shall not be less than _____ high, or less than _____ wide.
   a) 6¼′; 3′
   b) 6¼′; 3½′
   c) 6½′; 3′
   d) 6½′; 3½′
39. The _____ extend(s) from the service point to the service disconnecting means.
   a) service lateral
   b) service raceways
   c) service conductors
   d) service equipment

40. Where conductors of different insulations are run together, the maximum operating temperature is _____.
    a) the temperature of the lowest conductor temperature rating
    b) limited by the ampacity
    c) the ambient temperature
    d) the ultimate insulation temperature
41. Which of these may be connected to the supply side of the service disconnect?
    I. Load management devices
    II. Fire alarms
    III. Sprinkler alarms
    a) I only
    b) I and II
    c) II and III
    d) I, II, and III
42. Mobile homes are covered in Chapter _____ of the *NEC.*
    a) 4
    b) 5
    c) 7
    d) 8
43. _____ means that the implied action requires personal intervention for its control.
    a) Intermittent
    b) Nonautomatic
    c) Automatic
    d) Varying
44. Article 800 covers all but which of the following?
    a) Radio
    b) Telephone
    c) Outside wiring for fire alarm
    d) Outside wiring for burglar alarm
45. Factory-installed internal wiring need not be inspected at the time of installation of the equipment if the equipment has been listed by _____.
    a) a qualified electrical testing laboratory having the facilities required to test wiring equipment and materials
    b) government organizations
    c) National Fire Underwriters
    d) a qualified electrical testing laboratory that is recognized as having the facilities required to test wiring equipment and materials
46. Which of these is not covered in the *NEC?*
    a) X-ray equipment
    b) Portable X-ray equipment
    c) Radiation safety and performance
    d) Transportable X-ray equipment

47. Optional standby systems do not cover _____.
    a) life support
    b) refrigeration
    c) data processing
    d) private business
48. The words "thermally protected" appearing on the nameplate of a motor indicate that the motor is provided with a _____.
    a) breaker
    b) heat sensing element
    c) switch
    d) fuse
49. The ratio of the maximum demand of a system, or part of a system, to the total connected load of a system or the part of the system under consideration is called the _____.
    a) demand factor
    b) maximum demand
    c) demand ratio
    d) load diversity
50. An oil switch is a switch having contacts that operate under _____.
    I. oil
    II. askarel
    III. other suitable liquids
    a) I only
    b) II only
    c) I or II
    d) I, II, or III

Exam Number 14

Time Limit: 2 Hours

1. Class 2 or Class 3 power sources may have the output connections paralleled or otherwise interconnected where listed for such interconnection.
   a) True
   b) False
2. A grounded circuit conductor may be used for grounding non-current-carrying metal parts of _____ on existing branch-circuit installations.
   I. dryers
   II. ranges
   III. water heaters
   a) I and II
   b) I and III
   c) II and III
   d) I, II, and III
3. When installing office furnishings, convenience receptacles _____ be located in lighting accessories.
   a) duplex-type only can
   b) shall
   c) single-type only can
   d) shall not
4. Which of the following is not required on a motor nameplate?
   a) Watts
   b) Voltage
   c) Horsepower
   d) Manufacturer's identification
5. The ampacity of supply branch-circuit conductors for X-ray equipment shall be based on _____ percent of the momentary rating of the equipment.
   a) 50
   b) 70
   c) 40
   d) 80
6. Fluorescent luminaires (lighting fixtures) may be used as a raceway for circuit conductors if _____.
   a) wired so that conductors are not closer than 3″ from ballast
   b) listed and marked for use as a raceway
   c) connected by a conduit wiring method
   d) none of these

7. Each circuit leading to or through motor fuel-dispensing equipment shall be provided with a clearly identified and readily accessible switch or other acceptable means, located remote from the dispensing devices, to disconnect simultaneously from the source of supply, _____.

   I. all ungrounded conductors
   II. all grounded conductors
   III. all grounding conductors

   a) I only
   b) II only
   c) I and II
   d) I, II, and III

8. In commercial garages generally, the floor area up to a level of 18″ above the floor is designated as _____.

   a) Class II, Division 2
   b) Class II, Division 1
   c) Class I, Division 2
   d) Class I, Division 1

9. Where rigid nonmetallic conduit is used in a bulk storage plant, the raceway shall include _____.

   a) a grounded conductor
   b) sunlight resistant listing
   c) a bushing with double locknuts
   d) an equipment grounding conductor

10. Cablebus shall be permitted to be used for _____.

   I. feeders
   II. services
   III. branch circuits

   a) I and II
   b) I and III
   c) II and III
   d) I, II, and III

11. Which of the following is not a standard size fuse?

   a) 75 amperes
   b) 110 amperes
   c) 125 amperes
   d) 250 amperes

12. What is the minimum working clearance on a circuit 120 volts to ground, exposed live parts on one side and no live or grounded parts on the other side of the working space?

   a) 6 feet
   b) 4 feet
   c) 3 feet
   d) 3½ feet

13. All buildings or portions of buildings or structures designed or intended for gathering together of _____ or more persons are assembly occupancies.
    a) 50
    b) 100
    c) 250
    d) 500
14. Where located within 7 feet of the floor, power-limited fire alarm circuit cables shall be securely fastened in an approved manner at intervals of not more than _____ inches.
    a) 12
    b) 18
    c) 24
    d) 48
15. A main common-return conductor in the electromagnetic supply of an electrically operated pipe organ shall not be less than _____.
    a) 28 AWG
    b) 26 AWG
    c) 20 AWG
    d) 14 AWG
16. A cord connector that is supplied by a permanently connected cord pendant shall be considered a(n) _____.
    a) receptacle outlet
    b) individual outlet
    c) lighting outlet
    d) general-purpose outlet
17. _____ fixture wires shall be permitted to be tapped to the branch circuit conductors of a 30-ampere branch circuit.
    a) 16-AWG and larger
    b) 14-AWG and larger
    c) 12-AWG and larger
    d) 10-AWG and larger
18. Service-entrance conductors may be installed before the manufactured building is erected at the site.
    a) True
    b) False
19. Flexible cords and cables shall be connected to devices and to fittings so that ____ is not transmitted to joints or terminals.
    a) tension
    b) voltage
    c) shock
    d) heat

20. Health care low-voltage equipment that is frequently in contact with the bodies of persons shall not exceed _____.
   a) 115 volts
   b) 50 volts
   c) 10 volts
   d) 8 volts
21. Flexible metal conduit used as a fixture whip to bond a fixture enclosure to a circuit junction box must not exceed _____ feet in length.
   a) 10
   b) 8
   c) 4
   d) 6
22. The definition of a qualified person is a _____.
   a) Master Electrician
   b) person designated by the management of a building to supervise or work on equipment
   c) person who has skills and knowledge related to the construction and operation of the electrical equipment and installations and has received safety training on the hazards involved
   d) person with a degree or license qualifying them to install, maintain, or operate a specific type of equipment or system
23. Rigid metal conduit shall be permitted to be installed in concrete, in direct contact with the earth, or in areas subject to severe corrosive influences where protected by _____ and judged suitable for the condition.
   a) PVC
   b) Orangeburg
   c) corrosion protection
   d) ceramic
24. The rating of a standard lampholder on a circuit operating at less than 50 volts shall be at least _____.
   a) 660 watts
   b) 550 watts
   c) 330 watts
   d) 220 watts
25. Which of the following is not an approved wiring method in all Class II locations?
   a) Threaded bosses
   b) Dusttight boxes
   c) EMT
   d) Flexible connections
26. The largest stranded conductors permitted to be connected to terminals by means of upturned lugs are _____ conductors.
   a) 12-AWG
   b) 10-AWG
   c) 6-AWG
   d) 8-AWG

27. Optical fiber cables transmit light for _____ through an optical fiber.

    I. signaling

    II. communications

    III. control

    a) I only
    b) III only
    c) I and II
    d) I, II, and III

28. Ampacity is _____.

    a) the current-carrying capacity expressed in amperes
    b) the current-carrying capacity of conductors expressed in amperes
    c) the current-carrying capacity of conductors expressed in volt-amperes
    d) the current, in amperes, a conductor can carry continuously under the conditions of use without exceeding its temperature rating

29. A point on the wiring system at which current is taken to supply utilization equipment is called a(n) _____.

    a) tap
    b) outlet
    c) junction
    d) receptacle

30. Open conductors run individually as service-drop conductors shall be _____.

    I. insulated

    II. bare

    III. covered

    a) I only
    b) II only
    c) III only
    d) either I or III

31. A unit of an electrical system that is intended to carry but not utilize electrical energy is a(n) _____.

    a) device
    b) light bulb
    c) motor
    d) appliance

32. A(n) _____ is an assembly of a fuse support with either a fuseholder, fuse carrier, or disconnecting blade.

    a) fuse
    b) circuit breaker
    c) cutout
    d) oil switch

33. The service disconnecting means shall be located _____.
   I. inside
   II. outside
   III. near the meter
   a) I only
   b) II only
   c) III only
   d) I or II
34. Disconnects shall be installed in a(n) _____ position.
   a) diagonal
   b) vertical
   c) horizontal
   d) upright
35. Power outlet cords shall not be used on _____.
   a) sheds
   b) mobile homes
   c) recreational vehicles
   d) temporarily installed equipment
36. Where rigid nonmetallic conduit is feeding underwater audio equipment, a(n) _____ solid or stranded copper bonding jumper shall be installed in the conduit.
   a) 12-AWG insulated
   b) 12-AWG bare
   c) 8-AWG insulated
   d) 8-AWG bare
37. The definition of a branch circuit is _____.
   a) the circuit conductors prior to the final overcurrent device protecting the circuit
   b) the conductors between the service and the sub-panel
   c) the circuit conductors between the final overcurrent device protecting the circuit and the outlet(s)
   d) the circuit conductors between the final overload device protecting the circuit and the outlet(s)
38. Is it permissible to install direct current and alternating current conductors in one pull box?
   a) No, never
   b) Yes, if the ampacity is the same for both conductors
   c) Yes, in dry places
   d) Yes, if insulated for the maximum voltage of any conductor
39. _____ is a system or circuit conductor that is intentionally grounded.
   a) Grounded conductor
   b) Grounding conductor
   c) Grounding electrode conductor
   d) Neutral conductor

40. The definition of concealed is _____.
   a) attached to the surface
   b) not readily visible
   c) surrounded by walls
   d) made inaccessible by the structure or finish of the building

41. Wires, whether stranded or solid, will have the same ampacity if they have the same _____.
   a) diameter
   b) circumference
   c) cross-section area
   d) insulation

42. In a residence, receptacle outlets shall be installed so that no point along the floor line in any wall space is more than _____ feet from an outlet in that space.
   a) 6
   b) 6½
   c) 10
   d) 12

43. Type _____ cable is a factory assembly of one or more insulated circuit conductors with or without optical fiber members enclosed in an armor of interlocking metal tape, or a smooth or corrugated metallic sheath.
   a) MC
   b) MI
   c) MV
   d) none of these

44. The definition of a dry location says that it _____.
   a) may be temporarily subject to wetness
   b) is not normally subject to dampness
   c) is not normally subjected to wetness
   d) all of the above

45. Which of the following is not a type of optical fiber cable?
   a) Conductive
   b) Nonconductive
   c) Composite
   d) AC

46. Which of the following type(s) of fixtures shall be permitted to be installed inside a closet?
   I. A recessed incandescent fixture with a glass lens
   II. A recessed fluorescent fixture
   III. A pendant fixture
   a) I or II
   b) I or III
   c) II or III
   d) I, II, or III

47. The conductors and equipment for delivering electric energy from the serving utility to the wiring system of the premises served is called the _____.

a) service

b) primary

c) distribution

d) main supply feeder

48. The _____ is the area between the top of direct burial cable and the top surface of finished grade.

a) gap

b) notch

c) cover

d) none of these

49. A fitting is _____.

a) part of a wiring system that is intended primarily to perform a mechanical function

b) part of a wiring system that is intended primarily to perform an electrical function

c) used for pulling cable into a confined area

d) to be suitable or proper for protection of the conductors

50. *Code* requirements may be waived _____.

a) under no circumstances

b) by the utility company

c) by the fire department

d) by the authority having jurisdiction

Exam Number 15

Time Limit: 2 Hours

1. A blue lead wire in a heating cable means the circuit voltage is _____ volts.
   a) 277
   b) 240
   c) 208
   d) 120
2. Nonmetallic surface extensions shall be secured in place by approved means at intervals not exceeding _____ inches.
   a) 16
   b) 10
   c) 8
   d) 6
3. The grounded (neutral) conductors brought to the service shall not be smaller than the required grounding electrode conductor.
   a) True
   b) False
4. Steel cable trays shall not be used as equipment grounding conductors for circuits with ground-fault protection above _____ amperes.
   a) 150
   b) 300
   c) 600
   d) 1200
5. A _____ location is a location in which easily ignitable fibers are stored or handled other than in the process of manufacture.
   a) nonhazardous
   b) Class III, Division 1
   c) Class II, Division 2
   d) Class III, Division 2
6. All heating elements that are _____ shall be legibly marked with the ratings in volts and watts or in volts and amperes.
   I. rated over 1 ampere
   II. part of an electric heater
   III. replaceable in the field
   a) I and II
   b) I and III
   c) II and III
   d) I, II, and III

7. Aboveground storage tanks shall be classified _____ for the area between 5 feet and 10 feet from the open end of a vent, extending in all directions.
   a) Class I, Division 1
   b) Class I, Division 2
   c) Class II, Division 1
   d) Class II, Division 2
8. A(n) _____ branch circuit consists of two or more ungrounded conductors having a voltage between them, and a grounded conductor having equal voltage between it and each ungrounded conductor of the circuit, and that is connected to the neutral or grounded conductor of the system.
   a) multipurpose
   b) general-purpose
   c) multiwire
   d) individual
9. For garages and outbuildings on residential property, a _____ shall be permitted as the disconnecting means.
   a) single-pole switch
   b) set of 3-way or 4-way switches
   c) either a or b
   d) None of these are permitted.
10. The *National Electrical Code* prohibits all splices and taps in any type of extension cord.
    a) True
    b) False
11. 100-ampere fuses rated 600 volts can be used on a _____ system.
    I. 120-volt
    II. 240-volt
    III. 480-volt
    a) I or III
    b) II or III
    c) III only
    d) I, II, or III
12. Doorbell wiring of Class 2 circuits in a residence may be run in the same raceway with conductors of _____.
    a) electric light circuits
    b) electric power circuits
    c) nonpower-limited fire alarm circuits
    d) none of these
13. It is the intent of the *National Electrical Code* that _____ wiring need not be inspected at the time of installation of the equipment.
    a) factory-installed external
    b) outside
    c) factory-installed internal
    d) underground

14. The maximum voltage between conductors in a swimming pool luminaire (light fixture) installation shall not exceed _____.
   a) 150 volts
   b) 125 volts
   c) 24 volts
   d) 12 volts
15. Where the *Code* specifies that one equipment shall be "in sight from" another equipment, the specified equipment is to be visible and not more than _____ feet distant from the other.
   a) 25
   b) 50
   c) 75
   d) 100
16. Duct heaters installed within 4 feet of heat pump outlets may require _____ on the inlet side of the duct heater to ensure an even distribution of air over the face of the heater.
   I. turning vanes
   II. pressure plates
   III. other devices
   a) I only
   b) II only
   c) I or II
   d) I, II, or III
17. A branch circuit feeding a sign that supplies a combination of incandescent and fluorescent lamps shall be rated not to exceed _____.
   a) 15 amperes
   b) 20 amperes
   c) 30 amperes
   d) 50 amperes
18. In normal practice, the *NEC* discourages the practice of switching the grounded conductor. However, in one of the following cases, the *Code* makes an exception and actually requires that this be adopted. Select that case.
   a) Electrical circuits that lead into or pass through paint spray booths
   b) Electrical circuits that lead into or pass through wall-mounted fixtures mounted at a height of less than 5′ 6″ and located in pediatric wards
   c) Electrical circuits that lead into or pass through a gasoline dispensing pump
   d) Electrical circuits that lead into or pass through wet niche light fixtures located in swimming pools
19. Where connected to a branch circuit having a rating in excess of _____ amperes, lampholders shall be of the heavy-duty type.
   a) 20
   b) 25
   c) 30
   d) 35

20. Type MC cable installed outside of buildings or as aerial cable shall comply with ______.
    I. 225.10
    II. 396.10
    III. 396.12
    a) I only
    b) II only
    c) II and III
    d) I, II, and III
21. What is the difference in burial depth between PVC underground and PVC under two inches of concrete?
    a) 6 inches
    b) 12 inches
    c) 18 inches
    d) 24 inches
22. Nonmetallic auxiliary gutters shall be supported at intervals not to exceed ______ and at each end or joint.
    a) 3′
    b) 5′
    c) 6′
    d) 10′
23. ______ of insulating material shall be permitted to be used without boxes in exposed nonmetallic-sheathed cable wiring.
    I. Switch
    II. Outlet
    III. Tap devices
    a) I only
    b) II only
    c) III only
    d) I, II, and III
24. X-ray equipment mounted on a permanent base with wheels, casters, or a combination of both to facilitate moving the equipment while completely assembled is classified as ______.
    a) mobile
    b) portable
    c) transportable
    d) none of these
25. Storage batteries used as a source of power for emergency systems shall be ______.
    I. of the acid type
    II. of the alkali type
    III. provided with automatic battery charging means
    a) I only
    b) II only
    c) II or III
    d) I or II, and III

26. An appliance that is not easily moved from one place to another in normal use is a _____ appliance.
   a) stationary
   b) fixed
   c) dwelling-unit
   d) fastened-in-place
27. Thermal _____ is a protective device for assembly as an integral part of a motor-compressor that, when properly applied, protects the motor against dangerous overheating due to overload and failure to start.
   a) cutout
   b) protector
   c) heat anticipator
   d) overcurrent
28. Which chapter of the *National Electrical Code* covers power limited circuits?
   a) Chapter 7
   b) Chapter 5
   c) Chapter 8
   d) Chapter 6
29. Motor horsepower ratings are based on an observable safe temperature rise above ambient temperature. The ambient temperature is taken as _____.
   a) 60° C
   b) 50° C
   c) 45° C
   d) 40° C
30. An accessible conductor _____.
   a) can be reached without use of a ladder
   b) is not guarded by locked doors, elevation, or other methods
   c) admits close approach
   d) is not permanently enclosed by a structure
31. Continuous duty is _____.
   a) operation at a substantially constant load for an indefinitely long time
   b) intermittent operation in which the load conditions are regularly recurrent
   c) a load where the maximum current is expected to continue for 1 hour or more
   d) a load where the maximum current is expected to continue for 3 hours or more
32. _____ are used to provide explanatory material (referenced to other sections), precautions to be observed, and further mandatory requirements.
   a) Listings
   b) Exceptions
   c) Mandatory rules
   d) Fine print notes

33. Listed equipment must be installed or used in accordance with _____.
   a) any instructions provided by the manufacturer of the equipment
   b) the electrician
   c) the authority having jurisdiction
   d) any instructions included in the listing or labeling

34. Which of the following is not permitted for connecting conductors 6 AWG and larger to terminal parts?
   a) Wire-binding screws
   b) Splices to flexible leads
   c) Solder lugs
   d) Pressure connectors

35. Optional standby systems are intended to protect _____.
   I. private business
   II. refrigeration systems
   III. data processing
   a) I or II
   b) I or III
   c) II or III
   d) I, II, or III

36. The highest current at rated voltage that a device is intended to interrupt under standard test conditions is its _____ rating.
   a) maximum
   b) operating
   c) rated
   d) interrupting

37. A current-limiting overcurrent protective device is a device that _____ the current flowing in the faulted circuit.
   a) maintains
   b) reduces
   c) increases
   d) none of these

38. Connections of conductors to terminal parts shall ensure _____.
   a) proper bonding
   b) proper torque values
   c) proper crimping pressure
   d) a thoroughly good connection without damaging the conductors

39. Approved is _____.
   a) UL listed only
   b) listed and labeled equipment
   c) acceptable to the authority having jurisdiction
   d) tested and approved for the purpose by a qualified testing lab

40. The type letter for moisture- and heat-resistant rubber is _____.

a) THW
b) RHW
c) MRH
d) RUH

41. Flexible cords can be used to _____.

I. prevent the transmission of noise or vibration
II. wire cranes or hoists
III. run through holes in floors

a) I only
b) II only
c) I and II
d) I, II, and III

42. All wiring must be installed so that when completed it _____.

a) will withstand a hi-pot test
b) is acceptable to *Code* compliance authorities
c) is free of shorts and unintentional grounds
d) meets the current-carrying requirements of the load

43. Compliance with the provisions of the *Code* will result in _____.

a) an efficient system
b) convenient operation
c) good electrical service
d) freedom from hazard

44. Accessible (as applied to wiring methods) means _____.

I. capable of being removed or exposed without damaging the building structure or finish
II. not permanently closed in by the structure or finish of the building

a) I only
b) II only
c) I or II
d) neither I nor II

45. Covered; shielded; fenced or enclosed by means of suitable covers, casings, barriers, rails, screens, mats, or platforms is the definition of _____.

a) guarded
b) enclosed
c) isolated
d) protected

46. What type of transformer is used on a swimming pool wet-niche fixture?

a) An auto-transformer listed as a swimming pool transformer
b) A grounded auto-transformer
c) An isolated winding type transformer listed as a swimming pool transformer
d) None of these

47. The definition of overload is _____.

a) the heaters in a motor circuit

b) any current in excess of the rated current of equipment or the ampacity of a conductor

c) the operation of equipment in excess of normal full-load rating, or of a conductor in excess of rated ampacity

d) a thermal relay used to protect a motor

48. A requirement of service that demands operation for alternate intervals of (1) load and no load; (2) load and rest; or (3) load, no load, and rest is called _____ duty.

a) periodic

b) intermittent

c) variable

d) short-time

49. The connection between the grounded circuit conductor and the equipment grounding conductor at the service is called the _____ bonding jumper.

a) main

b) circuit

c) electrode

d) equipment

50. Equipment grounding conductors shall have the same ampacity as circuit conductors for 15-, 20-, and 30-ampere circuits.

a) True

b) False

Exam Number 16

Time Limit: 2 Hours

1. Service lateral is defined as the underground service conductors between _____.
   a) the service point and the first point of connection to the service entrance conductors in a terminal box, meter, or other enclosure
   b) the street main and the first point of connection to the service entrance conductors in a terminal box, meter, or other enclosure
   c) the street main and the service point
   d) the terminals of the service equipment and the point of connection
2. The sum of the cross-sectional areas of all contained conductors at any cross section of a sheet metal auxiliary gutter shall not exceed _____ percent of the interior cross-sectional area of the sheet metal auxiliary gutter.
   a) 20
   b) 25
   c) 33
   d) 75
3. A hermetic refrigerant motor-compressor branch-circuit short-circuit and ground-fault protective device shall have a rating or setting not exceeding _____ percent of the motor-compressor rated-load current or branch-circuit selection current, whichever is greater.
   a) 115
   b) 125
   c) 175
   d) 225
4. The installation and operation of optional standby systems are _____.
   a) not covered in the *NEC*
   b) covered under Article 701
   c) covered under Article 700: Emergency Systems
   d) covered in their own section of the *NEC*
5. The branch circuit conductors supplying one or more units of a data processing system shall have an ampacity not less than _____ of the total connected load.
   a) 50%
   b) 75%
   c) 80%
   d) 125%
6. Not more than _____ feet of secondary circuit conductors shall be permitted in nonmetallic conduit from a high-voltage terminal of a transformer to the first neon tube electrode.
   a) 15
   b) 20
   c) 50
   d) 100

7. Where ventilation is required in transformer vaults, ventilation openings shall be located as far away as possible from _____.

   I. fire escapes

   II. doors and windows

   III. combustible material

   a) II and III
   b) I and II
   c) III only
   d) I, II, and III

8. Mobile home service equipment shall be located in sight from and not more than _____ feet from the exterior wall of the mobile home it serves.

   a) 20
   b) 30
   c) 40
   d) 50

9. All transformer vault ventilation openings to the indoors shall be provided with automatic closing fire dampers that operate in response to a vault fire. Such dampers shall possess a standard fire rating of not less than _____.

   a) 1½ hours
   b) 2 hours
   c) 2½ hours
   d) 3 hours

10. Single conductors that are installed in a cable tray shall be securely bound in circuit groups to prevent _____ due to fault-current magnetic forces.

    a) voltage surges
    b) current unbalance
    c) excessive movement
    d) inductive reactance

11. Storage batteries used as sources of power for emergency systems shall be of suitable rating and capacity to supply and maintain the total load for a minimum period of _____.

    a) 1 hour
    b) 1½ hours
    c) 2 hours
    d) 3 hours

12. Secondary circuits of current and potential instrument transformers _____ where the primary windings are connected to circuits of 300 volts or more to ground.

    a) shall be grounded
    b) shall not be grounded
    c) shall be bonded
    d) none of the above

13. All boxes and enclosures (including transfer switches, generators, and power panels) for emergency circuits shall be permanently marked so they will be _____ as a component of an emergency circuit or system.

a) listed

b) classified

c) readily identified

d) easily recognized

14. A photo cell is mounted in an 88-cubic-inch box on a tree. The box is not mounted to the tree. One rigid conduit is threaded into the top of the box and one into the bottom. The conduits have straps within 18 inches of the box. What additional means of support for the box is required by the *NEC?*

a) Two #12 pan head sheet metal screws

b) Two #10 pan head sheet metal screws

c) One lag bolt

d) No additional support is required by the *NEC.*

15. A luminaire (lighting fixture) that weighs more than _____ pounds shall be supported independently of the outlet box unless the outlet box is listed for the weight to be supported.

a) 6

b) 20

c) 30

d) 50

16. Mobile X-ray equipment is mounted on a _____ base with wheels, casters, or a combination of both to facilitate moving the equipment while completely assembled.

a) permanent

b) temporary

c) transportable

d) portable

17. _____ shall be permitted to be installed in concrete, in direct contact with the earth, or in areas subject to severe corrosive influences where protected by corrosion protection and judged suitable for the condition.

a) PVC

b) Orangeburg

c) Rigid metal conduit

d) Ceramic conduit

18. Live parts of generators operated at more than _____ to ground shall not be exposed to accidental contact where accessible to unqualified persons.

a) 50 volts

b) 65 volts

c) 150 volts

d) 250 volts

19. In community antenna television and radio distribution systems, the grounding conductor shall not be smaller than 14 AWG. It shall have a current-carrying capacity approximately equal to that of the _____ of the coaxial cable.
   a) inner shield
   b) outer shield
   c) inner conductor
   d) outer conductor
20. The installation of underfloor raceways shall be permitted _____.
   I. beneath the surface of a concrete floor
   II. in areas where subject to corrosive vapors
   III. in office occupancies where laid flush with the concrete floor and covered with linoleum
   a) I and II
   b) I and III
   c) II and III
   d) I, II, and III
21. The feeder neutral conductor must be capable of carrying the _____ load.
   a) average unbalanced
   b) average demand
   c) maximum connected
   d) maximum unbalanced
22. All lights and any receptacles adjacent to the mirror(s) and above the dressing table counter(s) installed in theater dressing rooms shall be controlled by wall switches installed in the _____.
   a) stage office
   b) dressing rooms
   c) projection room
   d) control room
23. Expansion fittings and telescoping sections of metal raceways shall be made electrically continuous by equipment _____ or other means.
   a) grounding conductors
   b) bonding jumpers
   c) bonding conductors
   d) bonding straps
24. Data processing equipment shall have a rating of _____ of the ampacity of the branch circuit conductors.
   a) 50%
   b) 75%
   c) 80%
   d) 125%

25. Where a change occurs in the size of the ungrounded conductor, _____.
   a) the maximum unbalanced load shall be considered as 70 percent of the load on the ungrounded conductors
   b) the grounded conductor shall be permitted to be not more than two AWG sizes smaller than the ungrounded conductors
   c) a similar change shall be permitted to be made in the size of the grounded conductor
   d) the grounding conductor shall not be required to be larger than the ungrounded conductor
26. Cooling of electrical equipment within enclosures is _____.
   a) covered by the *National Electrical Code*
   b) covered by Article 440
   c) not required
   d) the responsibility of the equipment manufacturer
27. An outlet where one or more receptacles are installed is called a _____.
   a) duplex outlet
   b) tri-plex outlet
   c) receptacle outlet
   d) multi-outlet assembly
28. Article 800 of the *NEC* does not cover _____.
   a) telephone
   b) telegraph
   c) radio
   d) outside wiring for fire alarm
29. In copper-clad aluminum conductors, the copper forms a minimum of _____ percent of the cross-sectional area of a solid conductor or each strand of a stranded conductor.
   a) 25
   b) 20
   c) 15
   d) 10
30. Article 680 of the *National Electrical Code* covers all of the following except _____.
   a) bathtubs
   b) hot tubs
   c) fountains
   d) therapeutic pools
31. Wooden plugs are not allowed to anchor electrical equipment to concrete walls.
   a) True
   b) False
32. The term for equipment that has no live parts exposed to a person on the operating side of the equipment is _____.
   a) externally operable
   b) interrupted
   c) dead front
   d) isolated

33. In anesthetizing areas, all non-current-carrying conductive portions of fixed electric equipment operating at more than 10 volts between conductors shall be _____.
   a) shielded
   b) grounded
   c) unshielded
   d) ungrounded
34. The service disconnecting means for each set of service-entrance conductors shall consist of not more than _____ switches.
   a) two
   b) six
   c) ten
   d) twelve
35. Covers of outlet boxes having holes through which flexible cord pendants may pass shall be provided with _____.
   a) nonmetallic fiber ducts
   b) inserts
   c) nipples
   d) approved bushings
36. A luminaire (fixture) that weighs more than _____ pounds shall not be supported by the screw shell of a lampholder.
   a) 3
   b) 6
   c) 10
   d) 16
37. Where a permanently installed pool is installed at a dwelling unit, at least one 125-volt convenience receptacle shall be located a minimum of _____ from, and not more than _____ from the inside wall of the pool.
   a) 5 feet; 15 feet
   b) 10 feet; 20 feet
   c) 10 feet; 15 feet
   d) 15 feet; 20 feet
38. What thickness is required for the sealing compound in an explosionproof fitting that is fed by a 1″ conduit?
   a) 5⁄8″
   b) 3⁄4″
   c) 1″
   d) The *Code* does not specify.
39. A multifamily dwelling is a building containing _____ or more dwelling units.
   a) 1
   b) 2
   c) 3
   d) 4

40. Power-limited fire alarm circuits shall have overcurrent devices, where required, located _____.
  a) on the supply side of the transformer
  b) on the secondary side of the transformer
  c) on the load side of the conductor
  d) at the point where the conductor to be protected receives its supply
41. What is the maximum number of duplex receptacles allowed on a two-wire, 20-amp circuit?
  a) 7
  b) 9
  c) 11
  d) 13
42. Receptacles located around swimming pools shall be located at least _____ feet from the inside walls of the pool.
  a) 3
  b) 5
  c) 10
  d) 20
43. What size lampholder is required for a 1000-watt incandescent light bulb?
  a) Admedium
  b) Medium
  c) Edison
  d) Mogul
44. A hybrid system is a system comprising multiple power sources. These power sources may include all but which of the following?
  a) Wind
  b) Batteries
  c) Photovoltaic
  d) Micro-hydro generators
45. Where the conductor material is not specified, the material and the sizes given in the *Code* shall apply to _____ conductors.
  a) copper
  b) aluminum
  c) copper-clad aluminum
  d) AL CU
46. The dc resistance of a conductor is highest when rated _____.
  a) 60° C
  b) 75° C
  c) 80° C
  d) 90° C

47. Conductor sizes are expressed _____.

a) in diameter or area

b) only in circular mils

c) in AWG or millimeters

d) in AWG or in circular mils

48. Loads for structural additions to an existing dwelling unit or for a previously unwired portion of an existing dwelling unit, either of which exceeds _____ square feet, shall be calculated in accordance with 220.12 and 220.14.

a) 500

b) 300

c) 100

d) 46.5

49. Unless identified for use in the operating environment, no conductors or equipment shall be located in damp or wet locations where exposed to _____.

I. gases

II. fumes

III. vapors

a) I only

b) II only

c) II and III

d) I, II, and III

50. Conductors, splices, and taps within enclosures of switches or overcurrent devices shall not fill the wiring space at any cross section to more than _____ percent of the cross-sectional area of the space.

a) 30

b) 40

c) 60

d) 75

Highlight List

APPENDIX

A

Introduction

Taking an electrical exam with a highlighted *Code* book is a tremendous time saver and can be the difference between pass and fail. Many students have reported back after passing their exam, saying that although highlighting their *Code* book took time, it was invaluable. A great deal of time has been spent compiling and fine-tuning this highlighting list. Do not overlook this essential element when preparing for an electrical exam.

One potential problem can arise from using a highlighted *Code* book: nonhighlighted sections of the *Code* can fade into the background and become nearly invisible. Although this highlight list is quite extensive, not all answers will be found in the highlighted *Code* sections. The only way to guarantee that every answer will come from highlighted *Code* sections is to highlight all the text in the *Code,* and that would defeat the purpose of highlighting your book. Provided you are aware of this, a highlighted *Code* book should not be a distraction.

Highlighting your *Code* book will take approximately eight to ten hours. While a highlighted *Code* book can be purchased, it is not recommended. It is also not recommended to have someone else highlight your *Code* book. Highlighting your own book is part of the learning process. Set aside plenty of time to highlight your *Code* book. Don't try to finish this task in one day. As you are highlighting, take time to read the text you are highlighting. Having taken your time and having read the highlighted sections will be extremely helpful during your exam.

During an exam, if you are not familiar with a question, search in nonhighlighted and highlighted sections to find the answer. If you read the question and don't know the answer, but you know you have highlighted the section containing the answer, search in the highlighted sections. This is where you will benefit from reading what you have highlighted.

In a few jurisdictions, taking a highlighted *Code* book into an exam is not allowed. Therefore, before highlighting your *Code* book, make sure a highlighted *Code* book is permitted in the exam you will be taking.

Highlight List

90.1(A)
90.1(B)
90.1(C)
90.2(A)
90.2(A)(1) and (2)
90.2(B)(4)
90.4—(1st sentence)
90.4—(2nd paragraph)
90.5(C)—(1st paragraph)
90.7—(2nd paragraph)
Art. 100—Accessible (as applied to equipment)
Art. 100—Accessible (as applied to wiring methods)
Art. 100—Accessible, Readily
Art. 100—Ampacity
Art. 100—Appliance
Art. 100—Approved
Art. 100—Askarel
Art. 100—Automatic
Art. 100—Bonding (Bonded)
Art. 100—Bonding Jumper, Equipment
Art. 100—Bonding Jumper, Main
Art. 100—Branch Circuit
Art. 100—Branch Circuit, General-Purpose
Art. 100—Branch Circuit, Multiwire
Art. 100—Cabinet
Art. 100—Concealed
Art. 100—Conductor, Covered
Art. 100—Continuous Load
Art. 100—Controller
Art. 100—Copper-Clad Aluminum
Art. 100—Dead Front
Art. 100—Demand Factor
Art. 100—Device

Art. 100—Duty, Continuous
Art. 100—Duty, Intermittent
Art. 100—Duty, Varying
Art. 100—Dwelling, Multifamily
Art. 100—Equipment
Art. 100—Feeder
Art. 100—Festoon Lighting
Art. 100—Fitting
Art. 100—Grounded
Art. 100—Grounding Conductor
Art. 100—Guarded
Art. 100—Hoistway
Art. 100—Identified
Art. 100—In Sight From
Art. 100—Interrupting Rating
Art. 100—Location, Damp
Art. 100—Location, Dry
Art. 100—Location, Wet
Art. 100—Multioutlet Assembly
Art. 100—Nonautomatic
Art. 100—Outlet
Art. 100—Overcurrent
Art. 100—Overload
Art. 100—Power Outlet
Art. 100—Qualified Person
Art. 100—Rainproof
Art. 100—Receptacle
Art. 100—Receptacle Outlet
Art. 100—Remote-Control Circuit
Art. 100—Separately Derived System
Art. 100—Service
Art. 100—Service Conductors
Art. 100—Service Lateral (1st sentence)
Art. 100—Switch, Isolating
Art. 100—Thermally Protected (as applied to motors)
Art. 100—Thermal Protector (as applied to motors)
Art. 100—Thermal Protector (FPN)
Art. 100—Voltage (of a circuit)
Art. 100—Watertight
Art. 100—Weatherproof
Art. 100, Part II—Cutout
Art. 100, Part II—Fuse
Art. 100, Part II—Oil Switch
110.3(A)(2)
110.3(A)(4)
110.3(B)
110.5
110.6
110.7
110.9
110.11—(1st sentence)
110.12
110.12(A)
110.13(A)—(2nd sentence)
110.13(B)—Cooling
110.14—(2nd sentence)
110.14(A)
110.14(B)
110.15—orange in color
110.15—(last sentence)
110.19
Table 110.26(A)(1)
Table 110.26(A)(1)—Condition 2
110.26(A)(2)—(1st sentence)
110.26(E)
110.27(A)
110.32
110.34(C)
Table 110.34(E)—601-7500 and 9
110.54(B)—(1st sentence)
200.6(A)—(1st sentence)
200.6(A)(1)
200.10(A)
210.3—(2nd sentence)
210.4(B)
210.4(C)
210.4(C) Ex.1
210.4(C) Ex.2
210.6(D)(1) b
210.8(B)
210.8(B)(1)
210.9
210.12(B)
210.19(A)(1)—(2nd sentence)
210.19(A)(1) FPN No. 4
210.19(A)(3) Ex.1 (1st sentence)
210.19(A)(3) Ex. 2
210.21(A)—(1st sentence)
210.21(B)(1)
Table 210.21(B)(3)
210.23(A)(1)
210.23(A)(2)
Table 210.24—15-amp circuit = 15-amp receptacle
210.50(A)
210.50(C)
210.52—(1st paragraph)
210.52(A)(1)
210.52(A)(3)
210.62

210.70(A)(2)(b)
215.2(A)(2)
215.2(A)(3) FPN No. 2
215.4(A)
215.5—(1st sentence)
220.14(F)
220.14(I)
220.16(A)(1)
220.18(B)
220.54—5000 watts
220.55—twice the
220.61(A)—maximum unbalanced load
220.61(B)—70 percent
220.61(B)(1)
220.61(B)(2)
220.61(C)
225.4—(1st sentence)
225.6(A)(1)
225.6(B)
225.18—12 feet
225.19(B)
225.19(C)
225.22
225.36 Ex.
230.2—(1st sentence)
230.9(A) Ex.
230.22
230.23(B)
230.24(B)—12 feet
230.26—(last sentence)
230.28—(2nd sentence)
230.31(B)
230.40 Ex. 2
230.50(A)
230.51(A)
230.53
230.54(C) Ex.
230.62(B)
230.70(A)(1)
230.71(A)—(1st sentence)
230.82(4) and (5)
230.95
230.95(C) FPN No. 1—(1st sentence)
230.204(A)
230.204(D)—(1st sentence)
240.2—Current-Limiting Overcurrent Protective Device
240.4(C)
240.4(B)(2)
240.6(A)
240.21(b)(4)(2)
240.23
240.24(D)
240.33—(1st sentence)
240.40—(1st sentence)
240.51(A)
240.51(B)
240.61
240.83(B)
240.83(C)—(1st sentence)
240.100(B)
250.4(A)(1)
250.4(A)(2)
250.10
250.10(2)
250.20(A)(1)—exceeds 150
250.20(A)(2)—un (only "un" of ungrounded)
250.20(A)(3)
250.21—(last sentence)
250.22(1)
250.24(C)(1)—(1st sentence)
250.28
250.28(D)—(2nd sentence)
250.30(A)(1) Ex. 3
250.32(A)—(1st sentence)
250.32(B)(2)
250.34(A)
250.34(A)(1)
250.34(B)
250.52(A)(1)
250.52(A)(4)
250.52(A)(6)
250.53(B)—(1st sentence)
250.53(F)
250.53(H)
250.56
250.56 FPN
250.64(B)—(2nd sentence)
250.64(C)
250.64(C)(1)
250.64(E)—(1st sentence)
250.66(B)
250.70(3)
250.92(B)—(last paragraph)
250.97
250.97(2)
250.98
250.102(C)—(2nd sentence)
250.102(E)—(2nd sentence)
250.112(B)
250.116 FPN
250.118(5)(C)

250.119
250.122(A)—(1st sentence)
250.122(B)
250.122(C)
250.122(F)—(1st sentence)
250.140
250.140 Ex.
250.142(A)
250.142(A)(1)
250.162(A)
250.162(A) Ex. 3
250.166(A)
250.170
250.172 Ex.
250.174(C)—(1st sentence)
250.178
250.184(A)
250.184(C)(1)—(sentence and a., b., and c.)
280.2—(Surge Arrester)
280.3
280.4(A)—(1st sentence)
280.23
300.3(C)(1)—(1st and 2nd sentence)
300.4(A)(1)
300.4(F)—(1st sentence)
300.5(D)(1)
300.6(A)
300.6(A)(3)
300.6(B)
300.6(C)
300.6(D)
300.7(A)
300.11(A)
300.13(B)
300.14
Table 300.19(A)
300.20(B) FPN
300.21—(1st sentence)
300.22(A)
300.22(B)—(1st and 2nd sentences)
300.31
300.50(B)—(2nd sentence)
310.3
310.6—(1st sentence)
310.10
310.11(A)(1), (2), and (3)
310.11(C)(1)
310.12(B)
310.13 FPN—(2nd sentence)
310.13 FPN—(underline "normal temperatures")
Table 310.13—Type MTW
Table 310.13—Machine tool wiring
Table 310.13—Type RHH
Table 310.13—Type RHW
Table 310.13—Type THW
310.15(B)(2)(a)
310.15(B)(2)(a) Ex. 3
310.15(B)(3)
310.15(B)(4)(a)
312.6(C)
312.8
312.11(A)(3)
314.16(B)
314.16(B)(1) Ex.
314.16(B)(2)
314.16(B)(4)
314.16(C)(2)
314.23(F)
314.24
314.27(B)
314.27(D)
314.28(B)
314.40(B)—(1st sentence)
314.42—(1st sentence)
314.71(A)
320.30(D)(2)
322.10—(1st sentence)
322.10(2), (3), and (4)
322.56(B)—(last sentence)
324.10(B)(2)
324.10(H)
324.40(A)
324.40(D)
324.120(A)
324.120(A)(1), (2), and (3)
Table 326.80—Ampacity Type IGS
Table 326.80—500 kcmil & 168 amperes
328.2—Medium Voltage Cable
330.2
330.10(B)(4)
330.12—(1st sentence)
332.2
332.24(2)
332.104
334.40(B)—(1st sentence)
342.14
342.28
344.10(B)
344.14
344.28
Table 344.30(B)(2)—1¼–1½ and 14
344.42(A)

Table 348.22—(information above table)
Table 348.22—(footnote below table)
348.30(A)
350.22(B)
354.10—Uses Permitted
354.10(2)
354.10(3)
354.12—Uses Not Permitted
354.12(1)
356.12(3)
360.24(A)
Table 360.24(A)
362.20(B)
366.22(A)—(1st sentence)
366.22(A)—(2nd sentence)
366.23(A)
366.30(A)
366.30(B)—(1st sentence)
366.100(E)—(1st sentence)
368.17(A)
368.17(B) Ex.
368.30
368.238—(last sentence)
370.3—(last sentence of 1st paragraph)
372.2—(last sentence)
372.5—(1st sentence)
372.6
372.9—(3rd sentence)
374.2—(last sentence)
376.10(1), (2), and (3)
376.12(1) and (2)
376.22—(1st sentence)
376.56
380.2(A)
380.2(B)
380.2(B)(5)
380.3
382.15
382.30
386.12(2)
386.56—(3rd sentence)
390.2(A)
390.3(A)
390.6
390.7
392.2—Cable Tray System
392.4
392.5(E)
392.6(A)—(2nd sentence)
Table 392.7(B)—(last footnote)
392.8(D)—(2nd paragraph)
392.11(A)(1)
394.56—(1st sentence)
398.30(A)(1)
398.42
Table 400.4 Note 2
400.7(A)(5) and (7)
400.7(B)
400.8
400.8(1) and (2)
400.9
400.10
400.10 FPN
400.22(A)
400.31(B)
402.3 FPN
Table 402.3—RFH-1 and 167EF
Table 402.3—SF-1 Silicone Rubber
Table 402.3—Limited to 300 volts
402.6
404.2(B) Ex.
404.6(A)—(1st sentence)
404.6(B)—(last sentence)
404.9(A)
404.10(A)
404.13(A)
404.14(A)(1)
404.14(A)(2)
404.14(A)(3)—80 percent
404.14(B)
404.14(B)(2)—(1st sentence)
404.16
406.2(C)
406.2(B)
406.3(A)—(2nd sentence)
406.3(D)(a), (b), and (c)
406.8(B)(1)
408.3(C)—(1st sentence)
Table 408.5—Noninsulated busbars;10
408.35—(1st paragraph)
408.52
Table 408.56
410.4(D)
410.5
410.8(B)
410.8(C)
410.8(D)(3)
410.15(A)
410.16(A)
410.16(C)
410.23
410.24

410.27(B)
410.28(C)
410.28(E)
410.29(C)
410.29(C)(1), (2), and (3)
410.30(C)(2)
410.31
410.33
410.38(A)—(1st sentence)
410.38(C)—(1st paragraph, 2nd sentence)
410.42(A)—(last sentence)
410.46
410.47—(2nd sentence)
410.53
410.54(A)
410.65(A)
410.66(A)(1)
410.67(C)—(last sentence)
410.73(B)
410.75
410.76(A)
410.80(B)
410.80(C)
410.86
411.6
422.10(A)—(1st paragraph)
422.10(A)—(3rd paragraph)
422.11(A)—(1st sentence)
422.11(B)
422.11(C)
422.14—(1st sentence of 2nd paragraph)
422.16(B)(2)
422.31(A)
422.43(A)
422.46
422.61
424.3(A)—(1st sentence of 2nd paragraph)
424.19—(1st sentence)
424.29—underline "or"
424.34
424.35—(last paragraph)
424.35(1) through (5)
424.36—(1st sentence)
424.41(D)
424.44(A)
424.59 FPN
424.72(B)—(1st paragraph)
424.82—(1st sentence)
424.98(A)
426.2—Resistance Heating Element (FPN)
427.2—Resistance Heating Element (FPN)
430.2—Motor Control Circuit
430.7(A)
430.7(A)(1), (2), and (7)
430.7(B)(1)
430.12(A)
Table 430.12(C)(1)
430.14(B)
430.31
430.32(A)(4)
430.32(E)
430.37
Table 430.37
430.53(A)
430.55
430.62(B)
430.72(B)(2)—(2nd sentence)
Table 430.72(B)—Col. B, 120 (copper)
Table 430.72(B)—Col. B, 100 (aluminum)
430.82(B)—(1st sentence)
430.102(B) Ex.—(1st sentence)
430.109(A)(1)—(1st sentence)
430.109(A)(2)
430.109(A)(3)
430.109(F)—(2nd sentence)
430.110(A)
430.111(B)(3)—(1st sentence)
440.2 Branch-Circuit Selection Current—(last sentence)
440.2 Hermetic Refrigerant Motor-Compressor
440.6(A)—(1st sentence)
440.12(A)(1)
440.12(A)(2)—(1st and 2nd sentence)
440.22(A)—(1st and 2nd sentence)
440.41(B)
440.54(B)
440.55(B)
440.62(A)(1), (2), (3), and (4)
440.63—(first 5½ lines)
440.64
445.12(C)
445.13—(1st sentence)
445.14
Table 450.3(B) Note 3
450.4(A) Ex.
450.5(A)(1)
450.5(A)(3) FPN
450.6
450.6(A)(1)
450.11
450.21(B)
450.22—(2nd sentence)

450.25
450.42—(1st sentence)
450.45(A)
450.45(E)
460.2(A)
460.6(A)
460.6(B)
460.8(A)—(1st sentence)
460.8(B)—(2nd sentence)
460.24(A)
460.24(A)(1)
460.24(A)(2)
460.24(A)(3)—(underline "current")
460.24(A)(4)
460.24(C)
470.3
470.4
470.18(E)
480.2—Nominal Battery Voltage
480.6(B)—(last sentence)
480.10(A)
500.5(B)—(1st sentence)
500.5(C)—(1st sentence)
500.5(C)(1)(1)
500.5(D)—(1st sentence)
500.5(D)(1) FPN No. 2
500.5(D)(1) FPN No. 2—(underline "Spanish moss")
500.5(D)(2)
500.8(D)
500.8(D)(2)—(1st sentence)
501.10(A)(1)(a)
501.15—(2nd sentence)
501.15(A)(2)
501.15(C)(3)
501.105(B)(6)(3)
501.115(B)(2)
501.130(B)(3)
502.10(A)(1)(1)
502.10(A)(1)(4)—(1st sentence)
502.10(A)(2)
502.10(A)(2)(1)
502.10(B)(4)
502.100(A)(3)
504.2—Intrinsically Safe Circuit
505.9(E)
511.3(A)(1)
511.3(A)(2)
511.3(B)(3)(1)
511.7(B)(1)(b)
511.12
513.3(C)
Table 514.3(B)(1)—Pits; any pit
514.11(A)—(1st sentence)
514.8 Ex.2—(1st sentence)
Table 515.3—Tank–aboveground
Table 515.3—Vent
515.8(C)
516.4(D)
517.2—X-Ray Installations, Mobile
517.13(B)—(1st sentence)
517.14—(1st sentence)
517.18(B)
517.18(C)
517.25
517.30(B)(2)
517.31—(last paragraph)
517.60(A)(2)
517.61(A)(5)
517.62—(1st sentence)
517.64(A)(1)
517.72(A)
518.1
518.4(A)
520.21
520.24
520.73
540.1
540.13—(1st sentence)
545.6 Ex.
547.5(B)—(1st sentence)
550.2—Appliance, Portable
550.2—Appliance, Portable (FPN)
550.2—Appliance, Stationary
550.10(A)
550.10(D)
550.15(H)
550.15(H) Ex.
550.16(A)(1)—(2nd sentence)
550.32(A)—(2nd sentence)
550.32(C)—(1st sentence)
551.2—FPN under Appliance, Portable
551.55(F)
555.19(A)(4)
555.19(A)(4)(a)
590.3(B)
590.3(C)
590.4(C)
600.3
600.4(B)
600.5(A)
600.5(B)(1)

600.5(B)(2)
600.6—(1st sentence)
600.6 Ex.2
600.6(A)(1)—(1st sentence)
600.6(A)(2)(1)
600.6(B)
600.7(B)
600.7(D)
600.8—(2nd sentence)
600.8(B)
600.9(A)
600.9(C)
600.10(C)(2)
600.22(A) and (B)
600.23(C)
600.23(D)
600.32(B)
600.32(J)
600.32(J)(2)
605.5(C)
610.21(F)(4)
620.11(A)
620.12(A)(1)
620.12(A)(2)
620.13 FPN No. 1
620.21
620.21(C)(1)
620.21(C)(2)
620.22(A)
620.61(B)(1)—(1st sentence)
620.61(B)(2)
620.91(C)—(1st sentence)
630.12(A)—(last paragraph)
630.12(B)
630.31(B) FPN (1) and (2)
630.31(B) FPN (3)—(1st sentence)
640.1—(1st sentence)
640.9(C)—(1st sentence)
640.10(A)
640.21(A)
645.5(A)
645.10—(1st and 2nd sentence)
645.16
650.4
650.6(A)—(last sentence)
650.8—(1st sentence)
660.1—(1st sentence)
660.1 FPN No. 1
660.2—Mobile
660.5
660.6(A)
665.1—(1st sentence)
665.1—(underline "not covered")
668.2—Electrolytic Cell
669.9
670.3(A)(2)
670.3(B)
675.13—(2nd sentence)
680.1—(1st sentence)
680.2—Forming Shell
680.2—Storable Swimming Pool
680.6—(2nd sentence)
680.6(2)
680.6(5)
680.6(6)
680.8
680.8(A)
680.10—(1st sentence)
680.12
680.22(A)(1)
680.22(A)(2)
680.22(A)(3)
680.22(B)(1)
680.23(A)(2)—grounded metal barrier
680.23(A)(4)
680.23(A)(5)—(1st sentence)
680.23(B)(2)(b)
680.23(F)(2)—(1st paragraph)
680.24(A)(2)(c)(1)
680.24(D)
680.25(B)—(1st sentence)
680.26(B)(5)
680.26(B)(5)(1)
680.26(C)
680.27(A)(2)
680.32—(1st sentence)
680.42(A)(2)
685.12
690.2—Array
690.2—(Hybrid System)
700.4(A)
700.6(C)
700.7
700.7(A), (B), and (C)
700.9(A)
700.9(D)(2)
700.12—(1st sentence)
700.12—(4th paragraph)
700.12(A)—1½ hours
700.12(A)—87½ percent
700.12(A)—acid or alkali type
700.12(A)—(3rd paragraph)

700.12(A)—(last sentence)
700.12(B)(2)—(1st sentence)
700.25
701.9(A)
701.11(B)(1)
701.11(B)(2)
Art. 702—(main title)
702.1
702.2
702.2 FPN
720.5
725.3(C)
725.23
725.41(B)
725.55(A)
725.55(J)—(1st sentence)
760.51—(2nd sentence)
760.52(B)(1)—(last sentence)
760.55(A)
760.57
760.61(A)—(1st sentence)
770.6
770.9(A)—(title only)
770.9(B)—(title only)
770.9(C)—(title only)
800.1—except radio
800.3(C)
800.100(D)
Table 800.113
800.154(B)(1)
Table 800.154(B)(1)
800.179(E)
810.12
Table 810.16(A)—Copper-clad steel, and 17
810.20(A)
820.1
820.44(F)(3)
820.100(A)(2)
820.100(A)(3)
Chapter 9, Note 4
Chapter 9, Note 7—decimal of 0.8 or larger
Chapter 9, Tables 12(A) and (B)—
(3rd paragraph)

Answer Keys

APPENDIX

B

CHAPTER PROBLEMS ANSWER KEY

Chapter 1

Problems (page 2)

1. 499.44
2. 2232.1358
3. 0.1473
4. 115

End-of-Chapter Problems

1. c) 240
2. d) 18,650
3. c) 5000
4. b) 16
5. b) 0.5
6. a) 12
7. d) 8800
8. a) 3.6
9. b) 115
10. b) 7740
11. c) 100
12. a) 0.1473
13. b) 12.88384
14. a) 23.04
15. d) 25

Chapter 2

Problems (page 17)

1. 10 ohms
2. 12 amperes
3. 12 volts
4. 24 volts
5. 36 volts
6. 48 volts
7. 144 watts
8. 288 watts
9. 432 watts
10. 576 watts
11. 1440 watts

Problems (page 19)

1. 16 ohms
2. 0.5 ampere
3. 2 amperes
4. 1 ampere
5. 4 amperes
6. 7.5 amperes
7. 60 watts
8. 240 watts
9. 120 watts
10. 480 watts
11. 900 watts

Problems (page 22)

1. 30 ohms
2. 4 amperes
3. 20 volts
4. 4 amperes
5. 20 volts
6. 480 watts

End-of-Chapter Problems

1. a) 120 volts

   $R_T = (12 \div 3) + 20$

   $R_T = 24$ ohms

   $E = 5 \times 24 = 120$ volts

2. **d) 1, 2, and 3**

   1 is a wattmeter, 2 is an ammeter, and 3 is a voltmeter

3. **c) 12 amperes**

   R = 20 ÷ 2

   R = 10 ohms

   I = 120 ÷ 10 = 12 amperes

4. **d) 420 watts**

$$R = \frac{1}{\frac{1}{60}+\frac{1}{80}} = \frac{1}{0.0167+0.0125} =$$

$$\frac{1}{0.02917} = 34.2818 \text{ ohms}$$

   W = 120 × 120 ÷ 34.2818 = 420 watts

5. **d) 140 volts**

   R = 20 + 5 + 10 = 35 ohms

   E = 4 × 35 = 140 volts

6. **c) 30 ohms**

   R = 20 + 10 = 30 ohms

7. **d) 1, 2, 3, 4, and 5**

8. **a) 2 and 3**

9. **b) 6 volts**

   4 × 1.5 = 6 volts

10. **a) 0.0129 ohms**

    The resistance for 1000 feet of 500-kcmil uncoated copper from Table 8 in Chapter 9 is 0.0258 ohms.

$$R = \frac{1}{\frac{1}{0.0258}+\frac{1}{0.0258}} = \frac{1}{38.7597+38.7597} =$$

$$\frac{1}{77.5194} = 0.0129 \text{ ohms}$$

    Alternate Method 1. Since the two resistors are of equal value, divide the resistance of one by the number of resistors.

    0.0258 ÷ 2 = 0.0129 ohms

    Alternate Method 2. Since two 500-kcmil conductors are in parallel, the equivalent conductor size for a single conductor would be a 1000-kcmil conductor. The resistance for 1000 feet of 1000-kcmil uncoated copper from Table 8 in Chapter 9 is 0.0129 ohms.

11. **a) 90 amperes**

$$R = \frac{1}{\frac{1}{2}+\frac{1}{4}} = \frac{1}{0.5+0.25} = \frac{1}{0.75} = 1.3333 \text{ ohms}$$

    I = 120 ÷ 1.3333 = 90 amperes

12. **b) 2.03 ohms**

    R = 115 × 115 ÷ 6500 = 2.03 ohms

13. **b) Ammeter**

14. **b) 1.56 amperes**

    R = 120 × 120 ÷ 150 = 96 ohms (each heater)

    R_T = 96 × 4 = 384 ohms

    I = 600 ÷ 384 = 1.56 amperes

15. **a) 115 volts**

    E_{Drop} = 0.05 × 100 = 5 volts

    E_{Load} = 120 – 5 = 115 volts

16. **c) 1 ampere**

    1 + 6 + 3 = 10 ohms

    10 ÷ 2 = 5 ohms

    R_T = 2 + 5 + 3 = 10 ohms

    I = 10 ÷ 10 = 1 ampere

17. **a) 6530**

    Chapter 9, Table 8

18. **a) 114 volts**

    Total resistance of the conductors = 0.025 + 0.025 = 0.05 ohms

    Total current = 120 amperes

    Voltage drop across the conductors = 0.05 × 120 = 6 volts

    Voltage at the load = 120 – 6 = 114 volts

19. **c) 22 ohms**

$$R = \frac{1}{\frac{1}{20}+\frac{1}{30}} = \frac{1}{0.05+0.0333} = \frac{1}{0.0833} = 12$$

    12 + 10 = 22 ohms

20. **d) 56.52 amperes**

    6500 (W) ÷ 115 (E) = 56.52 amperes

21. **d) 13.3 amperes**

$I = W \div E = 200 \div 120 = 1.6667$ amperes each

$8 \times 1.6667 = 13.3$

22. **c) 23.3 ohms**

$$R = \frac{1}{\frac{1}{20} + \frac{1}{40}} = \frac{1}{0.05 + 0.025} = \frac{1}{0.075} = 13.3$$

$13.3 + 10 = 23.3$ ohms

23. **b) 360 watts**

$W = E^2 \div R$

$W = 120 \times 120 \div 40 = 360$ watts

Chapter 3

Problems (page 35)

1. **15 amperes**

   $I = W \div E = 1800 \div 120 = 15$ amperes

2. **3 amperes**

   $I = W \div E = 360 \div 120 = 3$ amperes

3. **8 ohms**

   $R = E \div I = 120 \div 15 = 8$ ohms

4. **40 ohms**

   $R = E \div I = 120 \div 3 = 40$ ohms

5. **120 volts**
6. **120 volts**

Problems (page 35)

1. **48 ohms**

   $R_1 + R_2 = 8 + 40 = 48$ ohms

2. **5 amperes**

   $I = E \div R = 240 \div 48 = 5$ amperes

3. **5 amperes**

   Current remains the same through a series circuit.

4. **40 volts**

   $E_{Drop} = I \times R = 5 \times 8 = 40$ volts

5. **200 volts**

   $E_{Drop} = I \times R = 5 \times 40 = 200$ volts

End-of-Chapter Problems

1. **a) maximum unbalance**

   [220.61(A)]

2. **d) any of these**

   [200.6(A)]

3. **b) II only (equipment grounding conductor)**

   [250.119]

4. **c) grounded conductor**

   [200.10(C)]

5. **a) grounded**

   [400.22(F)]

6. **c) both A and B (screw shell and identified blade)**

   [410.42(A)]

7. **b) carries the unbalanced current**

   [310.15(B)(4)(a)]

8. **c) in multiwire branch circuits**

   [300.13(B)]

9. **a) grounded**

   [200.6(A)]

10. **d) grounded conductor**

    [Art. 100 (Grounded Conductor)]

11. **c) multiwire branch circuit**

    [Art. 100 (Branch Circuit, Multiwire)]

12. **d) I, II, or III (by a metal or metal coating that is substantially white in color, by the word *white*, or by the letter W located adjacent to the identified terminal)**

    [200.10(B)(1)]

Chapter 4

End-of-Chapter Problems

1. **c) 11 kW**

   Column C for 2 units = 11 kW

2. **d) 41 kW**

   Column B = 26 × 6.75 × 24% = 42.12 kW

   Column C = 15 + 26 = 41 kW

   Compare Column B with Column C and choose the lowest kW.

3. c) 25.3 kW

13.5 = 14

<u>– 12</u>

2 × 5% = 10% = Increase Column C by 10%

Column C = 23 × 1.1 = 25.3 kW

4. d) 6.6 kW

The question is asking for the service demand load, not the branch-circuit demand load.

3.5 + 3.5 + 5 = 12 kW for 3 units

Column B for 3 units has a demand factor percent of 55%.

12 × 55% = 6.6 kW

5. c) 162.5 kW

Note that Problem 5 to Table 220.55 applies only to household cooking appliances that are used in instructional programs.

The appliances in Problem 5 are commercial ranges.

The demand factor percent, in Table 220.56, for 20 units is 65%.

20 × 12.5 × 65% = 162.5 kW

6. c) 38.5 kW

6 × 12 = 72

4 × ~~8~~ 12 = 48

10 × 15 = 150

Total number of units = 20 units

Total kW = 270 kW

Average kW per unit = 270 ÷ 20 = 13.5 kW

13.5 = 14

<u>– 12</u>

2 × 5% = 10% – Increase Column C by 10%

Column C = 35 × 1.1 = 38.5 kW

7. b) 58.3 kW

14

<u>– 12</u>

2 × 5% = 10% = Increase Column C by 10%

Column C = 15 + 38 = 53 × 1.1 = 58.3 kW

8. a) 25 kW

Column B = 10 × 8 × 34% = 27.2 kW

Column C = 25 kW

Compare Column B with Column C and choose the lowest kW.

9. c) 34.65 kW

6 × ~~8~~ 12 = 72

4 × 12 = 48

8 × 15 = 120

Total number of units = 18 units

Total kW = 240 kW

Average kW per unit = 240 ÷ 18 = 13.3 kW

13.3 = 13

<u>– 12</u>

1 × 5% = 5% = Increase Column C by 5%

Column C = 33 × 1.05 = 34.65 kW

10. b) 38 kW

Column B = 23 × 8 × 26% = 47.84 kW

Column C = 38 kW

Compare Column B with Column C and select the lowest kW.

11. c) 250 kW

5 (minimum) × 200 (units) × 25% [Table 220.54] = 250 kW

12. d) 160 amps

Column C = 15 + 40 = 55 kW

The neutral demand factor, found in 220.61(B)(1), is 70%.

55 × 70% = 38.5 kW

38.5 × 1000 = 38,500 watts

38,500 ÷ 240 = 160.4167 = 160 amps

13. d) 10-AWG THHN CU

8 kW × 1.1 = 8.8 = 8800 ÷ 240 = 36.7 amps

36.7 × 70% = 25.69 = 10 AWG THHN CU

14. a) 35 kW

Column B = 20 × 8 × 28% = 44.8 kW

Column C = 35 kW

Select the lowest kW of Columns B and C

15. a) 82 amperes

Table 220.55, Column C = 27 kW = 27,000 watts

27,000 × 70% = 18,900 watts on the neutral.

18,900 ÷ 230 = 82.17 = 82 amps

16. b) 28.7 kW

14 + 14 + 6 + 4 + 3 = 41

41 × 70% [Table 220.56] = 28.7 kW

17. a) 22.4 kW

12 + 6 + 8 + 3 + 3 = 32 × 70% [Table 220.56] = 22.4

18. **d) 27**

$(2 \times 5000) + (4 \times 3000) + (2 \times 6000) = 34{,}000$ watts

$34{,}000 \times 65\%$ [Table 220.56] $= 22{,}100$ watts

$$\frac{22{,}100}{480 \times 1.732} = 26.58 = 27 \text{ amperes}$$

19. **a) 5.2 kW**

$4 + 4 = 8$ kW for 2 units

Column B for two units has a demand factor percent of 65%.

$8 \times 65\% = 5.2$ kW

20. **c) 3.5 kVA**

$5 \times 70\%$ [220.61(B)(1)] $= 3.5$ kVA neutral demand load. Treat the clothes washer/dryer as a clothes dryer [Annex D, Example D2(b)].

21. **b) 22,500 watts**

Derating is not permissible in laundry facilities.

$6 \times 1500 = 9000$

$3 \times 4500 = 13{,}500$

$9000 + 13{,}500 = 22{,}500$ watts

22. **b) 12 kW**

Add the units together and treat as one range.

$8 + 7 + 8 = 23$ kW

$23 - 12 = 11$

$11 \times 5\% = 55\%$

One unit, Column C = 8 kW

$8 \times 55\% = 4.4$

$4.4 + 8 = 12.4 = 12$ kW

23. **b) 27,600 watts**

$12 \times 5000 = 60{,}000$

Demand factor percent for 12 units:
$47 - (12 - 11) = 46$

$60{,}000 \times 46\% = 27{,}600$ watts

24. **c) 15.6 kW**

Add the units together and treat as one range.

$10 + 9 + 12 = 31$ kW

$31 - 12 = 19$

$19 \times 5\% = 95\%$

One unit, Column C = 8 kW

$8 \times 95\% = 7.6$

$7.6 + 8 = 15.6$ kW

25. **b) 47 kW**

[Table 220.55]

Column C $= 15 + 32 = 47$ kW

Column B $= 8 \times 32 \times 22\% = 56.32$ kW

Compare the two and select the lowest.

26. **a) 57.6 kW**

[Table 220.84] (Use the nameplate rating.)

$4 \times 60 \times 24\% = 57.6$ kW

27. **c) 25.5 kW**

Derating is not permissible in laundry facilities.

$5 \times 1500 = 7500$

$4 \times 4500 = 18{,}000$

$7500 + 18{,}000 = 25{,}500$ watts

$25{,}500 \div 1000 = 25.5$ kW

Chapter 5

End-of-Chapter Problems

1. **c) 0.0129 ohm**

$0.0258 \div 2 = 0.0129$

2. **b) 105.5 volts**

$$V_D = \frac{2 \times 13.35659 \times 75 \times 60}{16{,}510} = 7.281 \text{ volts}$$

$$V_D = \frac{2 \times 21.20192 \times 75 \times 60}{26{,}240} = 7.272 \text{ volts}$$

$V_D = 7.281 + 7.272 = 14.553$ volts

$V_{Load} = 120 - 14.553 = 105.447$ volts

3. **b) 991**

An ohmmeter uses dc voltage to read resistance.

$$\frac{0.021}{0.0212} \times 1000 = 990.566 = 991$$

4. **c) 71′**

$0.035 \div 0.000491 = 71.28 = 71'$

5. **a) 0.016 ohm**

$$R = \frac{E}{I} = \frac{6.4}{400} = 0.016 \text{ ohms}$$

6. d) 0.0125 ohm

0.025 ÷ 1000 = 0.000025 × 500 = 0.0125 ohms

7. d) 400 kcmil THW

$$\text{Current} = \frac{120{,}000}{1.732 \times 208} = 333 \text{ amps}$$

$$\text{Cmil} = \frac{1.732 \times 12.9 \times 100 \times 333}{6.24} = 119{,}233 \text{ Cmils}$$

119,233 Cmils = 2/0 THW

The ampacity of a 2/0 THW conductor is only 175 amps. The only conductor in the answers that will carry 333 amps is a 400-kcmil THW conductor.

8. c) 10 AWG THW

A 5-hp, 230-volt, three-phase motor draws 15.2 amps.

[Table 430.250]

$$\text{Cmil} = \frac{1.732 \times 12.9 \times 150 \times 15.2}{6.9} = 7383 \text{ Cmils}$$

9. c) 0.11 ohm

[Chapter 9, Table 9]

10. d) 700 kcmil THW

900 ÷ 2 = 450 amps

Table 310.16 = 700-kcmil THW

11. a) 8.4

$$\frac{70 \times 120}{1000} = 8.4 \text{ kW}$$

12. b) 0.0212 ohm

0.0424 ÷ 2 = 0.0212

13. d) 0.0945 ohm

0.21 ÷ 1000 = 0.00021 × 450 = 0.0945

14. c) 0.0180 ohm

$$R = \frac{E}{I} = \frac{7.2}{400} = 0.0180 \text{ ohms}$$

15. b) 2 AWG THW

A 5-hp, 115-volt, single-phase motor draws 56 amps.

[Table 430.248]

$$\text{Cmil} = \frac{2 \times 12.9 \times 150 \times 56}{3.45} = 62{,}817 \text{ Cmils}$$

16. b) 61′

The drawing is only 120-volt, single phase because two of the three switches are open. The load is 60 amps. 7200 ÷ 120 = 60 amps

$$L = \frac{26{,}240 \times 3.6}{2 \times 12.88384 \times 60} = 61.099795 = 61 \text{ feet}$$

17. a) 0.34

$$L = \frac{41{,}740 \times 6.24}{1.732 \times 12.85592 \times 10.6} = 1103.5 \text{ feet}$$

$$\frac{1103.5 \times 0.308120}{1000} = 0.34 \text{ ohm}$$

Alternative Method. First, calculate the maximum voltage drop per conductor.

V_D per conductor = 3.6 volts
(208 ÷ 1.732 = 120 × 3%)

Next, find the full load current of the motor.

3-hp, 208-volt, 3-phase motor = 10.6 amps
[Table 430.250]

Finally, use the Ohm's law formula to find the resistance.

R = E ÷ I = 3.6 ÷ 10.6 = 0.3396 = 0.34 ohms

18. a) 5.8212 volts

A ½-hp, 115-volt, single-phase motor draws 9.8 amps.

[Table 430.248]

$$V_D = \frac{2 \times 12.9294 \times 150 \times 9.8}{6530} = 5.8212 \text{ volts}$$

19. b) 0.026 ohm

0.048 ÷ 1000 = 0.000048 × 545 = 0.02616

20. c) 0.0225 ohm

0.045 ÷ 1000 = 0.000045 × 500 = 0.0225

21. c) 1 AWG THW

$$\text{Cmil} = \frac{1.732 \times 21.2 \times 215 \times 61}{6.24} = 77{,}173 \text{ Cmils}$$

22. b) 82 amps

The drawing is only 120-volt, single phase, because two of the three switches are open.

$$I = \frac{52{,}620 \times 3.6}{2 \times 12.8919 \times 90} = 81.6 = 82 \text{ amps}$$

23. c) 0.063 ohm

Chapter 9, Table 9

24. d) Chapter 9, Table 8

Chapter 9, Note 8

25. b) 108 volts

$$V_D = \frac{2 \times 13.2 \times 75 \times 50}{16{,}510} = 6 \text{ volts}$$

$$V_D = \frac{2 \times 21.2 \times 75 \times 50}{26{,}240} = 6.059 \text{ volts}$$

$V_D = 6 + 6.059 = 12.059 = 12$ volts

$V_{Load} = 120 - 12 = 108$ volts

26. a) 116′

The drawing is only 120 volt, single phase, because two of the three switches are open.

The load is 100 amps. 12,000 ÷ 120 = 100 amps

$$L = \frac{83{,}690 \times 3.6}{2 \times 12.88826 \times 100} = 116.88 = 116 \text{ feet}$$

27. c) 6 AWG

$$\text{Cmil} = \frac{2 \times 12.9 \times 225 \times 25}{7.2} = 20{,}156 \text{ Cmils}$$

28. b) 1.167 ohms

0.778 × 1.5 = 1.167 ohms [Chapter 9, Table 8]

29. c) 11.3 volts

load = 6150 ÷ 240 = 25.625 amperes

$$V_D = \frac{2 \times 12.88384 \times 450 \times 25.625}{26{,}240} = 11.3 \text{ volts}$$

30. b) 82′

The resistance of the conductor is given (0.04 ohm).

The resistance for 1000 feet of 6-AWG copper is 0.491.

The resistance for 1 foot is 0.000491.

Divide the given resistance by the resistance per foot.

0.04 ÷ 0.000491 = 81.4664 = 82 feet

Chapter 6

End-of-Chapter Problems

1. **b) 80%**

   Every heater is connected to a separate grounded conductor. There are a total of six current-carrying conductors.

   6 conductors = 80%

2. **a) 50 amperes**

   XHHW in a wet location = 75° C [Table 310.13]

3. **b) 20 amperes**

   Do not derate nipples.

4. **c) 3 AWG**

   $$\frac{70}{1 \times 0.80} = 87.5 = 88 \text{ amperes} = 4 \text{ AWG}$$

   The minimum overcurrent protection for a 70-ampere continuous load is a 90-ampere fuse or circuit breaker (70 × 125% = 88 = 90).

   After derating, 4 AWG has an allowable ampacity of 76 amperes.

   95 × 0.80 = 76 amperes

   After derating, the maximum size fuse or circuit breaker for the conductor in this example is 80 amperes.

   After derating, a 3-AWG THHN copper conductor has an allowable ampacity of 88 amperes.

   110 × 0.80 = 88 amperes

5. **c) 8 AWG**

   $$\frac{24.2 \times 125\%}{0.67} = 45.15 = 45 \text{ amperes}$$

6. **d) 37 amperes**

   35 × 1.05 = 36.75 amperes

7. **d) 30 amperes**

   40 × 0.91 = 36.4 = 30 amperes (because of the asterisk)

8. **b) 30 amperes**

   35 × 1 = 35 = 30 amperes (because of the asterisk)

9. **c) 20 amperes**

   25 × 80% = 20 amperes

10. **d) Do not derate.**

    Do not derate nipples.

11. **c) 8 AWG**

    $$\frac{18.7 \times 125\%}{0.82 \times 0.70} = 40.72 = 41 \text{ amperes}$$

12. a) 20 amperes

$30 \times 0.80 = 24 = 20$ amperes (because of the asterisk)

13. b) 8 AWG

$\frac{8000}{240} = 33.3 \times 125\% = 41.63 = 42$ amperes

$= 8$ AWG [(422.13)]

14. a) 32 amperes

$40 \times 80\% = 32$ amperes

15. c) 45%

[Table 310.15(B)(2)(a)] (A nipple is 24″ or less.)

16. b) 36.4 amperes

$40 \times 0.91 = 36.4$

17. d) 82 amperes

$100 \times 0.82 = 82$ amperes

18. d) 66.12

Multiply the conductor ampacity by 80% because the load is a single motor.

$95 \times 0.87 \times 0.80 = 66.12$ amperes

19. c) 29.4 amperes

$35 \times 1.05 \times 80\% = 29.4$ amperes

20. b) 5–7

$70 \times 120 \times 80\% = 6720 \div 1000 = 6.72$ kW

21. b) 70%

Eight current-carrying conductors = 70%

22. d) Not at all

All of the heaters are connected to one common neutral. The load is a resistive load. The neutral does not count as a current-carrying conductor.

23. c) 50 amperes

XHHW in a wet location is a 75° C conductor [Table 310.13]. No derating is required; the conductors are in a nipple.

24. b) 8 AWG

Table 310.15(B)(2)(a): Six conductors = 80%

$40 \div 80\% = 50 = 8$ AWG

25. b) 40

$7500 \div 240 = 31.25$ amperes

$31.25 \times 125\% = 39.0625 = 40$ [422.13]

26. c) 26.32

$35 \times 0.94 \times 80\% = 26.32$

27. b) 30 amperes

$40 \times 0.87 = 34.8 = 30$ (because of the asterisk on Table 310.16)

28. c) 3/0

Divide the load because of the paralleled conductors.

$299 \div 2 = 149.5$ amperes on each set of conductors

All of the conductors are in one conduit. The ampacity must be derated because of the number of conductors in the conduit.

A 2/0-THW conductor is good only for 140 amps after derating the ampacity by 80%. ($175 \times 80\% = 140$)

A 3/0-THW conductor is good for 160 amps after derating the ampacity by 80%. ($200 \times 80\% = 160$)

29. b) 50 amperes

$50 \times 0.94 = 47 = 50$

30. b) 7.1–8 amperes

$20 \times 0.82 \times 45\% = 7.38$ amperes

31. c) 4-AWG

$44 \times 125\%$ [424.3(B)] $= 55$

$55 \div 80\%$ [Table 310.15(B)(2)(a)] $=$ $68.75 = 4$ AWG

32. d) 0%

There are only three current-carrying conductors.

33. c) 8 AWG

Table 310.15(B)(2)(a): Eight conductors = 70%

$30 \div 70\% = 43 = 8$ AWG

34. d) 24 amperes

Six current-carrying conductors = 80% derating [310.15(B)(2)(a)]

$30 \times 80\% = 24$ amperes

35. b) 25 amperes

$20 \times 125\% = 25$ amperes [422.10(A)]

36. b) 28 amperes

Because each cable contains a grounding conductor, there are only six current-carrying conductors.

The Table 310.15(B)(2)(a) adjustment factor for six current-carrying conductors is 80 percent.

$35 \times 80\% = 28$ amperes

Chapter 7

End-of-Chapter Problems

1. **d) 11**

   [Annex C, Table C4] (A nipple is 24″ or less.)

2. **c) 0.630″**

   [Chapter 9, Table 4]

3. **c) 12**

   60% column in Chapter 9, Table 4 = 0.532

   0.532 ÷ 0.0437 = 12.17 = 12

4. **b) 1″**

   4 × 0.0726 = 0.2904

   6 × 0.0333 = 0.1998

   0.2904 + 0.1998 = 0.4902

   The area in.^{2} required for these 10 conductors = 0.4902

   60% column in Chapter 9, Table 4 = 0.499 = 1″

5. **c) 65 amperes**

   65 × 1.00 = 65 amps

   Because of the conduit nipple, do not derate with the correction factors in Table 310-15(b)(2)(a).

6. **b) 0.3496**

   8 × 0.0437 = 0.3496

   Eight 8-AWG THW conductors will occupy 0.3496 area in.^{2} in any conduit or nipple.

7. **c) 0.743**

   [Chapter 9, Table 4]

8. **b) 0.026**

   [Chapter 9, Table 5]

9. **c) 0.722″**

   [Chapter 9, Table 4]

10. **a) 4**

    0.785 × 20% = 0.157

    0.157 ÷ 0.0353 = 4.448 = 4

11. **d) 14**

    60% – 20% = 40%

    40% column in Chapter 9, Table 4 = 0.829

    0.829 ÷ 0.059 = 14.05 = 14

    Table C8 (Annex C) allows fourteen 6 AWG XHHW conductors in 40 percent fill in rigid metal conduit.

12. **d) I, II, and III**

    This is a simple math problem. All three answers are the same: 2.139.

13. **c) 9**

    60% column in Chapter 9, Table 4 = 0.320

    0.320 ÷ 0.0353 = 9.065 = 9

14. **a) ¾″**

    2 × 0.0437 = 0.0874

    2 × 0.0181 = 0.0362

    3 × 0.0139 = 0.0417

    0.0874 + 0.0362 + 0.0417 = 0.1653

    60% column in Chapter 9, Table 4 = 0.272 = ¾″

15. **c) 30 amperes**

    35 × 0.88 = 30.8 = 30 amperes

    Because of the conduit nipple, do not derate with the correction factors in Table 310-15(b)(2)(a).

16. **b) 10**

    [Annex C, Table C7] (assume "with covering")

17. **a) 5.858**

    [Chapter 9, Table 4]

18. **b) 31%**

    [Chapter 9, Tables 1 and 4]

19. **d) 37**

    60% column in Chapter 9, Table 4 = 2.178

    2.178 ÷ 0.059 = 36.9 = 37

20. **b) 10**

    [Annex C, Table C5]

21. **a) 14**

    [Annex C, Table C1]

22. **b) 1.36″**

    [Chapter 9, Table 4]

23. **c) 0.814**

    [Chapter 9, Table 4]

24. **a) 18**

    Since 20 percent of a nipple is full, 40 percent remains open. All the conductors are the same size, and therefore the answer can be found in Annex C, Table C1. The answer could also be calculated by using Chapter 9, Tables 4 and 5.

25. a) 60%

Chapter 9, Note 4

26. d) 0.766

60% column in Chapter 9, Table 4

27. c) 2″

0.3904 + 0.3904 + 0.3904 + 0.0814 = 1.2526

2″ = 1.316

28. b) 1.0532

[Chapter 9, Table 5]

29. a) 20

60% column in Chapter 9, Table 4 = 0.519

0.519 ÷ 0.026 = 19.96 = 20

30. c) 32

[Annex C, Table C8]

31. c) 2

Nine 12-AWG THHNs are allowed in a ½″ rigid metal conduit. Two more (than the original seven) can be installed in the conduit. 9 – 7 = 2

NOTE: The key word is "contained."

32. b) 4.866

[Chapter 9, Table 4]

33. a) 13

[Annex C, Table C2]

34. a) 1.298

40% column in Chapter 9, Table 4

35. b) 2.013

60% column in Chapter 9, Table 4

Chapter 8

End-of-Chapter Problems

1. a) 12″

Since there is no raceway entering the top of the box, there is no straight pull for the X dimension. The raceway entering the bottom of the box must be calculated by the angle pull method.

6 × 2 = 12″

2. d) 21.5 cubic inches

Two 14-2s = 4 × 2 = 8 in.^{3} (do not include the grounds)

One 12-2 passing through the box = 2 × 2.25 = 4.5 in.^{3}

One 12-AWG ground = 2.25 in.^{3} (largest ground in box)

Three cable clamps = 2.25 in.^{3}
one hickey = 2.25 in.^{3}

Fixture wires = 0

8 + 4.5 + 2.25 + 2.25 + 2.25 = 19.25 in.^{3}

Minimum size box is 21.5 cubic inches. [314.16(B)]

3. b) 2

Volume allowance for one 12-AWG = 2.25

The receptacle counts as two 12-AWG conductors = 4.5 in.^{3}

18 – 4.5 = 13.5 13.5 ÷ 2.25 = 6

Two cables (not counting grounds) count as four conductors.

Two grounds count as one conductor.

Maximum cables permitted = 2

Although space remains for one conductor, there are two conductors in each cable.

4. b) 3

Two 12-2 cables count as five 12-AWG conductors (the two grounds count as one).

5 × 2.25 = 11.25 cubic inches

24 – 11.25 = 12.75 cubic inches remaining

12.75 ÷ 2 = 6 individual conductors permitted

Since the largest ground has already been counted, each cable will count as two conductors. 6 ÷ 2 = 3

5. a) 12 inches

[314.28(A)(2)] Raceway entries enclosing same conductors must not be less than six times the trade diameter of the largest raceway.

2 × 6 = 12 inches

6. c) 3

Volume allowance for one 14-AWG = 2

The switch counts as two 14-AWG conductors = 4 in.^{3}

18 – 4 = 14 14 ÷ 2 = 7

Three cables (not counting grounds) count as six conductors.

Three grounds count as one conductor.

Maximum cables permitted = 3

7. d) 30

One 14-AWG = 2 cubic inches 60 ÷ 2 = 30

8. **d) 24″**

   This time, the X dimension has both angle and straight pulls. The largest angle pull is the left side. 6 × 3 = 18 + 2 = 20″

   The straight pull method is the larger of the two methods. 8 × 3 = 24″

9. **c) 13**

   [Table 314.16(A)]

10. **b) 4**

    one switch = 2, three clamps = 1, three grounding conductors = 1

    2 + 1 + 1 = 4 [314.16(B)]

11. **b) 15 inches**

    [314.28(A)(2)] Raceway entries enclosing the same conductors must not be less than six times the trade diameter of the largest raceway.

    2.5 × 6 = 15 inches

12. **c) 2.25**

    [Table 314.16(B)]

13. **b) 4 × 1½ inch**

    [Table 314.16(A)]

14. **a) 5**

    One 12-AWG = 2.25 cubic inches

    6 × 2.25 = 13.5 cubic inches

    27 − 13.5 = 13.5 cubic inches remaining

    One 10-AWG = 2.5 cubic inches

    13.5 ÷ 2.5 = 5.4 = 5 conductors permitted

Chapter 9

End-of-Chapter Problems

1. **b) full voltage**
2. **d) Start is pushed, energizing the M relay coil; normally open M contacts close.**
3. **a) The overload is in series with the M relay coil.**
4. **d) Start is pushed, energizing the M relay coil; the auxiliary M contact is bypassed in the jog mode; the M relay coil remains energized only when the start button is in the closed position.**
5. **d) 3**
6. **d) 1**
7. **a) 2**
8. **c) 4**
9. **b) 5**
10. **b) resistor**
11. **c) all normally open CR contacts are closed**
12. **a) in parallel with the CR contacts**
13. **d) When the CR coil is energized, the motor will run at full speed.**
14. **b) parallel**
15. **c) full voltage**
16. **c) L2 and L3 are interchanged**
17. **a) F1 must close before For is released.**
18. **b) A is closed, F1 is closed, and F2 is open.**
19. **d) the reverse button may be pressed with the motor running forward**
20. **d) the control circuit**

Chapter 10

Problems (page 130)

1. **6 amperes**
2. **40 amperes**
3. **17 amperes**
4. **83 amperes**
5. **56 amperes**
6. **211 amperes**
7. **255 amperes**

End-of-Chapter Problems

1. **c) 150 amperes**

   59.4 × 175% = 103.95

   56 × 175% = 98

   103.95 is not a standard size fuse. A standard size fuse higher than 103.95 is 110 amperes. The largest branch circuit protection is 110 amperes.

   110 + 56 = 166

166 is not a standard size fuse.

The maximum feeder overcurrent protection is 150 amperes.

2. **a) 19.2 amperes; 21.7 amperes**

   16.7 × 115% = 19.2 minimum

   16.7 × 130% = 21.7 maximum

3. **b) 70 amperes**

   40 × 175% = 70

4. **c) 175 amperes**

   54 × 250% = 135

   135 is not a standard size fuse.

   A standard size fuse higher than 135 is 150 amperes.

   150 + 17 + 17 = 184

   The single-phase motors are 230-volt.

   When the motors are balanced between phases, each phase will have two 3-hp motors.

   184 is not a standard size fuse.

   The maximum feeder overcurrent protection is 175 amperes.

5. **c) 70 amperes**

   28 × 250% = 70

6. **a) 11,150**

   1.732 × 230 × 28 = 11,154

7. **c) 12-AWG THW**

   10 × 125% = 12.5 amperes = 14-AWG THW

   Because the length is given, this is also a voltage drop problem.

   $$\text{Cmil} = \frac{2 \times 12.9 \times 150 \times 10}{6.9} = 5608.7 \text{ Cmils}$$

   The length requires the use of a 12-AWG THW conductor.

8. **c) 23.4 amperes**

   18.7 × 125% = 23.375 = 23.4

9. **a) 4000 watts**

   12 kW = 12,000 watts

   There are three phases.

   12,000 ÷ 3 = 4000

10. **c) 797 watts**

    1.732 × 230 × 6 = 2390.16

    2390.16 ÷ 3 = 796.72 = 797

11. **d) 50 amperes**

    24 × 125% = 30

    The three single-phase, 120-volt motors are allowed to be balanced, as evenly as possible, between phases. Each 2-hp motor will be on a separate phase. Each phase will carry a 2-hp motor and one leg of the 5-hp motor.

    30 + 16.7 = 46.7 = 50

12. **c) 24.2 amperes**

    [Table 430.250]

13. **b) 60 amperes**

    24 × 250% = 60

14. **d) 20.9 amperes**

    16.7 × 125% = 20.875 = 20.9

15. **c) 70**

    To calculate feeder protection, the largest branch-circuit protection is needed. The 120-volt motor calculates to be the largest branch-circuit protection.

    24 × 250% = 60

    Because of the 120-volt motors, there are only two motors on each phase.

    60 + 16.7 = 76.7

    The feeder protection cannot exceed this number.

    The next standard size breaker less than 76.7 is 70 amperes.

16. **b) 8-AWG**

    24 × 125% = 30

    30 + 16.7 = 46.7

    8-AWG THW is rated for 50 amperes.

17. **a) 3000 watts**

    9 kW = 9000 watts

    There are three phases.

    9000 ÷ 3 = 3000 watts

18. **d) 6-AWG THW**

    20 × 125% = 25 = 12-AWG THW

    Because the length is given, this is also a voltage drop problem.

    $$\text{Cmil} = \frac{2 \times 12.9 \times 150 \times 20}{3.45} = 22{,}434.78 \text{ Cmils}$$

    The length requires the use of a 6-AWG THW conductor.

19. c) 22 amperes

[Table 430.250]

20. a) 3730 watts

The output of a motor is 746 watts per horsepower.

5 × 746 = 3730

21. b) 175 amperes

59.4 × 250% = 148.5

148.5 is not a standard size breaker.

A standard size breaker higher than 148.5 is 150 amperes.

The largest branch-circuit protection is 150 amperes.

150 + 34 = 184 (not a standard size breaker)

The maximum feeder overcurrent protection is 175 amperes.

22. c) 8 AWG

23 × 125% = 28.75 = 10 AWG

Because the length is given, this is also a voltage drop problem.

$$\text{Cmil} = \frac{2 \times 12.9 \times 175 \times 23}{7.2} = 14{,}423 \text{ circular mils}$$

The length requires the use of an 8-AWG THW conductor.

23. c) 6 AWG

[Table 250.122]

24. d) 30 ampere

10.6 × 250% = 26.5

26.5 is not a standard size breaker.

A standard size breaker higher than 26.5 is 30 amperes.

25. b) 150 amperes

59.4 × 250% = 148.5

148.5 is not a standard size breaker.

A standard size breaker higher than 148.5 is 150 amperes.

The largest branch circuit protection is 150 amperes.

150 + 24 = 174 (not a standard size breaker)

The maximum feeder overcurrent protection is 150 amperes.

26. a) 26 amperes

[Table 430.250]

27. b) 48

42 × 115% = 48.3 = 48

28. d) 27.83–31.46 amperes

24.2 × 115% = 27.83 minimum

24.2 × 130% = 31.46 maximum

29. b) 28 amperes

10.6 × 125% = 13.25 + 7.5 + 7.5 = 28.25

30. c) 16.1 amperes

All other motors = 115% [430.32(A)(1)]

14 (nameplate) × 115% = 16.1

31. c) 21

Find the full load current of each motor.

3 hp = 10.6 amperes

2 hp = 7.5 amperes

10.6 (largest motor) × 125% = 13.25

Add the other motor.

13.25 + 7.5 = 20.75 = 21 amperes

32. c) 125 amperes

Do not use the nameplate rating.

The full load current from Table 430.250 is 65 amperes.

65 × 175% = 113.75

113.75 is not a standard size fuse.

The next higher standard size = 125 amperes

33. b) 140 amperes

15 hp = 46.2 amps

7½ hp = 24.2 amps

3 hp = 10.6 amps

46.2 × 125% = 57.75

57.75 + 24.2 + 24.2 + 10.6 + 10.6 + 10.6 = 137.95 = 140 amperes

34. b) 2.68 volts

$$V_D = \frac{1.732 \times 12.88384 \times 150 \times 21}{26{,}240} = 2.68 \text{ volts}$$

35. d) 60 amperes

22 × 250% = 55

The next higher standard size = 60

36. a) 109 amperes

NOTE: These are synchronous-type motors.

25 hp = 26 amps

30 hp = 32 amps

40 hp = 41 amps

41 × 125% = 51.25

51.25 + 26 + 32 = 109.25 = 109 amperes

37. b) 8 AWG

[Table 250.122]

38. c) 6 AWG

[Table 430.250]

10 hp = 14

30 hp = 40

40 × 125% = 50

50 + 14 = 64 amps = 6 AWG

39. c) 150

59.4 × 125% = 74.25

74.25 + 59.4 = 133.65 amperes

The next higher answer above 133.65 amperes = 150

40. d) 250 amperes

80 × 300% = 240

The next higher standard size = 250 amperes

Chapter 11

End-of-Chapter Problems

1. c) 4000 watts

Optional Method [220.82(C)]

4000 × 100% = 4000 air conditioner

6000 × 65% = 3900 heat

Compare the heating and air-conditioning systems after the demand has been calculated, and then select the larger number.

2. d) None of these

In a dwelling unit all general-use receptacle outlets, 15 or 20 amperes, shall be considered as part of the general lighting load. No additional load calculations shall be required for these outlets. [220.14(J)]

3. b) 8 AWG

13.55 = 14

−12

2 × 5% = 10% = Increase Column C by 10%

Column C = 8 × 0.1 = 0.8 + 8 = 8.8 kW = 8800 watts

8800 ÷ 240 = 36.7 amperes = 8-AWG THW

4. b) 6 AWG

175 ampere service [Table 310.15(B)(6)] = 1/0-AWG copper

1/0-AWG copper [Table 250.66] = 6-AWG copper

5. a) 14.58

5000 ÷ 240 = 20.8333

20.8333 × 70% [220.61(B)] = 14.5833 = 14.58

6. d) 5.52 kW

4 (units) × 6 (amperes) × 230 (volts) = 5520 ÷ 1000 = 5.52 kW

5 (units) × 1 (kW) = 5 kW (omit this heating load)

7. a) 0 watts

In dwellings, receptacle outlets are included in the general lighting load calculations of 220.12. [220.14(J)]

8. b) 3900 watts

Optional Method [220.82(C)]

3000 × 100% = 3000 air conditioner

6000 × 65% = 3900 heat

Compare the heating and air-conditioning systems after the demand has been calculated, and then select the larger number.

9. c) 150 amperes

[220.82(B)] first 10,000 at 100% and the remainder at 40%

67,500 − 10,000 = 57,500 × 40% = 23,000

10,000 + 23,000 = 33,000

33,000 ÷ 240 = 137.5 = 150 amperes

10. **b) 4**

6000 ÷ 120 = 50

50 ÷ 15 = 3.33 = 4 circuits

11. **a) 24.35**

Column C = 8 kW = 8000 watts

8000 × 70% [220.61(B)] = 5600

5600 ÷ 230 = 24.3478 = 24.35

12. **d) 5.04 kW**

4 + 4 + 5.1 = 13.1 kW for three units

Column B for three units has a demand factor percent of 55%.

13.1 × 55% = 7.205 kW

7.205 × 70% [220.61(B)] = 5.0435 = 5.04 kW

Chapter 12

End-of-Chapter Problems

1. **c) 4**

General lighting will be on 15-ampere circuits.

small appliance = 2, laundry = 1, bathroom = 1

2 + 1 + 1 = 4

2. **b) 55 kW**

40 + 15 = 55 kW [Table 220.55]

3. **b) 225**

125 × 4 = 500 × 45% [Table 220.84] = 225

4. **b) 44.55 kW**

4.5 × 30 × 33% = 44.55 kW

5. **c) 18.7 kW**

16 + 14 + ~~10~~ 12 + ~~8~~ 12 = 54 ÷ 4 = 13.5

~~13.5~~ 14 – 12 = 2 × 5% = 10% = 1.1

Column C (4 units) = 17 × 1.1 = 18.7 kW

6. **c) 16.8 kW**

Column B = 3.5 × 15 × 32% = 16.8 kW

7. **a) 6**

3000 ÷ 120 = 25 amperes = 2 circuits

Small appliance = 2, laundry = 1, bathroom = 1

2 + 2 + 1 + 1 = 6

8. **b) 47.3 kW**

14 – 12 = 2 × 5% = 10% = 1.1

Column C (28 units) = 15 + 28 = 43 × 1.1 = 47.3 kW

9. **d) 45 kW**

10 × 6 = 60

60 × 75% [220.53] = 45

10. **d) 65 kW**

10 × 1500 = 15,000

10 × 5000 = 50,000

15,000 + 50,000 = 65,000 watts

65,000 ÷ 1000 = 65 kW

Derating is not permissible in laundry facilities.

11. **c) 182 amperes**

Fifty units from Column C = 25 kW plus ¾ kW for each range

25 + (0.75 × 50) = 25 + 37.5 = 62.5 kW = 62,500 watts

62,500 × 70% [220.61(B)(1)] = 43,750 watts on the neutral

43,750 ÷ 240 = 182.29 = 182 amperes

12. **d) 6**

4800 ÷ 120 = 40 amperes = 2 circuits

Small appliance = 2, laundry = 1, bathroom = 1

2 + 2 + 1 + 1 = 6

13. **c) 270 kW**

60 × 6 = 360

360 × 75% [220.53] = 270

14. **b) 82 amperes**

Table 220.55, Column C = 27 kW = 27,000 watts

27,000 × 70% [220.61(B)(1)] = 18,900 watts on the neutral

18,900 ÷ 230 = 82.17 = 82 amperes

15. c) 72.9 kW

Fifty-five units from Column C = 25 kW plus ¾ kW for each range

25 + (0.75 × 55) = 25 + 41.25 = 66.25 kW

Because each range exceeds 12 kW, Column C must be increased.

14 – 12 = 2 × 5% = 10% = Increase Column C by 10%

66.25 × 1.10 = 72.875 = 72.9 kW

16. a) 8.1

[220.84(C)(3)] use nameplate rating of 4.5 kVA

4 × 4.5 = 18

18 × 45% [Table 220.84] = 8.1

17. c) 64–72 kVA

800 × 3 × 12	= 28,800	
Small appliance × 12	= 36,000	
Laundry × 12	= 18,000	
	82,800	
	–3000 × 100% =	3000
	79,800 × 35% =	27,930
General lighting and receptacles:		30,930

Service

General lighting and receptacles:	30,930
Ranges, Column C:	27,000
Dishwasher 1200 × 12 × 75%:	10,800
Total:	68,730

68,730 ÷ 1000 = 68.73 kVA

18. d) 34.1 kW

5 × ~~8~~ 12 = 60

4 × 12 = 48

7 × 16 = 112

Total number of units = 16 units

Total kW = 220 kW

Average kW per unit = 220 ÷ 16 = 13.75 kW

13.75 = 14

– 12

2 × 5% = 10% = Increase Column C by 10%

Column C = 31 × 1.10 = 34.1 kW

Chapter 13

End-of-Chapter Problems

1. b) 17,000 VA

85 × 200 = 17,000 volt-amperes

2. c) 36–46 kVA

3 × 20 × 16 × 32 × 125%	=	38,400
2000 × 0.5 × 125%	=	1250
60 × 18 × 3.5 × 125%	=	4725
60 × 18 × 1	=	1080
		45,455

45,455 ÷ 1000 = 45.455

3. b) 12.2 kVA

80 × 180 = 14,400

–10,000	×	100%	=	10,000
4400	×	50%	=	2200
				12,200

12,200 ÷ 1000 = 12.2

4. b) 324 kVA

408,000 ÷ 24,000 = 17 volt-amperes per square foot

The first 3 at 100%

3 × 100% = 3.0

The remainder at 75%

14 × 75% = 10.5

3 + 10.5 = 13.5 VA per sq. ft.

13.5 × 24,000 = 324,000 ÷ 1000 = 324

5. a) 76 amperes

1 hp at 120 volts = 16 amps × 125% = 20

2 × 1920 × 125% = 4800 ÷ 120 = 40

16 (receptacles at 100%) = 16

20 + 40 + 16 = 76 amperes

6. d) 322.5 amperes

72,000 ÷ 240 = 300 × 125% = 375 amperes (continuous)

375

–200 × 100% = 200

175 × 70% = 122.5

322.5 amperes

7. b) 90

40,000 × 3.5 × 125% = 175,000 (continuous lighting)

40,000 × 1 = 40,000 (noncontinuous receptacles)

175,000 + 40,000 = 215,000 volt-amperes

215,000 ÷ 120 = 1792 amperes

1792 ÷ 20 = 89.6 = 90 circuits

8. c) 22,500 VA

50 × 15 × 30 × 2 = 45,000 volt-amps for rooms before derating

[Table 220.42]

45,000

<u>– 20,000</u> × 50% = 10,000

25,000 × 40% = 10,000

10,000 + 10,000 = 20,000 volt-amperes for rooms

4000 × 0.5 × 125% = 2500 volt-amperes for halls

20,000 + 2500 = 22,500 total volt-amperes

9. d) 135 kVA

160,000 ÷ 20,000 = 8 volt-amperes per square foot

The first 3 at 100%

3 × 100% = 3.0

The remainder at 75%

5 × 75% = 3.75

3 + 3.75 = 6.75 VA per sq. ft.

6.75 × 20,000 = 135,000 ÷ 1000 = 135 kVA

10. c) 1.25 kVA

2000 × 0.5 × 125% = 1250 volt-amperes

1250 ÷ 1000 = 1.25 kVA

11. c) 7

20 amperes × 80% = 16 amps maximum for continuous loads

250 watts ÷ 120 = 2.0833 amps each fixture

16 ÷ 2.0833 = 7.68 = 7 light fixtures maximum on each circuit

46 ÷ 7 = 6.57 = 7 circuits required

12. b) 25 amperes

6000 ÷ 240 = 25 amperes (no derating on commercial dryers)

13. b) 385 kVA

680,000 ÷ 20,000 = 34 volt-amperes per square foot

The first 3 at 100%

3 × 100% = 3.0

The next 17 at 75%

17 × 75% = 12.75

The remainder at 25%

14 × 25% = 3.5

3 + 12.75 + 3.5 = 19.25 VA per sq. ft.

19.25 × 20,000 = 385,000 ÷ 1000 = 385 kVA

14. b) 8 AWG

[Table 250.122]

15. d) One on each side

[210.62]

16. c) 7200 VA

180 × 40 = 7200 volt-amperes

17. b) 704 kVA

60 (amps) × 240 (volts) = 14,400 volt-amperes per trailer

Section 550.31 requires a minimum of 16,000 volt-amperes when calculating mobile home park services and feeders.

200 × 16,000 = 32,000,000 VA

Use Table 550.31 to derate.

32,000,000 × 22% = 704,000 ÷ 1000 = 704 kVA

18. b) 6 AWG

[Table 250.122]

19. c) 170 kVA

206,600 ÷ 20,000 = 10.33 volt-amperes per square foot

The first 3 at 100%

3 × 100% = 3.0

The remainder at 75%

7.33 × 75% = 5.4975

3 + 5.4975 = 8.4975 VA per sq. ft.

8.4975 × 20,000 = 169,950 ÷ 1000 = 169.95 = 170 kVA

20. **c) 8 AWG**

    [Table 250.122]

21. **b) 10,000–20,000 volt-amperes**

    80 × 100 × 1 [Table 220.12] × 125% = 10,000

    30 × 40 × 2 [Table 220.12] × 125% = 3000

    10,000 + 3000 = 13,000 volt-amperes

22. **b) 730 amperes**

    Receptacles supplying shore power for boats shall not be less than 30 amperes.

    [555.19(A)(4)]

    Find the total number of receptacles and amperes (per phase).

    20 × ~~20~~ 30 = 600

    17 × 30 = 510

    7 × 50 = 350

    44 rec. = 1460 amperes

    Multiply the total rating of the receptacles by the demand factor percent. [Table 555.12]

    1460 × 50% = 730 amperes

23. **c) 23 kVA**

    200 × 180 (VA per receptacle) = 36,000 volt-amperes

    Derate according to Table 220.44.

    The first 10,000 at 100%

    10,000 × 100% = 10,000

    The remainder at 50%

    26,000 × 50% = 13,000

    10,000 + 13,000 = 23,000 ÷ 1000 = 23 kVA

24. **d) 170 kVA**

    206,600 ÷ 20,000 = 10.33 volt-amperes per square foot

    The first 3 at 100%

    3 × 100% = 3.0

    The remainder at 75%

    7.33 × 75% = 5.4975

    3 + 5.4975 = 8.4975 VA per sq. ft.

    8.4975 × 20,000 = 169,950 ÷ 1000 = 169.95 = 170 kVA

25. **a) 2400 VA**

    One volt-ampere per sq. ft. for unknown receptacles in banks

    2400 × 1 = 2400 VA

26. **d) 78.75 amperes**

    39 × 1.25 = 48.75

    24 × 1.25 = 30

    48.75 + 30 = 78.75 amperes

27. **a) 3 AWG**

    [250.32(B)(2) and Table 250.122] There will not be much load on the grounded (neutral) conductor. For a grounded system at the separate building, the size of the grounded conductor shall not be smaller than the equipment grounding conductor required by 250.122. Minimum size equipment grounding conductor for a 400-ampere overcurrent device is 3 AWG.

28. **c) 4 AWG**

    [Table 310.16 and Table 250.66]

29. **b) 2**

    [210.62]

30. **a) 1/0 THW**

    [240.21(B)(1) and Table 310.16]

31. **a) 1/0 AWG**

    [250.24(C), 250.24(C)(1), and Table 250.66] The grounded (neutral) conductor shall not be smaller than the required grounding electrode conductor specified in Table 250.66. Minimum size grounding electrode conductor for 500 kcmil is 1/0 AWG.

32. **c) 300-kcmil THW**

    [240.21(B)(2) and Table 310.16] 800 ÷ 3 = 267 amperes minimum

33. **d) 8 inches**

    [300.34 and Chapter 9, Table 5] The approximate diameter of 500-kcmil THW is 1.003.

    8 × 1.003 = 8.024 = 8 inches

Chapter 14

End-of-Chapter Problems

1. **b) 0.96**

$$PF = \frac{12{,}442}{208 \times 1.732 \times 36} = 0.96$$

Alternative method: 208 × 1.732 × 36 = 12,969 VA

12,442 ÷ 12,969 = 0.959 = 0.96

2. **d) 77.09%**

$$Eff. = \frac{20 \times 746}{240 \times 1.732 \times 48 \times 0.97} = 0.7709 = 77.09\%$$

3. **d) Unity**

$$PF = \frac{14{,}965}{240 \times 1.732 \times 36} = 1.00 = \text{unity}$$

Alternative method: 240 × 1.732 × 36 = 14,964.5 VA

14,965 ÷ 14,964.5 = 1.00 = unity

4. **c) 87%**

$$PF = \frac{10{,}000}{230 \times 50} = 0.8696 = 0.87 = 87\%$$

Alternative method: 230 × 50 = 11,500 VA

10,000 ÷ 11,500 = 0.8696 = 0.87 = 87%

5. **a) 5.7 amperes**

$$I = \frac{hp \times 746}{E \times Eff.} = \frac{1.5 \times 746}{230 \times 0.85} = 5.7238 = 5.7$$

6. **d) Unity**

$$PF = \frac{12{,}960}{208 \times 1.732 \times 36} = 0.999 = 1.00 = \text{unity}$$

Alternative method: 208 × 1.732 × 36 = 12,969 VA

PF = 12,960 ÷ 12,969 = 0.999 = 1.00 = unity

7. **c) 42.5**

$$I = \frac{15{,}000}{208 \times 1.732 \times 0.98} = 42.4868 = 42.5 \text{ amperes}$$

8. **c) 43.37**

$$\frac{15{,}000}{208 \times 1.732} = 41.637 \div 0.96 = 43.37 \text{ amperes}$$

9. **b) 12.2 amperes**

$$I = \frac{hp \times 746}{E \times Eff. \times PF} = \frac{1.5 \times 746}{120 \times 0.90 \times 0.85} = 12.1895 = 12.2$$

10. **b) 12 AWG**

Do not calculate the actual current of the motor.

3-hp motor from Table 430.248 = 17 amperes

17 × 125% = 21.25 amperes

[Table 310.16] minimum of 21.25 amperes = 12 AWG

11. **c) 13.02 amperes**

$$I = \frac{hp \times 746}{E \times 1.73 \times Eff. \times PF}$$

$$= \frac{5 \times 746}{200 \times 1.73 \times 0.92 \times 0.90} = 13.02$$

12. **b) 91%**

$$PF = \frac{10{,}860}{208 \times 1.732 \times 33} = 0.91 = 91\%$$

Alternative method: 208 × 1.732 × 33 = 11,888 VA

10,860 ÷ 11,888 = 0.91 = 91%

13. **d) 6000–6100**

5-hp motor from Table 430.250 = 16.7 amperes

208 × 1.732 × 16.7 = 6016 VA

14. **a) 75%**

$$Eff. = \frac{4 \times 746}{4000} = 0.746 = 0.75 = 75\%$$

15. **b) 58%**

$$Eff. = \frac{5 \times 746}{6440} = 0.579 = 0.58 = 58\%$$

16. **b) 16,145**

15-hp motor from Table 430.250 = 46.2 amperes

208 × 1.732 × 46.2 × 0.97 = 16,144.512 = 16,145 VA

17. **b) 70%**

Since motor current is not given, use full-load current of motor in Table 430.250 = 59.4 amperes.

$$Eff. = \frac{20 \times 746}{208 \times 1.732 \times 59.4} = 0.6972 = 0.70 = 70\%$$

Chapter 15

End-of-Chapter Problems

1. **b) 208 volts**

   The secondary voltages are 208 and 120. Line voltage is higher than phase voltage. The secondary line voltage is 208 volts.

2. **b) 25 amperes**

$$I_L = \frac{VA}{E_L \times 1.732} = \frac{9000}{208 \times 1.732} = 24.98 = 25$$

3. **a) 1442 VA**

   The voltage across each coil is 208 volts. The resistance is 30 ohms. Ohm's law can be used to solve this problem.

   $W = E^2 \div R$

   $208 \times 208 \div 30 = 1442$

4. **a) 60 amperes**

   Assume the question is asking for primary if it does not specify primary or secondary.

$$I_L = \frac{VA}{E_L \times 1.732} = \frac{50{,}000}{480 \times 1.732} = 60 \text{ amperes}$$

5. **c) 67 amperes**

$$I_P = \frac{VA}{E_P \times 3} = \frac{24{,}000}{120 \times 3} = 66.6667 = 67 \text{ amperes}$$

6. **c) less than the secondary line voltage**

7. **d) 50 amperes**

   Use only the 120-volt loads. There are three 2000-watt loads on each 120-volt phase.

   $3 \times 2000 = 6000$ on each phase

   $6000 \div 120 = 50$

8. **b) 200**

   $1.732 \times 480 \times 240 = 199{,}526.4$ volt-amperes

   $199{,}526.4 \div 1000 = 199.5264 = 200$

9. **c) 25**

   $I = W \div E = 3000 \div 120 = 25$

10. **b) 5000 watts**

    $15{,}000 \div 3 = 5000$

11. **d) 160**

$$T_S = \frac{E_S \times T_P}{E_P} = \frac{480 \times 40}{120} = 160$$

12. **a) 79 amperes**

$$I_L = \frac{VA}{E_L \times 1.732} = \frac{28{,}460}{208 \times 1.732} = 78.999 = 79 \text{ amperes}$$

13. **c) 240 amperes**

    $10 \times 5000 = 50{,}000 \div 208 = 240$ amperes

    There are no derating factors for office machines.

14. **c) 30 × $\sqrt{3}$ 3 amps**

    First calculate the volt-amperes on the secondary.

    $208 \times 1.732 \times 120 = 43{,}230.72$

    Efficiency is not given, so assume 100%.

    Use 43,230.72 VA to calculate the primary line current.

$$I_L = \frac{VA}{E_L \times 1.732} = \frac{43{,}230.72}{480 \times 1.732} = 52 \text{ amperes}$$

    $30 \times 1.732 = 51.96 = 52$ amperes

15. **b) 199 volts**

    $115 \times 1.732 = 199.18 = 199$ volts

16. **c) 3600 watts**

    $1.732 \times 208 \times 10 = 3602.56 = 3600$

17. **d) 14.4 amperes**

    $1500 \div 208 = 7.2$ amperes each dryer

    $7.2 \times 2 = 14.4$ amperes (only two dryers are on each phase)

18. **c) 86.7**

$$\frac{36{,}000}{240 \times 1.732} = 86.6 \text{ amperes}$$

19. **b) 75**

    Three 3000-watt loads will be on each leg (phase).

    $9000 \div 120 = 75$

20. **b) 2 to 1**

    480 compared to 240 = 2 to 1

    The "120" in this delta transformer is ½ of the phase voltage.

21. **a) 60 amperes**

$$\frac{50{,}000}{480 \times 1.732} = 60.14 = 60 \text{ amperes}$$

22. **d) 30 amperes**

First calculate the volt-amperes on the secondary.

208 × 1.732 × 120 = 43,230.72

Efficiency is not given, so assume 100%.

Use 43,230.72 VA to calculate the primary phase current.

$$I_P = \frac{VA}{E \times 3} = \frac{43,230.72}{480 \times 3} = 30 \text{ amperes}$$

23. **b) 125 amperes**

$$\frac{45,000}{208 \times 1.732} = 124.9 = 125 \text{ amperes}$$

24. **a) I only**

The transformer kVA divided by (1.732 × 480)

25. **b) 417**

No neutral current is required for 240-volt loads.

150,000 – 50,000 = 100,000 volt-amperes

100,000 ÷ 240 = 416.6667 = 417 amperes

26. **c) 26.7**

$$\frac{9600}{208 \times 1.732} = 26.6477 = 26.7$$

27. **b) 333 amperes**

No neutral current is required for 240-volt loads.

100,000 – 20,000 = 80,000 volt-amperes

80,000 ÷ 240 = 333.3333 = 333 amperes

28. **c) 30 amperes**

No neutral is required for the pumps rated 240 volts.

The three lighting loads can be balanced as evenly as possible: two loads on one phase and one load on the other phase.

The maximum neutral current is two lighting loads energized on one phase and one load deenergized on the other phase.

1800 + 1800 = 3600 watts

3600 ÷ 120 = 30 amperes

29. **d) 75**

208 × 200 × 1.732 = 72,051.2 volt-amperes

72,051.4 ÷ 1000 = 72.051 = 75

30. **b) 25 amperes**

[Table 450.3(B) and Note 1]

$$\frac{15,000}{480 \times 1.732} = 18.04 \text{ amperes}$$

18.04 × 125% = 22.55

Since 22.55 is not a standard size overcurrent protective device, select the next higher standard rating = 25 amperes

31. **c) 50%**

[450.5(A)(3) FPN]

32. **b) 10 amperes**

[Table 450.3(B) and 240.6(A)] 8 × 167% = 13.36

The overcurrent protection rating cannot exceed 167%.

Next standard size fuse or breaker below 13.36 = 10 amperes

33. **c) 20 amperes**

[Table 450.3(B) and Note 1] 13 × 125% = 16.25

Since 16.25 is not a standard size overcurrent protective device, select the next higher standard rating = 20 amperes

34. **b) 125%**

[Table 450.3(B)]

GENERAL EXAM QUESTIONS ANSWER KEY

Answers to General Exam Number 1

1. d) electrochemistry
2. a) True
3. b) parallel
4. d) magneto electricity
5. c) eddy current loss at a minimum
6. c) 148.8 watts
7. c) volts
8. a) transformer
9. b) watt-hour
10. c) 120-volts
11. d) Electro
12. b) shorted
13. d) the lack of isolation between the primary and secondary
14. b) quench the arc
15. a) XL
16. d) expansion bolts
17. b) electrons to flow
18. c) B-C-A
19. c) $\frac{1}{120}$ sec.
20. c) has its power supplied by transformers and batteries
21. a) EMT [358.28(B)]
22. d) relay
23. b) impedance
24. c) I, II, and V
25. b) switch
26. b) inductance
27. b) one hot leg is shut off
28. d) multiply the number of strands by the circular-mil area of one strand
29. a) Polarity
30. c) oscilloscope
31. c) current
32. d) I, II, or III
33. d) I, II, or III
34. b) False
35. c) in a dc circuit is
36. b) toggle bolts
37. b) False
38. c) burn at full brightness
39. d) threads per inch
40. c) 746
41. a) Rubber
42. a) high
43. a) conductivity
44. c) auto
45. c) provide greater voltage than each individual battery
46. c) 9
47. b) 100 volts
48. d) 140 volts
49. b) 6 amperes
50. d) 1, 2, 3, and 4

Answers to General Exam Number 2

1. c) A tachometer
2. a) Less voltage drop
3. b) less
4. d) horsepower
5. b) electrolyte
6. d) One
7. a) Δ
8. b) 60 amperes
9. a) shunt
10. b) effective
11. c) National Fire Protection Association
12. a) plan
13. c) condenser
14. a) Paper
15. a) Condenser
16. b) series-parallel
17. a) turns-ratio
18. b) the conduit thread is tapered
19. c) even spacing, numerous lights
20. b) accurately only when clamped perpendicular to a conductor
21. c) Squirrel cage
22. a) cartridge
23. b) connect two sections of rigid conduit when one section cannot be turned
24. c) temperature
25. d) soft iron
26. c) both a and b
27. a) A 10-ohm resistor
28. c) is determined by load, distance, and size of conductors
29. b) 6″ ($1500 \div 1000 = 1.5 \div 0.25 = 6$)
30. c) three
31. b) all in parallel
32. b) continuity
33. b) Branch-circuit protection
34. a) hydrometer
35. c) make the blades dead when the switch is open
36. a) strip of sandpaper
37. b) largest
38. a) either a vacuum or gas
39. a) From the top of the slab on the first floor to the bottom of the slab on the second floor
40. c) $V \times C \times \cos \theta$
41. d) the insulation to deteriorate
42. c) the effective difference of potential between two conductors
43. b) lower-wattage bulb has the higher resistance
44. b) 14.4 ohms
45. c) EMF
46. b) parallel
47. c) Aluminum
48. b) 384 ohms; 1.56 amperes
49. d) prevent electrical shock, which might otherwise result from contact with the conduit
50. a) the phase current

Answers to General Exam Number 3

1. c) I and II
2. b) 1.56 amperes
3. c) I and III
4. c) whether the supply is 25-cycle or 60-cycle ac
5. b) 3Ø 208-volt, 1Ø 120-volt, and 1Ø 208-volt
6. d) commutator bar separators
7. d) a loose connection
8. c) at least two wires
9. c) Capacitor
10. a) XC
11. d) conductance
12. b) control the speed of the motor
13. c) three-phase
14. a) Ease of voltage variations
15. b) heater coils
16. c) RMS or effective value
17. c) accidentally energized at a higher potential than ground

18. a) in accordance with the local code when more stringent than the *NEC*
19. a) commutator
20. c) ½
21. b) a conductor moving inside a magnetic field
22. b) E is equal to I × R
23. d) frequency meter
24. c) wear leather gloves over rubber ones
25. b) 120
26. b) insulation resistance
27. c) use a template
28. a) volt
29. b) Rectifier
30. c) use a hacksaw and remove the burrs
31. c) ⅔ ohm
32. a) expansion joints
33. a) A
34. c) consider the circuit is always alive
35. b) one megavolt
36. a) an explosion
37. c) 160 watts
38. c) an interconnected system
39. c) can be stored in chemical form
40. d) where the ratio of transformation is low
41. b) R/Z
42. b) real power
43. a) broken
44. b) is the output of the motor
45. d) 1000 amp turns
46. b) 0.002
47. a) protect the end of the wire
48. b) Three-phase, 4-wire
49. b) the fuse clips are too loose
50. c) 120

Answers to General Exam Number 4

1. b) heat
2. c) direct
3. d) I, II, and III
4. c) henry
5. b) protect the user of the tool should the winding break down to the case
6. b) the same ampacity [Table 310.16]
7. d) DC
8. c) decrease
9. a) low resistance and low melting point
10. a) 90-degree
11. a) apparent
12. d) resistor
13. c) farad
14. d) 1,000,000 watts
15. a) Perpendicular
16. d) Safe for stepping down higher voltages
17. c) remains the same
18. a) remain constant regardless of voltage or current change
19. d) One
20. d) compensate for voltage drop
21. b) that it cannot be used in electroplating
22. a) short circuited
23. b) 6.9 amperes
24. c) 12 amperes
25. a) 1442 VA
26. c) 4326 VA
27. d) the feeder
28. c) voltmeter
29. c) both ac and dc
30. c) Oil
31. d) the temperature of the area surrounding the conductor
32. a) 50 [314.27(B)]
33. b) in series
34. b) chemical activity
35. c) unbalanced [220.61(A)]
36. c) capacitance
37. b) average
38. c) thermocouple
39. b) in a dc circuit
40. c) raintight and arranged to drain [225.22]
41. a) effective value
42. c) capacitive reactance exceeds the inductive reactance in the circuit

43. b) 70.7 volts
44. d) I, II, and III
45. c) the contact resistance
46. b) high resistance
47. a) primary
48. d) zero
49. d) Conductance
50. b) False

Answers to General Exam Number 5

1. c) annunciating
2. b) its total watt output decreases
3. a) 1⁄80 sec.
4. c) 3 amperes
5. c) one wattmeter
6. d) shorted
7. b) TW [Table 310.13]
8. c) 1600 watts
9. c) lead sheath
10. b) use more power
11. c) 1000 ($W = I^2R = 10 \times 10 \times 10$)
12. b) film-disc cutout
13. d) voltmeter, ohmmeter, and ammeter
14. d) 30 amperes
15. b) power factor
16. c) will give longer service
17. c) 2500
18. b) 114 volts
19. c) for the purpose of strengthening a splice
20. c) inductive reactance exceeds the capacitive reactance in the circuit
21. a) the inside
22. d) silver
23. b) voltage difference
24. c) 120
25. a) oil
26. c) parallel
27. c) hanging fixture
28. b) longer bulb life
29. d) resistance
30. a) reverse any two of the three leads
31. c) specific gravity
32. c) III only
33. b) 35
34. a) Edison
35. a) True
36. d) one strand of the conductor
37. b) maximum [Art. 100]
38. d) 10,800 watts
39. b) False
40. b) zero
41. c) 1⁄240 sec.
42. b) shock or injury
43. d) shunt
44. a) thermal cutout
45. b) Vacuum
46. d) Oil
47. b) current
48. c) ammeter, wattmeter, and voltmeter
49. d) ohms
50. c) Conductor

Answers to General Exam Number 6

1. c) 480 volts
2. a) magnetic effect
3. a) rheostat
4. d) Primary
5. a) some sort of ballast
6. c) Optical
7. a) ohms
8. c) dc
9. a) cable terminal
10. b) tin and lead
11. a) Current
12. c) universal
13. a) True
14. a) 6 amperes
15. a) Mica
16. c) 25

17. d) High leg
18. d) either I or II
19. c) 168 volts
20. d) wires do not support the light
21. c) highest when bulb is on
22. d) reduce pitting of the contacts
23. a) induction
24. d) prevent the connection from loosening under vibration
25. a) the insulation should not be used as the only protective measure
26. b) a percentage of the applied voltage
27. d) 120
28. c) because of inductance
29. c) prevent corrosion
30. b) less than any one resistor
31. b) direct
32. a) good conductors
33. b) electrons passing a point per second
34. b) Kirchoff's law
35. c) Power factor meter
36. b) chemical
37. d) commutator
38. d) magnetism
39. a) for moisture proofing
40. d) 1, 2, and 3
41. a) electron flow
42. c) iron
43. a) current flow develops heat
44. c) the relationship among voltage, current, and resistance
45. b) pressure
46. d) it is a constant value throughout the U.S. for distribution systems
47. a) ohms
48. a) Zinc
49. b) II only
50. c) 110 volts

Answers to General Exam Number 7

1. c) use a hacksaw and ream the ends
2. a) Delta-wye
3. a) 1⁄60 sec.
4. c) $R = \frac{V^2}{W}$
5. a) an increase in the voltage drop across the connection
6. b) rectifier
7. a) True
8. c) a tool used to bend small sizes of rigid conduit
9. b) 600
10. a) at the higher voltage
11. d) delta
12. a) commutator or slip rings
13. a) oil
14. c) 1¼″
15. b) 11,612.9 watts
16. c) CO_2 fire extinguisher
17. d) two 3-ways and three 4-ways
18. b) Aluminum
19. a) mica
20. d) I, II, and III
21. b) Standard
22. b) diameter
23. c) a megger
24. d) 12 [210.52(A)(1)]
25. d) Shall [90.5(A)]
26. b) larger than
27. a) greater than
28. a) Wattmeter
29. c) Aluminum Sheathed Cable
30. b) False
31. d) DC generator driven by a dc motor
32. a) inversely proportional to its cross-sectional area
33. b) rawl plugs
34. a) current flowing through each load will be the same
35. c) may transmit shock to the user
36. d) direct
37. b) short circuits

38. b) False
39. d) 6000 watts
40. a) piezoelectricity
41. d) a loose connection
42. c) series
43. b) cost
44. b) hydrometer
45. a) damaging the outside finish
46. d) load
47. b) voltage
48. c) cancel the magnetic fields generated around the armature coils
49. a) 1
50. d) Rectifiers and filters

NEC EXAM QUESTIONS ANSWER KEY

Answers to *NEC* Exam Number 1

1. **d) grounding conductor** [Art. 100 (Grounding Conductor)]
2. **a) main bonding jumper** [408.3(C)]
3. **d) I, II, or III** [520.21 and 520.24]
4. **d) I, II, and III** [410.24]
5. **b) 20 AWG** [620.12(A)(2)]
6. **c) of the locking and grounding type** [555.19(A)(4)(a)]
7. **a) 14 AWG** [600.7(D)]
8. **d) 3.356″** [Chapter 9, Table 4]
9. **d) where exposed to destructive corrosive conditions, unless the metallic sheath is suitable for the conditions or is protected by material suitable for the conditions** [330.12]
10. **c) I and II** [600.6]
11. **d) the dimension does not matter** [680.2]
12. **d) I, II, or III** [460.24(C)]
13. **b) III only** [110.15]
14. **c) CO/ALR** [406.2(C)]
15. **a) terminals** [200.10(A)]
16. **d) any of these** [820.100(A)(2)]
17. **a) I, II, and III** [398.30(A), 398.42, 404.10(A)]
18. **d) installed with the grounded circuit conductor insulated from the grounding conductor and equipment enclosures** [550.16(A)(1)]
19. **a) electrolytic cell** [668.2]
20. **b) 600** [250.184(A)(1)]
21. **c) 9′** [Table 110.34(E)]
22. **b) 10** [230.26]
23. **b) 1⁄4″** [300.6(C)]
24. **a) 6 feet** [250.53(B), 250.56]
25. **c) ampere rating of the ballast** [220.18(B)]
26. **b) False** [240.51(B)]
27. **d) I, II, or III** [300.6(A)]
28. **c) 3′; 8′** [410.4(D)]
29. **c) a and b** [210.4(C) Ex. 1 and Ex. 2]
30. **a) 25 ohms or less** [250.56]
31. **d) 0.864″** [Chapter 9, Table 4]
32. **c) main bonding jumper** [Art. 100, 250.28]
33. **c) I and II** [380.2(B) and 380.3]
34. **b) 12 AWG** [225.6(B)]
35. **b) 3⁄4** [Table 408.56]
36. **a) the length of a standard pipe thread** [342.28, 344.28]
37. **c) 8** [410.4(D)]
38. **c) 30** [366.22(A)]
39. **b) 4** [314.16(B)]
40. **a) 3** [460.2(A)]
41. **b) flame arrester** [480.10(A)]
42. **d) Nonlinear loads** [220.61(C)]
43. **b) 8** [547.5(B)]
44. **c) can be started** [430.7(B)(1)]
45. **b) 100** [550.32(C)]
46. **d) 70 pounds** [314.27(D)]
47. **b) 3 percent** [210.19(A)(1) FPN No. 4]
48. **c) 30 inches** [110.26(A)(2)]
49. **a) 0** [250.64(C)]
50. **a) flame-retardant type** [620.21(C)(2)]

Answers to *NEC* Exam Number 2

1. **b) shall be permitted with a minimum size jumper 6-AWG copper** [800.100(D)]
2. **d) 3½** [410.38(C)]
3. **a) 1⁄4** [600.4(B)]
4. **b) header** [372.2, 374.2]
5. **a) 20** [210.19(A)(3) Ex.1]
6. **c) 115%** [440.12(A)(1)]
7. **c) substantially increased** [300.21]
8. **a) 15 amperes** [322.56(B)]
9. **b) Two** [210.70(A)(2)(b)]
10. **d) the manual controls are readily accessible and located within 6′ of the floor** [440.63]

11. **d) I, II, or III** [250.178]
12. **c) exposed ballasts or transformers** [410.76(A)]
13. **b) I, II, and III** [551.2 (Appliance, Portable)]
14. **c) Grounding conductor** [Art. 100 (grounding conductor)]
15. **c) intended for isolating an electrical circuit from its source of power** [Art. 100 (Switch, Isolating)]
16. **b) 90** [410.5, 410.65(A)]
17. **c) 1¼** [300.4(A)(1)]
18. **b) four** [517.18(B)]
19. **a) 20-AWG** [620.12(A)(1)]
20. **d) 30 volts** [650.4]
21. **c) electrical continuity** [392.6(A)]
22. **a) 2** [600.9(C)]
23. **d) 6** [Table 450.3(B) Note 3]
24. **a) 2 hours' full-demand operation of the system** [700.12(B)(2)]
25. **b) 0.030** [250.162(A) Ex. 3]
26. **c) This does include the emergency power service.** [620.91(C)]
27. **d) b and c** [620.11(A)]
28. **b) I and II only** [110.3(A)]
29. **b) 50 feet** [368.17(B) Ex.]
30. **a) True** [410.54(A)]
31. **c) 4-AWG** [250.66(B)]
32. **c) 300** [410.73(B)]
33. **b) False** [422.46]
34. **b) 12½** [250.28(D), 250.102(C)]
35. **c) listed tamper resistant** [517.18(C)]
36. **a) 14 AWG** [600.7(D)]
37. **b) 10 AWG** [517.14]
38. **d) I, II, or III** [422.14]
39. **c) 75** [700.12]
40. **a) clothes closets** [240.24(D)]
41. **c) I and II** [392.5(E)]
42. **c) root-mean-square current** [620.13 FPN No. 1]
43. **b) visible break** [230.204(A)]
44. **b) listed** [600.3]
45. **c) 5 kVA** [220.54]
46. **d) 5** [500.8(D) and 505.9(E)]
47. **c) electrically operated and mechanically held** [700.6(C)]
48. **d) 8¾ inches** [332.24(2)]

    $0.875 \times 10 = 8.75$
49. **d) 8 feet** [300.50(B)]
50. **d) Never** [725.3(C) and 300.22(A)]

Answers to *NEC* Exam Number 3

1. **c) covered in their own section of the *Code*** [540.1]
2. **c) to both a and b if both conditions exist** [310.15(B)(2)(a), 210.19(A)(1)]
3. **d) both I and II** [390.6]
4. **b) compensate for voltage drop** [210.19(A)(1) FPN No. 4]
5. **c) the type of insulation is important**
6. **d) At least one more than the number of conduit entries** [680.24(D)]
7. **a) at least 50 percent of the input required for the momentary rating** [660.5, 517.72(A)]
8. **a) equal to aluminum** [Tables 310.16 through 310.20]
9. **b) 600** [450.6]
10. **d) all of these** [511.12]
11. **b) II only** [324.10(H)]
12. **a) provided with factory-installed ground-fault circuit-interrupter protection** [600.10(C)(2)]
13. **d) I or III** [240.100(B)]
14. **c) 14′** [Table 344.30(B)(2)]
15. **a) I only** [398.42]
16. **c) either I or II** [600.8(B)]
17. **d) either I or II** [314.16(C)(2)]
18. **b) II only** [250.20(A)]
19. **a) 20** [376.22]
20. **d) I, II, or III** [430.109(A)(1), (2), and (3)]
21. **a) True** [250.34(A)]
22. **b) False** [430.7(A)(7)]
23. **b) 18** [210.6(D)(1)]
24. **d) are not part of the emergency system** [517.25, 517.30(B)]
25. **d) 100** [430.72(B)(2)]
26. **a) D** [310.11(C)(1)]
27. **b) II only** [504.2 (Intrinsically Safe Circuit)]
28. **c) 1″** [312.11(A)(3)]

29. **d) I, II, and III** [701.11(B)(1) and (2)]
30. **d) 0.027** [Chapter 9, Table 8]
31. **c) multioutlet assembly** [Art. 100 (Multioutlet Assembly)]
32. **b) the wall of the equipment** [110.12(A)]
33. **d) I, II, and III** [800.179(E)]
34. **b) 40** [225.6(B)]
35. **b) 55 amperes** [215.2(A)(2)]
36. **d) accessible to authorized personnel only** [700.25]
37. **c) equipped with an identified temperature-limiting means** [422.46]
38. **a) 0.0293** [Chapter 9, Table 5]
39. **c) I and II** [426.2, 427.2]
40. **d) I, II, or III** [230.62(B)]
41. **b) I or II** [280.4(A)]
42. **a) 70** [220.61(B)]
43. **d) I, II, and III** [410.29(C)]
44. **c) no induced currents** [300.20(B) FPN]
45. **d) All of the above** [810.12]
46. **a) 24** [Chapter 9, Note 4]
47. **b) arc-fault protected** [210.12(B)]
48. **c) 78 inches** [110.26(E)]
49. **a) for environmental air** [300.22(A) and (B)]
50. **a) 15** [Table 210.21(B)(3)]

Answers to *NEC* Exam Number 4

1. **b) 8** [Table 314.16(A)]
2. **b) False** [410.16(A)]
3. **d) all of these** [Art. 100 (Hoistway)]
4. **b) 20** [Table 220.55]
5. **c) average** [Art. 100 (Voltage)]
6. **a) ½″** [314.24]
7. **b) 6 kW** [220.55, Note 4]
8. **a) 10** [680.8]
9. **c) II or III** [440.2 (Branch-Circuit Selection)]
10. **b) 6** [820.44(F)(3)]
11. **a) 300** [410.75]
12. **d) I, II, and III** [551.55(F)]
13. **a) 18** [Annex C, Table C2]
14. **b) warm; colder** [300.7(A)]
15. **b) A dc system operating at 50 volts or less between conductors** [250.162(A)]
16. **d) I, II, and III** [400.10 FPN]
17. **c) 1200** [220.14(F)]
18. **a) with ground-fault protection above 600 amperes** [Table 392.7(B) footnote b]
19. **c) does not exceed 600 volts** [430.111(B)(3)]
20. **a) the starting current of the motor** [440.22(A)]
21. **d) Class I, Group C type receptacle** [517.61(A)(5)]
22. **d) time delay** [440.54(B)]
23. **c) 48; 60** [424.72(B)]
24. **a) one service** [230.2]
25. **a) excluding moisture** [501.15]
26. **b) 6 feet** [250.102(E)]
27. **c) in accordance with the conductor ampacity** [725.23]
28. **b) III only** [250.4(A)(2)]
29. **b) False** [250.142(A)(1)]
30. **c) three, one in each phase wire** [Table 430.37]
31. **d) I, II, and III** [430.32(E)]
32. **c) 18 inches** [300.5(D)(1)]
33. **a) 3′** [225.19(B)]
34. **d) I, II, and III** [460.24(A)]
35. **b) 115** [430.110(A)]
36. **d) I, II, or III** [640.9(C)]
37. **c) I or II** [440.12(A)(2)]
38. **c) I and III** [372.6]
39. **b) 8-AWG insulated** [680.23(B)(2)(b)]
40. **c) transition assembly** [324.40(D)]
41. **a) 3′** [225.19(B)]
42. **c) 18** [511.3(B)(3)(1)]
43. **d) I, II, and III** [390.3(C)]
44. **c) 8 AWG** [Table 250.122]
45. **a) 1000** [240.4(C), 240.6(A)]
46. **c) 5 percent** [215.2(A)(3) FPN No. 2]
47. **d) 36½ feet** [550.10(D)]
48. **b) 20 amperes** [600.6(B)]
    $10 \times 2 = 20$ amperes
49. **c) NM cable** [300.22(B)]
50. **b) It must be sized proportionately according to circular mil area of the ungrounded conductors.** [250.122(B)]

Answers to *NEC* Exam Number 5

1. a) I only [402.3 FPN]
2. c) I and III [354.10]
3. a) 50% [210.23(A)(2)]
4. b) be permitted where buried under not less than 2 feet of cover [514.8 Ex.2]
5. a) True [110.19]
6. c) 112½ [450.21(B)]
7. d) I, II, or III [332.104]
8. a) 2′ [320.30(D)(2)]
9. a) Class I, Division 1 [Table 514.3(B)(1)]
10. b) hoistways [392.4]
11. d) ¾″ taper per foot [344.28]
12. a) thermal protector [Art. 100 (Thermal Protector, FPN)]
13. c) 200% [600.6(B)]
14. c) the grounding terminal of the service equipment [680.25(B)]
15. b) False [110.14(B)]
16. a) 12 [470.3]
17. b) effectively insulated from ground and the motor driving it [250.112(B)]
18. c) II or III [250.64(C)(1)]
19. a) is covered in the *Code* under Article 640 [640.1]
20. d) 18 [Table 2, Chapter 9]
21. a) 300 volts [386.12(2)]
22. c) 25 amperes [210.3]
23. b) shall be removed from the raceway [374.7]
24. b) overload [430.55]
25. c) 50 [422.11(C)]
26. d) 90° C [410.33]
27. b) if less than 600 volts and if insulated for at least the maximum circuit voltage applied to any conductor within the enclosure, cable, or raceway [300.3(C)(1)]
28. c) 600 [110.54(B)]
29. c) 150° C [Table 402.3]
30. d) sized not less than the largest supply conductor [675.13]
31. a) 0 [Table 220.12]
32. d) electrically continuous [250.64(E)]
33. d) I, II, and III [250.184(C)(1)]
34. d) those permitted for the adjacent insulated conductors [310.15(B)(3)]
35. c) 4 [424.59 FPN]
36. c) 16 [410.15(A)]
37. b) II only [250.119]
38. b) At intervals not exceeding 4½ feet [348.30(A)]
39. b) the same
40. a) 24 [Table 300.5]
41. c) 5000 [240.83(C)]
42. d) 300 [600.23(D)]
43. d) I, II, and III [322.10]
44. c) shall not [110.14]
45. c) I and III [645.10]
46. b) ⅓ hp [430.109(F)]
47. b) the total ampere rating of each fixture [220.18(B)]
48. d) RHH [Table 310.13]
49. c) within sight of the pool equipment [680.12]
50. a) 1 [700.9(D)(2)]

Answers to *NEC* Exam Number 6

1. b) soldered [394.56]
2. b) must be the same size as the insulated circuit conductors [Table 250)122]
3. a) 65 [445.12(C)]
4. d) 90 [590.3(B)]
5. c) shall be used only as isolating switches [404.13(A)]
6. d) 14-AWG copper [250.30(A) Ex.3]
7. d) I, II, or III [410.23 and 47]
8. c) I, II, or III [669.9]
9. b) 5 [368.30]
10. b) conductors and equipment on certain public or private premises [90.2(A)(1)]
11. c) 8 [408.35]
12. d) 18 [314.23(F)]
13. b) 2000 amperes [366.23(A)]
14. a) shielded [400.31(B)]
15. c) 100 [240.83(B)]
16. b) freedom from hazard [90.1(A)]
17. a) 10 [225.4]

18. a) True [300.3(C)(1)]
19. b) 10 seconds [517.31]
20. c) 6 [410.67(C)]
21. b) I, II, and III [702.2 FPN]
22. d) from induced circulating currents [470.18(E)]
23. c) 50% [450.5(A)(3) FPN]
24. b) 2′ [230.54(C) Ex.]
25. a) distinctive marking at its terminations [200.6(A)(1)]
26. a) 150 [250.172 Ex.]
27. b) 3-inch [424.41(D)]
28. d) 200 [630.12(A)]
29. c) 30 [324.10(B)(2)]
30. d) 125% [424.82]
31. a) a connection on the header [372.9]
32. d) 100′ [240.21(B)(4)(2)]
33. c) 115 [445.13]
34. d) I, II, or III [406.3(D)(3)]
35. c) are not part of the emergency system [517.30(B)]
36. a) as low as practicable [460.8(B)]
37. d) none of these [410.8(D)(3)]
38. b) I and III [460.24(A) (current, not voltage)]
39. c) I, II, and III [312.8]
40. c) either a or b [424.29]
41. c) in bathrooms of guest rooms [210.8(B)]
42. a) higher [Chapter 9, Table 8]
43. b) service entrance [701.9(A)]
44. d) 135 [460.8(A)]
45. d) I, II, and III [600.32(B)]
46. b) 50 pounds [314.27(B)]
47. b) 2 AWG [250.122(F) and Table 250.122]
48. a) 5 [513.3(C)]
49. c) must be Type FPLP [760.61(A)]
50. d) 12 feet [680.22(B)(1)]

Answers to *NEC* Exam Number 7

1. d) integral mechanical means [404.6(B)]
2. a) 200 [404.16]
3. d) combustible dust [500.5(C)]
4. a) thermal protector [Art. 100 (Thermally Protected)]
5. d) I, II, and III [310.11(A)]
6. c) 5′ [366.30(A)]
7. a) 6 [314.28(B)]
8. c) 12 [230.24(B), 225.18]
9. b) 12 [210.62]
10. a) Class II, Division 1 [500.5(C)(1)]
11. a) 50 [422.43(A)]
12. d) I and II, or III [520.24]
13. b) 17 AWG [Table 810.16(A)]
14. a) cabinet [Art. 100 (Cabinet)]
15. b) I and III [610.21(F)(4)]
16. a) Under certain conditions, equipment grounding conductors may be required to be larger than circuit conductors. [250.95]
17. c) 200% [630.12(B)]
18. b) 14 AWG [410.27(B)]
19. a) not classified [511.3(A)(1)]
20. d) regardless of distance apart [Art. 100 (Festoon Lighting)]
21. c) the ampacity of the feeder conductors [430.62(B)]
22. b) I and III [406.3(A)]
23. a) 2″ [382.15]
24. c) the installation of the feeders [215.5]
25. a) 1⁄8 [422.31(A)]
26. c) 167°F [Table 402.3]
27. d) both a and b [440.41(B)]
28. b) shall not be used in a spray area during spray operations [516.4(D)]
29. a) 5 [640.10(A)]
30. d) I, II, and III [645.16]
31. b) 0.8 [Chapter 9, Note (7)]
32. c) 18″ [680.23(A)(5)]
33. d) 0.040 [410.38(A)]
34. b) 50 [110.27(A)]
35. d) 12″ [450.22]
36. c) 48 times [314.71(A)]
37. b) 8 AWG [230.31(B)]
38. c) 10 [700.12]
39. a) 35 [Table 310.15(B)(2)(a)]
40. b) surge arrester [280.2 (Surge Arrester)]
41. b) 1⁄2

42. a) I or III [400.22(A)]
43. d) not require an additional enclosure [600.8]
44. b) 10′ [440.64]
45. a) 2½′ [230.51(A)]
46. b) 20 [760.51]
47. a) 2 [725.55(J)]
48. b) 3 feet [225.19(C)]
49. b) 125% [Table 450.3(B)]
50. c) 3 [Table 430.37]

Answers to *NEC* Exam Number 8

1. b) discharging [280.2]
2. a) ungrounded [685.12]
3. d) I, II, or III [640.21(A)]
4. b) shall not prohibit the installation of these motors on wooden floors [430.14(B) Ex.]
5. b) 15-20-25-30 [424.3(A)]
6. a) I only [501.10(A)(1)]
7. b) 30 [322.10(1)]
8. c) 7500 [600.23(C)]
9. b) 2 [250.52(A)(6)]
10. b) ½″ [410.46]
11. c) cable tray system [392.2 (Cable Tray System)]
12. c) 1″ [366.100(E)]
13. b) ½″ [410.66(A)(1)]
14. c) 30 amperes [555.19(A)(4)]
15. d) I, II, and III [324.40(A)]
16. d) be smaller than 14 AWG [820.100(A)(3)]
17. d) I, II, or III [300.6(B)]
18. b) ¾ inch [Table 408.56]
19. d) I, II, or III [800.3(C), 300.21]
20. b) II only [90.7]
21. b) 300 volts [410.75]
22. c) be permitted to be installed in general-purpose enclosures [501.115(B)(2)]
23. a) 250; 125 [440.55(B)]
24. d) 15,000 volts [600.23(C)]
25. c) 36 [324.10(H)]
26. b) ozone-resistant [310.6]
27. c) I only [665.1]
28. b) Class I, Division 1 [517.60(A)(2)]
29. a) I only [410.86]
30. a) 1½′ [210.52(A)(3)]
31. d) 1000 [700.12]
32. d) 33 watts [424.98(A)]
33. a) 6″ [Table 300.5]
34. c) 24″ [Chapter 9, Note 4]
35. b) fitting [Art. 100 (Fitting)]
36. d) I, II, and III [410.24]
37. d) nonautomatic [Art. 100 (Nonautomatic)]
38. c) outlet [600.5(A)]
39. a) antenna discharge unit [810.20(A)]
40. d) unless they are durably and legibly marked by the manufacturer with their volume [314.16(C)(2)]
41. b) attachment plug [400.7(B)]
42. c) service disconnecting means [Art. 100 (Service Conductors)]
43. b) 250 [810.12]
44. c) I and III [620.22(A)]
45. b) I or III [517.18(C)]
46. a) 14 AWG [250.30(A)(1) Ex. 3]
47. b) 5 [250.52(A)(1)]
48. c) 5 feet [680.10]
49. a) 26 AWG [760.57]
50. d) 75 pounds [314.27(D)]

Answers to *NEC* Exam Number 9

1. a) II only [392.4]
2. b) 3 [Table 314.16(A)]
3. d) I, II, and III [430.82(B)]
4. b) II or III [700.12(A)]
5. c) a grounded metal barrier [680.23(A)(2)]
6. c) I and II [Table 210.21(B)(3)]
7. b) 10 AWG [210.19(A)(3) Ex.2]
8. b) II only [518.4(A)]
9. a) I and II [404.14(A)]
10. d) I, II, or III [511.3(A)(2)]
11. b) False

    IP = IL

12. **c) 4 AWG** [250.64(B)]
13. **c) 75%** [376.56]
14. **a) sight** [600.6(A)(1)]
15. **b) shall be identified** [410.42(A)]
16. **d) I, II, or III** [250.10]
17. **b) 14′** [600.9(A)]
18. **c) 12-AWG copper** [250.178]
19. **a) ¼″** [410.86]
20. **d) I, II, or III** [422.61]
21. **c) 50** [460.6(A)]
22. **b) 168** [Table 326.80]
23. **d) I, II, or III** [702.2]
24. **d) I, II, or III** [342.14, 344.14]
25. **c) conduct or witness** [700.4(A)]
26. **b) 40** [312.8]
27. **b) the total ampere rating of such units** [220.18(B)]
28. **d) array** [690.2 (Array)]
29. **c) neither a nor b** [90.1(C)]
30. **d) I, II, or III** [110.27(A)]
31. **c) 90°** [Table 310.13]
32. **b) feeder** [Art. 100 (Feeder)]
33. **d) Video systems** [640.1]
34. **c) metallic; conductive** [Art. 100 (Bonding)]
35. **c) MV** [328.2]
36. **b) 6** [650.8]
37. **a) True** [110.34(C)]
38. **a) True** [110.14(B)]
39. **d) intended for cutting off an electrical circuit from its source of power** [Art. 100 (Switch, Isolating)]
40. **c) capable of being reached quickly for operation** [Art. 100 (Accessible, Readily)]
41. **b) carries the unbalanced current** [310.15(B)(4)(a)]
42. **d) antenna systems** [820.1]
43. **a) Weatherproof** [Art. 100 (Weatherproof)]
44. **d) a load where the maximum current is expected to continue for 3 hours or more** [Art. 100 (Continuous Load)]
45. **c) 180 VA** [220.14(I)]
46. **d) NMC** [620.21 and 620.21(C)(1)]
47. **a) 6-AWG** [280.23]
48. **d) 250 volts** [250.97]
49. **a) 50 percent** [404.14(B)(2)]
50. **b) 115 percent** [430.32(A)(1)]

Answers to *NEC* Exam Number 10

1. **d) adequate bonding and grounding** [250.116 FPN]
2. **c) 6530** [Chapter 9, Table 8]
3. **b) in multiwire branch circuits** [300.13(B)]
4. **c) 100%** [210.21(B)(1)]
5. **b) MI** [332.2]
6. **a) 67%** [450.6(A)(1)]
7. **c) 10** [Table 408.10]
8. **c) 90° C** [470.4]
9. **d) I, II, and III** [250.174(C)]
10. **c) 30** [600.5(B)(2)]
11. **d) I, II, and III** [680.6]
12. **d) I, II, or III** [450.25]
13. **a) not be smaller than the neutral conductor** [250.166(A)]
14. **b) continuous** [620.61(B)(2)]
15. **b) 50** [550.10(A)]
16. **a) True** [210.9]
17. **d) none of these** [760.55(A)]
18. **a) metal** [430.12(A)]
19. **d) at all** [422.10(A)]
20. **a) 5** [Table 314.16(A)]
21. **b) 1** [Table 348.22]
22. **c) 8** [250.56 FPN]
23. **c) 90° C** [600.9(C)]
24. **d) I, II, or III** [680.26(C)]
25. **d) I, II, and III** [450.11]
26. **c) 125%** [Table 450.3(B)]
27. **d) 5** [680.22(A)(1)]
28. **b) MI** [332.2]
29. **d) Automatic** [Art. 100 (Automatic)]

30. d) **I, II, or III** [Art. 100 (Duty, Intermittent)]
31. b) **II only** [240.83(C)]
32. a) **Varying** [Art. 100 (Duty, Varying)]
33. a) **downward** [404.6(A)]
34. b) **1000 volts** [410.80(B)]
35. d) **Neither a nor b** [Art. 100 (Overload)]
36. a) **CMX** [Table 800.154 and 800.154(B)(1)]
37. b) **Rigid metal conduit** [344.10(B)]
38. c) **conductors and equipment on certain public or private premises** [90.2(A)(1)]
39. d) **the grounding terminal of the applicable service equipment** [680.25(B)]
40. a) **grounded** [200.6(A)]
41. d) **covered** [Art. 100 (Conductor, Covered)]
42. b) **fitting** [Art. 100 (Fitting)]
43. d) **I, II, and III** [Art. 100 (Automatic)]
44. b) **will not** [230.95(C) FPN No. 1]
45. c) **3142** [366.23(A)]
46. d) **22.5 feet** [Table 680.8]
47. b) **125 percent** [410.30(C)(2)]
48. c) **An independent means of secure support must be provided.** [300.11(A)(1)]
49. a) **removed from the raceway** [390.7]
50. a) **12** [501.130(B)(3)]

Answers to *NEC* Exam Number 11

1. b) **30 amperes** [240.51(A)]
2. d) **considered out of reach** [230.9(A) Ex.]
3. d) **6-AWG** [280.23]
4. c) **I or II** [700.12]
5. d) **All of these** [590.4(C)]
6. a) **I or II** [550.15(H) Ex.]
7. d) **I, II, and III** [324.120(A)]
8. b) **intended to protect persons and property** [90.1(A)]
9. c) **10-AWG Cu** [Table 250.122]
10. a) **True** [250.92(B)]
11. b) **galvanic action** [342.14, 344.14]
12. b) **identified** [230.28]
13. c) **III only** [460.6(B)]
14. b) **6′** [210.50(C)]
15. c) **arms or stems** [410.28(C)]
16. c) **Three** [350.22(B), Table 348.22]
17. d) **I, III, and IV** [424.19]
18. b) **False** [250.70(3)]
19. b) **False** [404.9(A)]
20. d) **25** [Table 310.16]
21. c) **5½′** [210.52]
22. a) **16½** [424.44(A)]
23. b) **18 AWG** [402.6]
24. a) **8** [300.5(D)(1)]
25. c) **red** [424.35]
26. c) **not counted** [314.16(B)(1) Ex.]
27. b) **2.0 volts per cell for the lead-acid type and 1.2 volts per cell for the alkali type** [480.2 (Nominal Battery Voltage)]
28. c) **100 amperes** [550.32(C)]
29. d) **6′ 6″** [110.26(E)]
30. d) **I, II, and III** [90.2(A)]
31. b) **damp locations** [Art. 100 (Locations, Damp)]
32. c) **Askarel** [Art. 100 (Askarel)]
33. b) **Stranded** [410.28(E)]
34. b) **equipment bonding jumper** [Art. 100 (Bonding Jumper Equipment)]
35. b) **forming shell** [680.2 (Forming Shell)]
36. d) **fitting** [Art. 100 (Fitting)]
37. b) **1″** [300.4(A)(1)]
38. b) **UF cable** [550.15(H)]
39. c) **8 AWG** [Table 250.122]
40. d) **both I and II** [90.4]
41. a) **fuse** [Art. 100, Part II (Fuse)]
42. b) **Watertight** [Art. 100 (Watertight)]
43. a) **5** [680.26(B)(5)]
44. b) **Overcurrent** [Art. 100 (Overcurrent)]
45. a) **equipment** [Art. 100 (Equipment)]
46. c) **17½ inches** [Table 360.24(A)]
47. a) **10 amperes** [725.23]
48. a) **easily ignitable** [500.5(D)(1) FPN No. 2]
49. d) **167 percent** [450.4(A) Ex.]
50. c) **6** [620.21(C)(1)]

Answers to *NEC* Exam Number 12

1. **b) 8** [225.6(A)(1)]
2. **d) 8 AWG** [540.13]
3. **c) hermetic refrigerant motor-compressor** [440.2]
4. **b) 60** [422.11(B)]
5. **c) rated at not less than 15 amperes, 125 volts** [406.2(B)]
6. **b) wet** [Art. 100 (Location, Wet)]
7. **b) machine tool wiring** [Table 310.13]
8. **c) 7** [424.34]
9. **b) II only** [Art. 100]
10. **b) False** [110.12]
11. **d) general-purpose** [Art. 100]
12. **b) locked** [430.102(B) Ex.]
13. **c) fixed** [400.8(1)]
14. **c) 2½** [250.53(F)]
15. **a) 50° C** [424.36]
16. **a) energized** [300.31]
17. **b) grounded** [250.22(1)]
18. **c) 8′** [Table 400.4 Note 2]
19. **b) ⅜″** [Table 430.12(C)(1)]
20. **c) 10** [680.22(A)(2)]
21. **a) 3 hours** [450.42]
22. **d) I, II, or III** [215.4(A)]
23. **c) in dry locations** [380.2(A)]
24. **c) I and II** [250.4(A)(1)]
25. **b) carries the electric signals directing the performance of the controller but does not carry the main power current** [430.2 (Motor Control Circuit)]
26. **b) II only** [210.4(C)]
27. **d) Overload** [Art. 100 (Overload)]
28. **d) I through V** [90.1(B)]
29. **b) concealed** [Art. 100 (Concealed)]
30. **d) I, II, and III** [Art. 100 (Location, Dry)]
31. **d) I, II, and III** [410.16(C)]
32. **a) 50** [314.27(B)]
33. **c) ¾″** [501.15(C)(3)]
34. **b) Class I, Division 1** [514.3]
35. **c) interrupting** [110.9]
36. **c) grounded** [Table 110.26(A), Condition 2]
37. **c) 18 AWG** [402-6]
38. **a) ten** [408.35]
39. **c) grounding conductor** [Art. 100 (Grounding Conductor)]
40. **d) at all** [422.11(A)]
41. **d) Equipment** [Art. 100 (Equipment)]
42. **d) Yes, if all circuit conductors are disconnected simultaneously.** [404.2(B) Ex.]
43. **c) an approved sealing compound** [501.15]
44. **a) remote-control** [Art. 100 (Remote-Control Circuit)]
45. **c) electrical energy only** [Art. 100 (Appliance)]
46. **d) raintight and arranged to drain** [225.22 and 230.53]
47. **b) rainproof** [Art. 100 (Rainproof)]
48. **b) bonding jumpers** [250.92(B)]
49. **b) II and III** [225.36 Ex.]
50. **c) telephones** [90.2(B)(4)]

Answers to *NEC* Exam Number 13

1. **b) 6** [356.12(3)]
2. **d) All of the above** [440.62(A)]
3. **c) are not permitted** [502.100(A)(3)]
4. **a) shielded** [310.6]
5. **a) 0.0625″** [3140.40(B)]
6. **b) 20** [480.6(B)]
7. **d) I, II, or III** [230.50(A)]
8. **a) 60′** [Table 300.19(A)]
9. **d) 12 AWG** [680.23(F)(2)]
10. **c) within sight of the controller or in the same enclosure with the controller** [600.6(A)(2)(1)]
11. **b) False** [701.1, 702.1]
12. **c) 2″** [362.20(B)]
13. **a) Right angle** [372.5]
14. **b) 4 AWG** [312.6(C), 300.4(F)]
15. **b) warm air to a colder** [300.7(A)]
16. **b) 2.0; 1.2** [480.2 (Nominal Battery Voltage)]
17. **a) GFCIs** [680.32]
18. **c) 8 AWG** [310.3]
19. **b) 300** [Table 402.3]
20. **a) 50%** [210.23(A)(2)]
21. **c) I and III** [670.3(A) and 670.3(B)]

22. c) 3/4 [342.28 and 344.28]
23. d) I, II, and III [Art. 100 (Weatherproof)]
24. d) 6 [300.14]
25. d) I, II, and III [368.238]
26. b) the rated kVA multiplied by 1000 and divided by the rated primary voltage, using values given on the name plate [630.31(B) FPN (1)]
27. d) I, II, III, and IV [Art. 100 (Equipment)]
28. c) both I and III [520.73]
29. a) True [230.95]
30. d) recognizable as suitable for the specific purpose, function, use, environment, application, and so forth, where described in a particular *Code* requirement [Art. 100 (Identified)]
31. b) Class II, Division 1 [500.5(C)(1)]
32. b) 100 [518.1]
33. c) Arresters [230.82(4)]
34. d) RHH [Table 310.16]
35. a) True [680.1]
36. a) single [Art. 100 (Receptacle)]
37. b) controller [Art. 100 (Controller)]
38. c) 6½′; 3′ [110.32]
39. c) service conductors [Art. 100 (Service Conductors)]
40. a) the temperature of the lowest conductor temperature rating [310.10]
41. d) I, II, and III [230.82(4) and (5)]
42. b) 5 [Article 550]
43. b) Nonautomatic [Art. 100 (Nonautomatic)]
44. a) Radio [800.1]
45. d) a qualified electrical testing laboratory that is recognized as having the facilities required to test wiring equipment and materials [90.7]
46. c) Radiation safety and performance [660.1 and FPN No. 1]
47. a) life support [702.2 and FPN]
48. b) heat sensing element [Art. 100 (Thermally Protected)]
49. a) demand factor [Art. 100 (Demand Factor)]
50. d) I, II, or III [Art. 100, Part II (Oil Switch)]

Answers to *NEC* Exam Number 14

1. a) True [725.41(B)]
2. a) I and II [250.140 Ex.]
3. d) shall not [605.5(C)]
4. a) Watts [430.7(A)]
5. a) 50 [660.6(A)]
6. b) listed and marked for use as a raceway [410.31]
7. c) I and II [514.11(A)]
8. c) Class I, Division 2 [511.3(B)(2) and (3)]
9. d) an equipment grounding conductor [515.8(C)]
10. d) I, II, and III [370.3]
11. a) 75 amperes [240.6(A)]
12. c) 3 feet [Table 110.26(A)]
13. b) 100 [518.1]
14. b) 18 [760.52(B)(1)]
15. d) 14 AWG [650.6(A)]
16. a) receptacle outlet [210.50(A)]
17. b) 14-AWG and larger [240.5(B)(2)]
18. a) True [545.6 Ex.]
19. a) tension [400.10]
20. c) 10 volts [517.64(A)(1)]
21. d) 6 [250.118(5)(C)]
22. c) person who has skills and knowledge related to the construction and operation of the electrical equipment and installations and has received safety training on the hazards involved [Art. 100 (Qualified Person)]
23. c) corrosion protection [344.10(B)]
24. a) 660 watts [720.5]
25. c) EMT [502.10]
26. b) 10-AWG [110.14(A)]
27. d) I, II, and III [770.6]
28. d) the current, in amperes, a conductor can carry continuously under the conditions of use without exceeding its temperature rating [Art. 100 (Ampacity)]
29. b) outlet [Art. 100 (Outlet)]
30. d) either I or III [230.22]
31. a) device [Art. 100 (Device)]
32. c) cutout [Art. 100, Part II (Cutout)]
33. d) I or II [230.70(A)(1)]
34. b) vertical [240.33]

35. a) **sheds** [Art. 100 (Power Outlet)]
36. c) **8-AWG insulated** [680.27(A)(2)]
37. c) **the circuit conductors between the final overcurrent device protecting the circuit and the outlet(s)** [Art. 100 (Branch Circuit)]
38. d) **Yes, if insulated for the maximum voltage of any conductor** [300.3(C)(1)]
39. a) **Grounded conductor** [Art. 100 (Grounded Conductor)]
40. d) **made inaccessible by the structure or finish of the building** [Art. 100 (Concealed)]
41. c) **cross-section area** [310.16 and Chapter 9, Table 8]
42. a) **6** [210.52(A)(1)]
43. a) **MC** [330.2]
44. d) **all of the above** [Art. 100 (Location, Dry)]
45. d) **AC** [770.9]
46. a) **I or II** [410.8(B)]
47. a) **service** [Art. 100 (Service)]
48. c) **cover** [Table 300.5]
49. a) **part of a wiring system that is intended primarily to perform a mechanical function** [Art. 100 (Fitting)]
50. d) **by the authority having jurisdiction** [90.4]

Answers to *NEC* Exam Number 15

1. c) **208** [424.35]
2. c) **8** [382.30]
3. a) **True** [250.24(C)(1)]
4. c) **600** [Table 392.7(B) footnote b]
5. d) **Class III, Division 2** [500.5(D)(2)]
6. c) **II and III** [424.29]
7. b) **Class I, Division 2** [Table 515.3]
8. c) **multiwire** [Art. 100 (Branch Circuit, Multiwire)]
9. c) **either a or b** [225.36 Ex.]
10. b) **False** [400.9, 530.12]
11. d) **I, II, or III** [240.61]
12. d) **none of these** [725.55(A)]
13. c) **factory-installed internal** [90.7]
14. a) **150 volts** [680.23(A)(4)]
15. b) **50** [Art. 100 (In Sight From)]
16. d) **I, II, or III** [424.59 FPN]
17. b) **20 amperes** [600.5(B)(1)]
18. c) **Electrical circuits that lead into or pass through a gasoline dispensing pump** [514.11(A)]
19. a) **20** [210.21(A)]
20. d) **I, II, and III** [330.10(B)(4)]
21. a) **6 inches** [Table 300.5]
22. a) **3′** [366.30(B)]
23. d) **I, II, and III** [334.40(B)]
24. a) **mobile** [517.2, 660.2]
25. d) **I or II, and III** [700.12(A)]
26. a) **stationary** [550.2 (Appliance, Stationary)]
27. b) **protector** [Art. 100 (Thermal Protector)]
28. a) **Chapter 7** [Articles 725 and 760]
29. d) **40° C** [430.32(A)(4)]
30. d) **is not permanently enclosed by a structure** [Art. 100]
31. a) **operation at a substantially constant load for an indefinitely long time** [Art. 100 (Duty, Continuous)]
32. d) **Fine print notes** [90.5(C)]
33. d) **any instructions included in the listing or labeling** [110.3(B)]
34. a) **Wire-binding screws** [110.14(A)]
35. d) **I, II, or III** [702.2]
36. d) **interrupting** [Art. 100 (Interrupting Rating)]
37. b) **reduces** [240.2]
38. d) **a thoroughly good connection without damaging the conductors** [110.14(A)]
39. c) **acceptable to the authority having jurisdiction** [Art. 100 (Approved)]
40. b) **RHW** [Table 310.13]
41. c) **I and II** [400.7 and 400.8]
42. c) **is free of shorts and unintentional grounds** [110.7]
43. d) **freedom from hazard** [90.1(B)]
44. c) **I or II** [Art. 100]
45. a) **guarded** [Art. 100 (Guarded)]

46. c) **An isolated winding type transformer listed as a swimming pool transformer** [680.23(A)(2)]
47. c) **the operation of equipment in excess of normal full-load rating, or of a conductor in excess of rated ampacity** [Art. 100 (Overload)]
48. b) **intermittent** [Art. 100 (Duty, Intermittent)]
49. a) **main** [Art. 100 (Bonding Jumper, Main)]
50. a) **True** [Table 250.95]

Answers to *NEC* Exam Number 16

1. b) **the street main and the first point of connection to the service entrance conductors in a terminal box, meter, or other enclosure** [Art. 100 (Service Lateral)]
2. a) **20** [366.6(A)]
3. c) **175** [440.22(A)]
4. d) **covered in their own section of the *NEC*** [702.1]
5. d) **125%** [645.5(A)]
6. c) **50** [600.32(J)]
7. d) **I, II, and III** [450.45(A)]
8. b) **30** [550.32(A)]
9. a) **1½ hours** [450.45(E)]
10. c) **excessive movement** [392.8(D)]
11. b) **1½ hours** [700.12(A)]
12. a) **shall be grounded** [250.170]
13. c) **readily identified** [700.9(A)]
14. d) **No additional support is required by the *NEC*.** [314.23(F)]
15. d) **50** [314.27(B)]
16. a) **permanent** [517.2, 660.2]
17. c) **Rigid metal conduit** [344.10(B)]
18. a) **50 volts** [445.14]
19. d) **outer conductor** [820.100(A)(3)]
20. b) **I and III** [390.2(A)]
21. d) **maximum unbalanced** [220.61(A)]
22. b) **dressing rooms** [520.73]
23. c) **bonding conductors** [250.98]
24. c) **80%** [645.5(A)]
25. c) **a similar change shall be permitted to be made in the size of the grounded conductor** [240.23]
26. a) **covered by the *National Electrical Code*** [110.13(B)]
27. c) **receptacle outlet** [Art. 100 (Receptacle Outlet)]
28. c) **radio** [800.1]
29. d) **10** [Art. 100]
30. a) **bathtubs** [680.1]
31. a) **True** [110.13(A)]
32. c) **dead front** [Art. 100 (Dead Front)]
33. b) **grounded** [517.62]
34. b) **six** [230.71(A)]
35. d) **approved bushings** [314.42]
36. b) **6** [410.15(A)]
37. b) **10 feet; 20 feet** [680.22(A)(3)]
38. c) **1″** [501.15(C)(3)]
39. c) **3** [Art. 100 (Dwelling, Multifamily)]
40. d) **at the point where the conductor to be protected receives its supply** [Chapter 9, Table 12 (3rd paragraph)]
41. d) **13** [220.14(I)]
42. c) **10** [680.22(A)(2)]
43. d) **Mogul** [410.53]
44. b) **Batteries** [690.2 (Hybrid System)]
45. a) **copper** [110.5]
46. d) **90° C** [Chapter 9, Table 8]
47. d) **in AWG or in circular mils** [110.6]
48. a) **500** [220.16(A)(1)]
49. d) **I, II, and III** [110.11]
50. d) **75** [312.8]

Index